AWS
Certified Solutions Architect Study Guide

Associate (SAA-C02) Exam

Third Edition

Ben Piper

David Clinton

Copyright © 2021 by John Wiley & Sons, Inc., Indianapolis, Indiana

Published simultaneously in Canada

ISBN: 978-1-119-71308-1
ISBN: 978-1-119-71309-8 (ebk.)
ISBN: 978-1-119-71310-4 (ebk.)

No part of this publication may be reproduced, stored in a retrieval system or transmitted in any form or by any means, electronic, mechanical, photocopying, recording, scanning or otherwise, except as permitted under Sections 107 or 108 of the 1976 United States Copyright Act, without either the prior written permission of the Publisher, or authorization through payment of the appropriate per-copy fee to the Copyright Clearance Center, 222 Rosewood Drive, Danvers, MA 01923, (978) 750-8400, fax (978) 646-8600. Requests to the Publisher for permission should be addressed to the Permissions Department, John Wiley & Sons, Inc., 111 River Street, Hoboken, NJ 07030, (201) 748-6011, fax (201) 748-6008, or online at www.wiley.com/go/permissions.

Limit of Liability/Disclaimer of Warranty: The publisher and the author make no representations or warranties with respect to the accuracy or completeness of the contents of this work and specifically disclaim all warranties, including without limitation warranties of fitness for a particular purpose. No warranty may be created or extended by sales or promotional materials. The advice and strategies contained herein may not be suitable for every situation. This work is sold with the understanding that the publisher is not engaged in rendering legal, accounting, or other professional services. If professional assistance is required, the services of a competent professional person should be sought. Neither the publisher nor the author shall be liable for damages arising herefrom. The fact that an organization or Web site is referred to in this work as a citation and/or a potential source of further information does not mean that the author or the publisher endorses the information the organization or Web site may provide or recommendations it may make. Further, readers should be aware that Internet Web sites listed in this work may have changed or disappeared between when this work was written and when it is read.

For general information on our other products and services or to obtain technical support, please contact our Customer Care Department within the U.S. at (877) 762-2974, outside the U.S. at (317) 572-3993 or fax (317) 572-4002.

Wiley publishes in a variety of print and electronic formats and by print-on-demand. Some material included with standard print versions of this book may not be included in e-books or in print-on-demand. If this book refers to media such as a CD or DVD that is not included in the version you purchased, you may download this material at booksupport.wiley.com. For more information about Wiley products, visit www.wiley.com.

Library of Congress Control Number: 2020947039

TRADEMARKS: Wiley, the Wiley logo, and the Sybex logo are trademarks or registered trademarks of John Wiley & Sons, Inc. and/or its affiliates, in the United States and other countries, and may not be used without written permission. AWS is a registered trademark of Amazon Technologies, Inc. All other trademarks are the property of their respective owners. John Wiley & Sons, Inc. is not associated with any product or vendor mentioned in this book.

SKY10022621_111720

Acknowledgments

We would like to thank the following people who helped us create *AWS Certified Solutions Architect Study Guide: Associate SAA-C02 Exam, Third Edition*.

First, a special thanks to our friends at Wiley. Kenyon Brown, senior acquisitions editor, got the ball rolling on this project and pushed to get this book published quickly. His experience and guidance throughout the project was critical. Stephanie Barton, project editor, helped push this book forward by keeping us accountable to our deadlines. Her edits made many of the technical parts of this book more readable.

Todd Montgomery reviewed the chapters and questions for technical accuracy. Not only did his comments and suggestions make this book more accurate, he also provided additional ideas for the chapter review questions to make them more challenging and relevant to the exam.

Lastly, the authors would like to thank each other!

About the Authors

Ben Piper is a networking and cloud consultant who has authored multiple books, including the *AWS Certified Cloud Practitioner Study Guide: Foundational CLF-C01 Exam* (Sybex, 2019) and *Learn Cisco Network Administration in a Month of Lunches* (Manning, 2017). You can contact Ben by visiting his website: benpiper.com.

David Clinton is a Linux server admin and AWS solutions architect who has worked with IT infrastructure in both academic and enterprise environments. He has authored books—including (with Ben Piper) the *AWS Certified Cloud Practitioner Study Guide: Foundational CLF-C01 Exam* (Sybex, 2019) and *Linux in Action* (Manning Publications, 2018)—and created more than two dozen video courses teaching Amazon Web Services and Linux administration, server virtualization, and IT security for Pluralsight.

In a "previous life," David spent 20 years as a high school teacher. He currently lives in Toronto, Canada, with his wife and family and can be reached through his website: bootstrap-it.com.

Contents at a Glance

Introduction	*xxi*
Assessment Test	*xxvii*

Part I		**The Core AWS Services**	**1**
Chapter	1	Introduction to Cloud Computing and AWS	3
Chapter	2	Amazon Elastic Compute Cloud and Amazon Elastic Block Store	21
Chapter	3	AWS Storage	59
Chapter	4	Amazon Virtual Private Cloud	83
Chapter	5	Database Services	133
Chapter	6	Authentication and Authorization— AWS Identity and Access Management	165
Chapter	7	CloudTrail, CloudWatch, and AWS Config	183
Chapter	8	The Domain Name System and Network Routing: Amazon Route 53 and Amazon CloudFront	211
Chapter	9	Simple Queue Service and Kinesis	233

Part II		**The Well-Architected Framework**	**245**
Chapter	10	The Reliability Pillar	247
Chapter	11	The Performance Efficiency Pillar	273
Chapter	12	The Security Pillar	301
Chapter	13	The Cost Optimization Pillar	335
Chapter	14	The Operational Excellence Pillar	353
Appendix		Answers to Review Questions	385

Index — *415*

Contents

Introduction   *xxi*

Assessment Test   *xxvii*

Part I		**The Core AWS Services**	**1**
Chapter	**1**	**Introduction to Cloud Computing and AWS**	**3**
		Cloud Computing and Virtualization	4
		Cloud Computing Architecture	4
		Cloud Computing Optimization	5
		The AWS Cloud	6
		AWS Platform Architecture	10
		AWS Reliability and Compliance	12
		The AWS Shared Responsibility Model	12
		The AWS Service Level Agreement	13
		Working with AWS	13
		The AWS CLI	14
		AWS SDKs	14
		Technical Support and Online Resources	14
		Support Plans	14
		Other Support Resources	15
		Summary	15
		Exam Essentials	16
		Review Questions	17
Chapter	**2**	**Amazon Elastic Compute Cloud and Amazon Elastic Block Store**	**21**
		Introduction	22
		EC2 Instances	22
		Provisioning Your Instance	23
		Configuring Instance Behavior	28
		Placement Groups	28
		Instance Pricing	29
		Instance Lifecycle	30
		Resource Tags	30
		Service Limits	31
		EC2 Storage Volumes	32
		Elastic Block Store Volumes	32
		Instance Store Volumes	34
		Accessing Your EC2 Instance	35

	Securing Your EC2 Instance	36
	Security Groups	36
	IAM Roles	37
	NAT Devices	37
	Key Pairs	38
	EC2 Auto Scaling	38
	Launch Configurations	39
	Launch Templates	39
	Auto Scaling Groups	40
	Auto Scaling Options	42
	AWS Systems Manager	46
	Actions	47
	Insights	49
	AWS CLI Example	51
	Summary	52
	Exam Essentials	53
	Review Questions	54
Chapter 3	**AWS Storage**	**59**
	Introduction	60
	S3 Service Architecture	61
	Prefixes and Delimiters	61
	Working with Large Objects	61
	Encryption	62
	Logging	63
	S3 Durability and Availability	64
	Durability	64
	Availability	65
	Eventually Consistent Data	65
	S3 Object Lifecycle	66
	Versioning	66
	Lifecycle Management	66
	Accessing S3 Objects	67
	Access Control	67
	Presigned URLs	69
	Static Website Hosting	69
	Amazon S3 Glacier	71
	Storage Pricing	72
	Other Storage-Related Services	73
	Amazon Elastic File System	73
	Amazon FSx	73
	AWS Storage Gateway	73
	AWS Snowball	74
	AWS DataSync	74
	AWS CLI Example	75

		Summary	76
		Exam Essentials	77
		Review Questions	78
Chapter	**4**	**Amazon Virtual Private Cloud**	**83**
		Introduction	84
		VPC CIDR Blocks	84
		Secondary CIDR Blocks	85
		IPv6 CIDR Blocks	85
		Subnets	87
		Subnet CIDR Blocks	87
		Availability Zones	88
		IPv6 CIDR Blocks	91
		Elastic Network Interfaces	91
		Primary and Secondary Private IP Addresses	91
		Attaching Elastic Network Interfaces	91
		Enhanced Networking	93
		Internet Gateways	93
		Route Tables	94
		Routes	94
		The Default Route	95
		Security Groups	98
		Inbound Rules	98
		Outbound Rules	99
		Sources and Destinations	99
		Stateful Firewall	99
		Default Security Group	100
		Network Access Control Lists	101
		Inbound Rules	102
		Outbound Rules	105
		Using Network Access Control Lists and Security Groups Together	106
		Public IP Addresses	106
		Elastic IP Addresses	107
		AWS Global Accelerator	109
		Network Address Translation	109
		Network Address Translation Devices	110
		Configuring Route Tables to Use NAT Devices	112
		NAT Gateway	113
		NAT Instance	113
		VPC Peering	114
		Hybrid Cloud Networking	115
		Virtual Private Networks	115
		AWS Transit Gateway	115
		AWS Direct Connect	123

		High-Performance Computing	125
		Elastic Fabric Adapter	125
		AWS ParallelCluster	126
		Summary	126
		Exam Essentials	127
		Review Questions	129
Chapter	**5**	**Database Services**	**133**
		Introduction	134
		Relational Databases	134
		Columns and Attributes	135
		Using Multiple Tables	135
		Structured Query Language	137
		Online Transaction Processing vs. Online Analytic Processing	137
		Amazon Relational Database Service	138
		Database Engines	138
		Licensing Considerations	139
		Database Option Groups	140
		Database Instance Classes	140
		Storage	141
		Read Replicas	145
		High Availability (Multi-AZ)	146
		Single-Master	147
		Multi-Master	147
		Backup and Recovery	148
		Automated Snapshots	148
		Maintenance Items	149
		Amazon Redshift	149
		Compute Nodes	149
		Data Distribution Styles	150
		Redshift Spectrum	150
		AWS Database Migration Service	150
		Nonrelational (NoSQL) Databases	151
		Storing Data	151
		Querying Data	152
		Types of Nonrelational Databases	152
		DynamoDB	153
		Partition and Hash Keys	153
		Attributes and Items	154
		Throughput Capacity	155
		Reading Data	157
		Global Tables	158
		Backups	158
		Summary	158

		Exam Essentials	159
		Review Questions	161
Chapter	**6**	**Authentication and Authorization—AWS Identity and Access Management**	**165**
		Introduction	166
		IAM Identities	166
		IAM Policies	167
		User and Root Accounts	168
		Access Keys	170
		Groups	172
		Roles	173
		Authentication Tools	173
		Amazon Cognito	174
		AWS Managed Microsoft AD	174
		AWS Single Sign-On	174
		AWS Key Management Service	175
		AWS Secrets Manager	175
		AWS CloudHSM	175
		AWS CLI Example	176
		Summary	177
		Exam Essentials	177
		Review Questions	179
Chapter	**7**	**CloudTrail, CloudWatch, and AWS Config**	**183**
		Introduction	184
		CloudTrail	185
		Management Events	185
		Data Events	186
		Event History	186
		Trails	186
		Log File Integrity Validation	189
		CloudWatch	189
		CloudWatch Metrics	190
		Graphing Metrics	192
		Metric Math	194
		CloudWatch Logs	195
		CloudWatch Alarms	198
		Amazon EventBridge	201
		AWS Config	202
		The Configuration Recorder	203
		Configuration Items	203
		Configuration History	203
		Configuration Snapshots	203
		Monitoring Changes	204

		Summary	206
		Exam Essentials	206
		Review Questions	207
Chapter	**8**	**The Domain Name System and Network Routing: Amazon Route 53 and Amazon CloudFront**	**211**
		Introduction	212
		The Domain Name System	212
		Namespaces	212
		Name Servers	213
		Domains and Domain Names	213
		Domain Registration	214
		Domain Layers	214
		Fully Qualified Domain Names	214
		Zones and Zone Files	215
		Record Types	215
		Alias Records	216
		Amazon Route 53	216
		Domain Registration	217
		DNS Management	217
		Availability Monitoring	219
		Routing Policies	220
		Traffic Flow	222
		Route 53 Resolver	223
		Amazon CloudFront	223
		AWS CLI Example	225
		Summary	226
		Exam Essentials	226
		Review Questions	228
Chapter	**9**	**Simple Queue Service and Kinesis**	**233**
		Introduction	234
		Simple Queue Service	234
		Queues	234
		Queue Types	235
		Polling	236
		Dead-Letter Queues	237
		Kinesis	237
		Kinesis Video Streams	237
		Kinesis Data Streams	238
		Kinesis Data Firehose	239
		Kinesis Data Firehose vs. Kinesis Data Streams	239
		Summary	240
		Exam Essentials	240
		Review Questions	241

Part II			**The Well-Architected Framework**	**245**
Chapter	**10**		**The Reliability Pillar**	**247**

    Introduction   248
    Calculating Availability   248
        Availability Differences in Traditional vs.
            Cloud-Native Applications   249
        Know Your Limits   252
        Increasing Availability   252
    EC2 Auto Scaling   253
        Launch Configurations   253
        Launch Templates   254
        Auto Scaling Groups   255
        Auto Scaling Options   256
    Data Backup and Recovery   261
        S3   261
        Elastic File System   261
        Elastic Block Storage   261
        Database Resiliency   262
    Creating a Resilient Network   263
        VPC Design Considerations   263
        External Connectivity   263
    Designing for Availability   264
        Designing for 99 Percent Availability   264
        Designing for 99.9 Percent Availability   265
        Designing for 99.99 Percent Availability   266
    Summary   267
    Exam Essentials   268
    Review Questions   269

Chapter	**11**		**The Performance Efficiency Pillar**	**273**

    Introduction   274
    Optimizing Performance for the Core AWS Services   274
        Compute   275
        Storage   279
        Database   282
        Network Optimization and Load Balancing   284
    Infrastructure Automation   286
        CloudFormation   286
        Third-Party Automation Solutions   288
    Reviewing and Optimizing Infrastructure Configurations   289
        Load Testing   289
        Visualization   290

		Optimizing Data Operations	291
		Caching	291
		Partitioning/Sharding	293
		Compression	294
		Summary	294
		Exam Essentials	295
		Review Questions	297
Chapter	**12**	**The Security Pillar**	**301**
		Introduction	302
		Identity and Access Management	302
		Protecting AWS Credentials	303
		Fine-Grained Authorization	303
		Permissions Boundaries	305
		Roles	306
		Enforcing Service-Level Protection	313
		Detective Controls	313
		CloudTrail	313
		CloudWatch Logs	314
		Searching Logs with Athena	315
		Auditing Resource Configurations with AWS Config	317
		Amazon GuardDuty	318
		Amazon Inspector	321
		Amazon Detective	322
		Security Hub	323
		Protecting Network Boundaries	323
		Network Access Control Lists and Security Groups	323
		AWS Web Application Firewall	323
		AWS Shield	324
		Data Encryption	324
		Data at Rest	325
		Data in Transit	326
		Macie	327
		Summary	327
		Exam Essentials	328
		Review Questions	329
Chapter	**13**	**The Cost Optimization Pillar**	**335**
		Introduction	336
		Planning, Tracking, and Controlling Costs	336
		AWS Budgets	337
		Monitoring Tools	338
		AWS Organizations	339

	AWS Trusted Advisor	340
	Online Calculator Tools	340
	Cost-Optimizing Compute	342
	Maximizing Server Density	343
	EC2 Reserved Instances	343
	EC2 Spot Instances	344
	Auto Scaling	347
	Elastic Block Store Lifecycle Manager	347
	Summary	347
	Exam Essentials	348
	Review Questions	349
Chapter 14	**The Operational Excellence Pillar**	**353**
	Introduction	354
	CloudFormation	354
	Creating Stacks	355
	Deleting Stacks	356
	Using Multiple Stacks	356
	Stack Updates	359
	Preventing Updates to Specific Resources	360
	Overriding Stack Policies	361
	CodeCommit	361
	Creating a Repository	362
	Repository Security	362
	Interacting with a Repository Using Git	363
	CodeDeploy	365
	The CodeDeploy Agent	366
	Deployments	366
	Deployment Groups	366
	Deployment Types	366
	Deployment Configurations	367
	Lifecycle Events	368
	The Application Specification File	369
	Triggers and Alarms	370
	Rollbacks	370
	CodePipeline	371
	Continuous Integration	371
	Continuous Delivery	371
	Creating the Pipeline	372
	Artifacts	373
	AWS Systems Manager	374
	Actions	374
	Insights	377

	AWS Landing Zone	378
	Summary	379
	Exam Essentials	379
	Review Questions	381
Appendix	**Answers to Review Questions**	**385**
	Chapter 1: Introduction to Cloud Computing and AWS	386
	Chapter 2: Amazon Elastic Compute Cloud and Amazon Elastic Block Store	387
	Chapter 3: AWS Storage	389
	Chapter 4: Amazon Virtual Private Cloud	391
	Chapter 5: Database Services	393
	Chapter 6: Authentication and Authorization—AWS Identity and Access Management	395
	Chapter 7: CloudTrail, CloudWatch, and AWS Config	397
	Chapter 8: The Domain Name System and Network Routing: Amazon Route 53 and Amazon CloudFront	399
	Chapter 9: Simple Queue Service and Kinesis	401
	Chapter 10: The Reliability Pillar	403
	Chapter 11: The Performance Efficiency Pillar	405
	Chapter 12: The Security Pillar	407
	Chapter 13: The Cost Optimization Pillar	409
	Chapter 14: The Operational Excellence Pillar	411
Index		*415*

Table of Exercises

Exercise	1.1	Use the AWS CLI	16
Exercise	2.1	Launch an EC2 Linux Instance and Log in Using SSH	27
Exercise	2.2	Assess the Free Capacity of a Running Instance and Change Its Instance Type	27
Exercise	2.3	Assess Which Pricing Model Will Best Meet the Needs of a Deployment	30
Exercise	2.4	Create and Launch an AMI Based on an Existing Instance Storage Volume	34
Exercise	2.5	Create a Launch Template	40
Exercise	2.6	Install the AWS CLI and Use It to Launch an EC2 Instance	51
Exercise	2.7	Clean Up Unused EC2 Resources	52
Exercise	3.1	Create a New S3 Bucket and Upload a File	62
Exercise	3.2	Enable Versioning and Lifecycle Management for an S3 Bucket	67
Exercise	3.3	Generate and Use a Presigned URL	69
Exercise	3.4	Enable Static Website Hosting for an S3 Bucket	70
Exercise	3.5	Calculate the Total Lifecycle Costs for Your Data	73
Exercise	4.1	Create a New VPC	85
Exercise	4.2	Create a New Subnet	89
Exercise	4.3	Create and Attach a Primary ENI	92
Exercise	4.4	Create an Internet Gateway and Default Route	96
Exercise	4.5	Create a Custom Security Group	100
Exercise	4.6	Create an Inbound Rule to Allow Remote Access from Any IP Address	103
Exercise	4.7	Allocate and Use an Elastic IP Address	107
Exercise	4.8	Create a Transit Gateway	117
Exercise	4.9	Create a Blackhole Route	122
Exercise	5.1	Create an RDS Database Instance	143
Exercise	5.2	Create a Read Replica	145
Exercise	5.3	Promote the Read Replica to a Master	146
Exercise	5.4	Create a Table in DynamoDB Using Provisioned Mode	156
Exercise	6.1	Lock Down the Root User	169
Exercise	6.2	Assign and Implement an IAM Policy	169
Exercise	6.3	Create, Use, and Delete an AWS Access Key	171
Exercise	6.4	Create and Configure an IAM Group	172

Exercise	7.1	Create a Trail	187
Exercise	7.2	Create a Graph Using Metric Math	194
Exercise	7.3	Deliver CloudTrail Logs to CloudWatch Logs	197
Exercise	8.1	Create a Hosted Zone on Route 53 for an EC2 Web Server	218
Exercise	8.2	Set Up a Health Check	219
Exercise	8.3	Configure a Route 53 Routing Policy	221
Exercise	8.4	Create a CloudFront Distribution for Your S3-Based Static Website	224
Exercise	10.1	Create a Launch Template	254
Exercise	11.1	Configure and Launch an Application Using Auto Scaling	277
Exercise	11.2	Sync Two S3 Buckets as Cross-Region Replicas	281
Exercise	11.3	Upload to an S3 Bucket Using Transfer Acceleration	282
Exercise	11.4	Create and Deploy an EC2 Load Balancer	285
Exercise	11.5	Launch a Simple CloudFormation Template	287
Exercise	11.6	Create a CloudWatch Dashboard	290
Exercise	12.1	Create a Limited Administrative User	305
Exercise	12.2	Create and Assume a Role as an IAM User	311
Exercise	12.3	Configure VPC Flow Logging	315
Exercise	12.4	Encrypt an EBS Volume	325
Exercise	13.1	Create an AWS Budget to Send an Alert	338
Exercise	13.2	Build Your Own Stack in Simple Monthly Calculator	341
Exercise	13.3	Request a Spot Fleet Using the AWS CLI	345
Exercise	14.1	Create a Nested Stack	357
Exercise	14.2	Create and Interact with a CodeCommit Repository	363

Introduction

Studying for any certification always involves deciding how much of your studying should be practical hands-on experience and how much should be simply memorizing facts and figures. Between the two of us, we've taken dozens of IT certification exams, so we know how important it is to use your study time wisely. We've designed this book to help you discover your strengths and weaknesses on the AWS platform so that you can focus your efforts properly. Whether you've been working with AWS for a long time or whether you're relatively new to it, we encourage you to carefully read this book from cover to cover.

Passing the AWS Certified Solutions Architect – Associate exam requires understanding the components and operation of the core AWS services as well as how those services interact with each other. Read through the official documentation for the various AWS services. Amazon offers HTML, PDF, and Kindle documentation for many of them. Use this book as a guide to help you identify your strengths and weaknesses so that you can focus your study efforts properly.

You should have at least six months of hands-on experience with AWS before taking the AWS Certified Solutions Architect – Associate exam. If you're relatively new to AWS, we strongly recommend our own *AWS Certified Cloud Practitioner Study Guide: CLF-C01 Exam* (Sybex, 2019) as a primer.

Even though this book is designed specifically for the AWS Certified Solutions Architect – Associate exam, some of your fellow readers have found it useful for preparing for the SysOps Administrator and DevOps Engineer exams.

Hands-on experience is crucial for exam success. Each chapter in this *AWS Certified Solutions Architect Study Guide: Associate SAA-C02 Exam, Third Edition* contains hands-on exercises that you should strive to complete during or immediately after you read the chapter. It's vital to understand that the exercises don't cover every possible scenario for every AWS service. In fact, it's quite the opposite. The exercises provide you with a foundation to build on. Use them as your starting point, but don't be afraid to venture out on your own. Feel free to modify them to match the variables and scenarios you might encounter in your own organization. Keep in mind that some of the exercises and figures use the AWS web console, which is in constant flux. As such, screenshots and step-by-step details of exercises may change. Use these eventualities as excuses to dig into the AWS online documentation and browse around the web console on your own. Also remember that although you can complete many of the exercises within the bounds of the AWS Free Tier, getting enough practice to pass the exam will likely require you to spend some money. But it's money well spent, as getting certified is an investment in your career and your future.

Each chapter contains review questions to thoroughly test your understanding of the services and concepts covered in that chapter. They also test your ability to integrate the concepts with information from preceding chapters. Although the difficulty of the questions varies, rest assured that they are not "fluff." We've designed the questions to help you realistically gauge your understanding and readiness for the exam. Avoid the temptation to rush through the questions to just get to the answers. Once you complete the assessment in each chapter, referring to the answer key will give you not only the correct answers but a detailed explanation as to why they're correct. It will also explain why the other answers are incorrect.

The book also contains a self-assessment exam with 39 questions, two practice exams with 50 questions each to help you gauge your readiness to take the exam, and flashcards to help you learn and retain key facts needed to prepare for the exam.

This *AWS Certified Solutions Architect Study Guide: Associate SAA-C02 Exam, Third Edition* is divided into two parts: "The Core AWS Services" and "The Well-Architected Framework."

Part I, "The Core AWS Services"

The first part of the book dives deep into each of the core AWS services. These services include ones you probably already have at least a passing familiarity with: Elastic Compute Cloud (EC2), Virtual Private Cloud (VPC), Identity and Access Management (IAM), Route 53, and Simple Storage Service (S3), to name just a few.

Some AWS services seem to serve similar or even nearly identical purposes. You'll learn about the subtle but important differences between seemingly similar services and, most importantly, when to use each.

Part II, "The Well-Architected Framework"

The second part of the book is a set of best practices and principles aimed at helping you design, implement, and operate systems in the cloud. Part II focuses on the following five pillars of good design:

- Reliability
- Performance efficiency
- Security
- Cost optimization
- Operational excellence

Each chapter of Part II revisits the core AWS services in light of a different pillar. Also, because not every AWS service is large enough to warrant its own chapter, Part II simultaneously introduces other services that, although less well known, may still show up on the exam.

Achieving the right balance among these pillars is a key skill you need to develop as a solutions architect. Prior to beginning Part II, we encourage you to peruse the Well-Architected Framework white paper, which is available for download at d0.awsstatic.com/whitepapers/architecture/AWS_Well-Architected_Framework.pdf.

What Does This Book Cover?

This book covers topics you need to know to prepare for the Amazon Web Services (AWS) Certified Solutions Architect – Associate exam:

Chapter 1: Introduction to Cloud Computing and AWS   This chapter provides an overview of the AWS Cloud computing platform and its core services and concepts.

Chapter 2: Amazon Elastic Compute Cloud and Amazon Elastic Block Store   This chapter covers EC2 instances—the virtual machines that you can use to run Linux and Windows workloads on AWS. It also covers the Elastic Block Store service that EC2 instances depend on for persistent data storage.

Chapter 3: AWS Storage   In this chapter, you'll learn about Simple Storage Service (S3) and Glacier, which provide unlimited data storage and retrieval for AWS services, your applications, and the Internet.

Chapter 4: Amazon Virtual Private Cloud   This chapter explains Amazon Virtual Private Cloud (Amazon VPC), a virtual network that contains network resources for AWS services.

Chapter 5: Database Services   In this chapter, you will learn about some different managed database services offered by AWS, including Relational Database Service (RDS), DynamoDB, and Redshift.

Chapter 6: Authentication and Authorization—AWS Identity and Access Management   This chapter covers AWS Identity and Access Management (IAM), which provides the primary means for protecting the AWS resources in your account.

Chapter 7: CloudTrail, CloudWatch, and AWS Config   In this chapter, you'll learn how to log, monitor, and audit your AWS resources.

Chapter 8: The Domain Name System and Network Routing: Amazon Route 53 and Amazon CloudFront   This chapter focuses on the Domain Name System (DNS) and Route 53, the service that provides public and private DNS hosting for both internal AWS resources and the Internet. It also covers CloudFront, Amazon's global content delivery network.

Chapter 9: Simple Queue Service and Kinesis   This chapter explains how to use the principle of loose coupling to create scalable and highly available applications. You'll learn how Simple Queue Service (SQS) and Kinesis fit into the picture.

Chapter 10: The Reliability Pillar   This chapter will show you how to architect and integrate AWS services to achieve a high level of reliability for your applications. You'll learn how to plan around and recover from inevitable outages to keep your systems up and running.

Chapter 11: The Performance Efficiency Pillar This chapter covers how to build highly performing systems and use the AWS elastic infrastructure to rapidly scale up and out to meet peak demand.

Chapter 12: The Security Pillar In this chapter, you'll learn how to use encryption and security controls to protect the confidentiality, integrity, and availability of your data and systems on AWS. You'll also learn about the various security services such as GuardDuty, Inspector, Shield, and Web Application Firewall.

Chapter 13: The Cost Optimization Pillar This chapter will show you how to estimate and control your costs in the cloud.

Chapter 14: The Operational Excellence Pillar In this chapter, you'll learn how to keep your systems running smoothly on AWS. You'll learn how to implement a DevOps mind-set using CloudFormation, Systems Manager, and the AWS Developer Tools.

Interactive Online Learning Environment and Test Bank

The authors have worked hard to provide some really great tools to help you with your certification process. The interactive online learning environment that accompanies the *AWS Certified Solutions Architect Study Guide: Associate SAA-C02 Exam, Third Edition* provides a test bank with study tools to help you prepare for the certification exam—and increase your chances of passing it the first time! The test bank includes the following:

Sample Tests All the questions in this book are provided, including the assessment test at the end of this Introduction and the chapter tests that include the review questions at the end of each chapter. In addition, there are two practice exams with 50 questions each. Use these questions to test your knowledge of the study guide material. The online test bank runs on multiple devices.

Flashcards The online text banks include 100 flashcards specifically written to hit you hard, so don't get discouraged if you don't ace your way through them at first. They're there to ensure that you're really ready for the exam. And no worries—armed with the review questions, practice exams, and flashcards, you'll be more than prepared when exam day comes. Questions are provided in digital flashcard format (a question followed by a single correct answer). You can use the flashcards to reinforce your learning and provide last-minute test prep before the exam.

Resources You'll find some AWS CLI and other code examples from the book for you to cut and paste for use in your own environment. A glossary of key terms from this book is also available as a fully searchable PDF.

> **Note:** Go to wiley.com/go/sybextestprep to register and gain access to this interactive online learning environment and test bank with study tools.

Exam Objectives

The AWS Certified Solutions Architect – Associate exam is intended for people who have experience in designing distributed applications and systems on the AWS platform. In general, you should have the following before taking the exam:

- A minimum of one year of hands-on experience designing systems on AWS
- Hands-on experience using the AWS services that provide compute, networking, storage, and databases
- Ability to define a solution using architectural design principles based on customer requirements
- Ability to provide implementation guidance
- Ability to identify which AWS services meet a given technical requirement
- An understanding of the five pillars of the Well-Architected Framework
- An understanding of the AWS global infrastructure, including the network technologies used to connect them
- An understanding of AWS security services and how they integrate with traditional on-premises security infrastructure

The exam covers five different domains, with each domain broken down into objectives.

Objective Map

The following table lists each domain and its weighting in the exam, along with the chapters in the book where that domain's objectives are covered.

Domain	Percentage of Exam	Chapters
Domain 1: Design Resilient Architectures	30%	
1.1 Design a multi-tier architecture solution		2, 3, 5, 8, 9, 10, 11
1.2 Design highly available and/or fault-tolerant architectures		2, 3, 5, 7, 8, 9, 10, 11, 14

Domain	Percentage of Exam	Chapters
1.3 Design decoupling mechanisms using AWS services		4, 5, 9, 10, 11, 14
1.4 Choose appropriate resilient storage		2, 3, 5, 9, 10, 11
Domain 2: Design High-Performing Architectures	28%	
2.1 Identify elastic and scalable compute solutions for a workload		2, 3, 5, 7, 8, 9, 11
2.2 Select high-performing and scalable storage solutions for a workload		2, 3, 9, 11
2.3 Select high-performing networking solutions for a workload		5, 8, 9, 11
2.4 Choose high-performing database solutions for a workload		5, 11
Domain 3: Design Secure Applications and Architectures	24%	
3.1 Design secure access to AWS resources		2, 3, 4, 6, 7, 12
3.2 Design secure application tiers		3, 6, 12
3.3 Select appropriate data security options		3, 4, 6, 7, 12
Domain 4: Design Cost-Optimized Architectures	18%	
4.1 Identify cost-effective storage solutions		2, 3, 13
4.2 Identify cost-effective compute and database services		2, 13
4.3 Design cost-optimized network architectures		8, 13

Assessment Test

1. True/false: The Developer Support plan provides access to a support application programming interface (API).
   A. True
   B. False

2. True/false: AWS is responsible for managing the network configuration of your EC2 instances.
   A. True
   B. False

3. Which of the following services is most useful for decoupling the components of a monolithic application?
   A. SNS
   B. KMS
   C. SQS
   D. Glacier

4. An application you want to run on EC2 requires you to license it based on the number of physical CPU sockets and cores on the hardware you plan to run the application on. Which of the following tenancy models should you specify?
   A. Dedicated host
   B. Dedicated instance
   C. Shared tenancy
   D. Bring your own license

5. True/false: Changing the instance type of an EC2 instance will change its elastic IP address.
   A. True
   B. False

6. True/false: You can use a Quick Start Amazon Machine Image (AMI) to create any instance type.
   A. True
   B. False

7. Which S3 encryption option does *not* require AWS persistently storing the encryption keys it uses to decrypt data?
   A. Client-side encryption
   B. SSE-KMS
   C. SSE-S3
   D. SSE-C

8. True/false: Durability measures the percentage of likelihood that a given object will not be inadvertently lost by AWS over the course of a year.
   A. True
   B. False

9. True/false: After uploading a new object to S3, there will be a slight delay (one to two seconds) before the object is available.
   A. True
   B. False

10. You created a Virtual Private Cloud (VPC) using the Classless Inter-Domain Routing (CIDR) block 10.0.0.0/24. You need to connect to this VPC from your internal network, but the IP addresses in use on your internal network overlap with the CIDR. Which of the following is a valid way to address this problem?
    A. Remove the CIDR and use IPv6 instead.
    B. Change the VPC's CIDR.
    C. Create a new VPC with a different CIDR.
    D. Create a secondary CIDR for the VPC.

11. True/false: An EC2 instance must be in a public subnet to access the Internet.
    A. True
    B. False

12. True/false: The route table for a public subnet must have a default route pointing to an Internet gateway as a target.
    A. True
    B. False

13. Which of the following use cases is well suited for DynamoDB?
    A. Running a MongoDB database on AWS
    B. Storing large binary files exceeding 1 GB in size
    C. Storing JSON documents that have a consistent structure
    D. Storing image assets for a website

14. True/false: You can create a DynamoDB global secondary index for an existing table at any time.
    A. True
    B. False

15. True/false: Enabling point-in-time RDS snapshots is sufficient to give you a recovery point objective (RPO) of less than 10 minutes.
    A. True
    B. False

16. Which of the following steps does the most to protect your AWS account?
   A. Deleting unused Identity and Access Management (IAM) policies
   B. Revoking unnecessary access for IAM users
   C. Rotating root access keys
   D. Restricting access to S3 buckets
   E. Rotating Secure Shell (SSH) key pairs

17. Which of the following can be used to encrypt the operating system of an EC2 instance?
   A. AWS Secrets Manager
   B. CloudHSM
   C. AWS Key Management Service (KMS)
   D. AWS Security Token Service (STS)

18. What is a difference between a token generated by the AWS Security Token Service (STS) and an IAM access key?
   A. The token generated by STS can't be used by an IAM principal.
   B. An IAM access key is unique.
   C. The token generated by STS can be used only once.
   D. The token generated by STS expires.

19. True/false: EC2 sends instance memory utilization metrics to CloudWatch every five minutes.
   A. True
   B. False

20. You configured a CloudWatch alarm to monitor CPU utilization for an EC2 instance. The alarm began in the `INSUFFICIENT_DATA` state and then entered the `ALARM` state. What can you conclude from this?
   A. The instance recently rebooted.
   B. CPU utilization is too high.
   C. The CPU utilization metric crossed the alarm threshold.
   D. The instance is stopped.

21. Where do AWS Config and CloudTrail store their logs?
   A. S3 buckets
   B. CloudWatch Logs
   C. CloudTrail Events
   D. DynamoDB
   E. Amazon Athena

22. True/false: An EC2 instance in a private subnet can resolve an "A" resource record for a public hosted zone hosted in Route 53.
    A. True
    B. False

23. You want to use Route 53 to send users to the application load balancer closest to them. Which of the following routing policies lets you do this with the least effort?
    A. Latency routing
    B. Geolocation routing
    C. Geoproximity routing
    D. Edge routing

24. True/false: You can use an existing domain name with Route 53 without switching its registration to AWS.
    A. True
    B. False

25. You're designing an application that takes multiple image files and combines them into a video file that users on the Internet can download. Which of the following can help you quickly implement your application in the fastest, most highly available, and most cost-effective manner?
    A. EC2 spot fleet
    B. Lambda
    C. Relational Database Service (RDS)
    D. Auto Scaling

26. You're using EC2 Auto Scaling and want to implement a scaling policy that adds one extra instance only when the average CPU utilization of each instance exceeds 90 percent. However, you don't want it to add more than one instance every five minutes. Which of the following scaling policies should you use?
    A. Simple
    B. Step
    C. Target tracking
    D. PercentChangeInCapacity

27. True/false: EC2 Auto Scaling automatically replaces group instances directly terminated by the root user.
    A. True
    B. False

28. Which ElastiCache engine can persistently store data?
    A. MySQL
    B. Memcached

C. MongoDB
D. Redis

29. Which of the following is *not* an AWS service?
    A. CloudFormation
    B. Puppet
    C. OpsWorks
    D. Snowball

30. True/false: S3 cross-region replication uses transfer acceleration.
    A. True
    B. False

31. Which of the following services can you deactivate on your account?
    A. Security Token Service (STS)
    B. CloudWatch
    C. Virtual Private Cloud (VPC)
    D. Lambda

32. Which of the following services can alert you to malware on an EC2 instance?
    A. AWS GuardDuty
    B. AWS Inspector
    C. AWS Shield
    D. AWS Web Application Firewall

33. True/false: If versioning is enabled on an S3 bucket, applying encryption to an unencrypted object in that bucket will create a new, encrypted version of that object.
    A. True
    B. False

34. Which instance type will, if left running, continue to incur costs?
    A. Spot
    B. Standard reserved
    C. On-demand
    D. Convertible reserved

35. True/false: The EBS Lifecycle Manager can take snapshots of volumes that were once attached to terminated instances.
    A. True
    B. False

36. Which of the following lets you spin up new web servers the quickest?
    A. Lambda
    B. Auto Scaling
    C. Elastic Container Service
    D. CloudFront

37. True/false: CloudFormation stack names are case-sensitive.
    A. True
    B. False

38. Where might CodeDeploy look for the `appspec.yml` file? (Choose two.)
    A. GitHub
    B. CodeCommit
    C. S3
    D. CloudFormation

39. True/false: You can use either CodeDeploy or an AWS Systems Manager command document to deploy a Lambda application.
    A. True
    B. False

Answers to Assessment Test

1. B. The Business plan offers access to a support API, but the Developer plan does not. See Chapter 1 for more information.

2. B. Customers are responsible for managing the network configuration of EC2 instances. AWS is responsible for the physical network infrastructure. See Chapter 1 for more information.

3. C. Simple Queue Service (SQS) allows for event-driven messaging within distributed systems that can decouple while coordinating the discrete steps of a larger process. See Chapter 1 for more information.

4. A. The dedicated host option lets you see the number of physical CPU sockets and cores on a host. See Chapter 2 for more information.

5. B. An elastic IP address will not change. A public IP address attached to an instance will change if the instance is stopped, as would happen when changing the instance type. See Chapter 2 for more information.

6. A. A Quick Start AMI is independent of the instance type. See Chapter 2 for more information.

7. D. With SSE-C you provide your own keys for Amazon to use to decrypt and encrypt your data. AWS doesn't persistently store the keys. See Chapter 3 for more information.

8. A. Durability corresponds to an average annual expected loss of objects stored on S3, not including objects you delete. Availability measures the amount of time S3 will be available to let you retrieve those objects. See Chapter 3 for more information.

9. B. S3 uses a read-after-write consistency model for new objects, so once you upload an object to S3, it's immediately available. See Chapter 3 for more information.

10. C. You can't change the primary CIDR for a VPC, so you must create a new one to connect it to your internal network. See Chapter 4 for more information.

11. B. An EC2 instance can access the Internet from a private subnet provided it uses a NAT gateway or NAT instance. See Chapter 4 for more information.

12. A. The definition of a public subnet is a subnet that has a default route pointing to an Internet gateway as a target. Otherwise, it's a private subnet. See Chapter 4 for more information.

13. C. DynamoDB is a key-value store that can be used to store items up to 400 KB in size. See Chapter 5 for more information.

14. A. You can create a global secondary index for an existing table at any time. You can create a local secondary index only when you create the table. See Chapter 5 for more information.

15. A. Enabling point-in-time recovery gives you an RPO of about five minutes. The recovery time objective (RTO) depends on the amount of data to restore. See Chapter 5 for more information.

16. B. Revoking unnecessary access for IAM users is the most effective of the listed measures for protecting your AWS account. See Chapter 6 for more information.

17. C. KMS can be used to encrypt Elastic Block Store (EBS) volumes that store an instance's operating system. See Chapter 6 for more information.

18. D. STS tokens expire and IAM access keys do not. An STS token can be used more than once. IAM access keys and STS tokens are both unique. An IAM principal can use an STS token. See Chapter 6 for more information.

19. B. EC2 doesn't track instance memory utilization. See Chapter 7 for more information.

20. C. The transition to the ALARM state simply implies that the metric crossed a threshold but doesn't tell you what the threshold is. Newly created alarms start out in the INSUFFICIENT_DATA state. See Chapter 7 for more information.

21. A. Both store their logs in S3 buckets. See Chapter 7 for more information.

22. A. An EC2 instance in a private subnet still has access to Amazon's private DNS servers, which can resolve records stored in public hosted zones. See Chapter 8 for more information.

23. C. Geoproximity routing routes users to the location closest to them. Geolocation routing requires you to create records for specific locations or create a default record. See Chapter 8 for more information.

24. A. Route 53 is a true DNS service in that it can host zones for any domain name. You can also register domain names with or transfer them to Route 53. See Chapter 8 for more information.

25. B. Lambda is a highly available, reliable, "serverless" compute platform that runs functions as needed and scales elastically to meet demand. EC2 spot instances can be shut down on short notice. See Chapter 10 for more information.

26. A. A simple scaling policy changes the group size and then has a cooldown period before doing so again. Step scaling policies don't have cooldown periods. Target tracking policies attempt to keep a metric at a set value. PercentChangeInCapacity is a simple scaling adjustment type, not a scaling policy. See Chapter 10 for more information.

27. A. Auto Scaling always attempts to maintain the minimum group size or, if set, the desired capacity. See Chapter 10 for more information.

28. D. ElastiCache supports Memcached and Redis, but only the latter can store data persistently. See Chapter 11 for more information.

29. B. Puppet is a configuration management platform that AWS offers via OpsWorks but is not itself an AWS service. See Chapter 11 for more information.

30. B. S3 cross-region replication transfers objects between different buckets. Transfer acceleration uses a CloudFront edge location to speed up transfers between S3 and the Internet. See Chapter 11 for more information.

31. A. You can deactivate STS for all regions except US East. See Chapter 12 for more information.

32. A. GuardDuty looks for potentially malicious activity. Inspector looks for vulnerabilities that may result in compromise. Shield and Web Application Firewall protect applications from attack. See Chapter 12 for more information.

33. A. Applying encryption to an unencrypted object will create a new, encrypted version of that object. Previous versions remain unencrypted. See Chapter 12 for more information.

34. C. On-demand instances will continue to run and incur costs. Reserved instances cost the same whether they're running or stopped. Spot instances will be terminated when the spot price exceeds your bid price. See Chapter 13 for more information.

35. A. The EBS Lifecycle Manager can take scheduled snapshots of any EBS volume, regardless of attachment state. See Chapter 13 for more information.

36. C. Elastic Container Service lets you run containers that can launch in a matter of seconds. EC2 instances take longer. Lambda is "serverless," so you can't use it to run a web server. CloudFront provides caching but isn't a web server. See Chapter 13 for more information.

37. A. Almost everything in CloudFormation is case sensitive. See Chapter 14 for more information.

38. A, C. CodeDeploy looks for the `appspec.yml` file with the application files it is to deploy, which can be stored in S3 or on GitHub. See Chapter 14 for more information.

39. B. You can use CodeDeploy to deploy an application to Lambda or EC2 instances. But an AWS Systems Manager command document works only on EC2 instances. See Chapter 14 for more information.

The Core AWS Services

PART I

Chapter 1

Introduction to Cloud Computing and AWS

The cloud is where much of the serious technology innovation and growth happens these days, and Amazon Web Services (AWS), more than any other, is the platform of choice for business and institutional workloads. If you want to be successful as an AWS solutions architect, you'll first need to understand what the cloud really is and how Amazon's end of it works.

TO MAKE SURE YOU'VE GOT THE BIG PICTURE, THIS CHAPTER WILL EXPLORE THE BASICS:

- ✓ What makes cloud computing different from other applications and client-server models
- ✓ How the AWS platform provides secure and flexible virtual networked environments for your resources
- ✓ How AWS provides such a high level of service reliability
- ✓ How to access and manage your AWS-based resources
- ✓ Where you can go for documentation and help with your AWS deployments

Cloud Computing and Virtualization

The technology that lies at the core of all cloud operations is virtualization. As illustrated in Figure 1.1, virtualization lets you divide the hardware resources of a single physical server into smaller units. That physical server could therefore host multiple virtual machines (VMs) running their own complete operating systems, each with its own memory, storage, and network access.

FIGURE 1.1 A virtual machine host

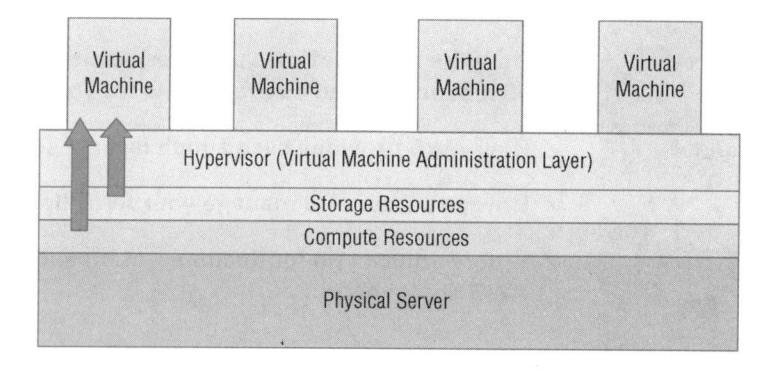

Virtualization's flexibility makes it possible to provision a virtual server in a matter of seconds, run it for exactly the time your project requires, and then shut it down. The resources released will become instantly available to other workloads. The usage density you can achieve lets you squeeze the greatest value from your hardware and makes it easy to generate experimental and sandboxed environments.

Cloud Computing Architecture

Major cloud providers like AWS have enormous server farms where hundreds of thousands of servers and disk drives are maintained along with the network cabling necessary to connect them. A well-built virtualized environment could provide a virtual server using storage, memory, compute cycles, and network bandwidth collected from the most efficient mix of available sources it can find.

A cloud computing platform offers on-demand, self-service access to pooled compute resources where your usage is metered and billed according to the volume you consume. Cloud computing systems allow for precise billing models, sometimes involving fractions of a penny for an hour of consumption.

Cloud Computing Optimization

The cloud is a great choice for so many serious workloads because it's scalable, elastic, and, often, a lot cheaper than traditional alternatives. Effective deployment provisioning will require some insight into those three features.

Scalability

A scalable infrastructure can efficiently meet unexpected increases in demand for your application by automatically adding resources. As Figure 1.2 shows, this most often means dynamically increasing the number of virtual machines (or *instances* as AWS calls them) you've got running.

FIGURE 1.2 Copies of a machine image are added to new VMs as they're launched.

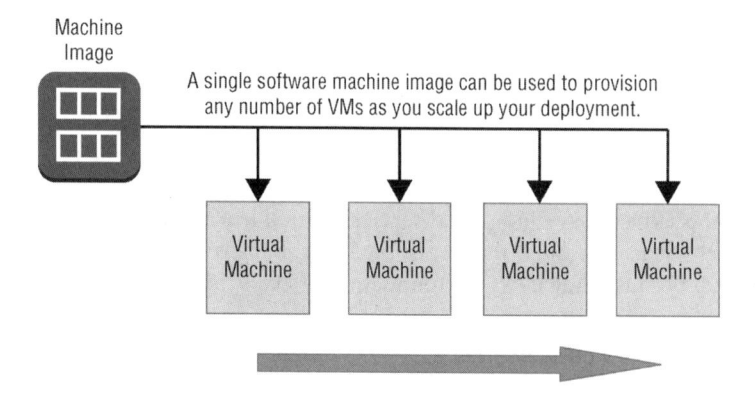

AWS offers its autoscaling service through which you define a machine image that can be instantly and automatically replicated and launched into multiple instances to meet demand.

Elasticity

The principle of elasticity covers some of the same ground as scalability—both address how the system manages changing demand. However, though the images used in a scalable environment let you ramp up capacity to meet rising demand, an elastic infrastructure will automatically *reduce* capacity when demand drops. This makes it possible to control costs, since you'll run resources only when they're needed.

Cost Management

Besides the ability to control expenses by closely managing the resources you use, cloud computing transitions your IT spending from a capital expenditure (capex) framework into something closer to operational expenditure (opex).

In practical terms, this means you no longer have to spend $10,000 up front for every new server you deploy—along with associated electricity, cooling, security, and rack space costs. Instead, you're billed much smaller incremental amounts for as long as your application runs.

That doesn't necessarily mean your long-term cloud-based opex costs will always be less than you'd pay over the lifetime of a comparable data center deployment. But it does mean you won't have to expose yourself to risky speculation about your long-term needs. If, sometime in the future, changing demand calls for new hardware, AWS will be able to deliver it within a minute or two.

To help you understand the full implications of cloud compute spending, AWS provides a free Total Cost of Ownership (TCO) Calculator at aws.amazon.com/tco-calculator. This calculator helps you perform proper "apples-to-apples" comparisons between your current data center costs and what an identical operation would cost you on AWS.

The AWS Cloud

Keeping up with the steady stream of new services showing up on the AWS Console can be frustrating. But as a solutions architect, your main focus should be on the core service categories. This section briefly summarizes each of the core categories (as shown in Table 1.1) and then does the same for key individual services. You'll learn much more about all of these (and more) services through the rest of the book, but it's worth focusing on these short definitions, because they lie at the foundation of everything else you're going to learn.

TABLE 1.1 AWS service categories

Category	Function
Compute	Services replicating the traditional role of local physical servers for the cloud, offering advanced configurations including autoscaling, load balancing, and even serverless architectures (a method for delivering server functionality with a very small footprint)
Networking	Application connectivity, access control, and enhanced remote connections
Storage	Various kinds of storage platforms designed to fit a range of both immediate accessibility and long-term backup needs

Category	Function
Database	Managed data solutions for use cases requiring multiple data formats: relational, NoSQL, or caching
Application management	Monitoring, auditing, and configuring AWS account services and running resources
Security and identity	Services for managing authentication and authorization, data and connection encryption, and integration with third-party authentication management systems

Table 1.2 describes the functions of some core AWS services, organized by category.

TABLE 1.2 Core AWS services (by category)

Category	Service	Function
Compute	Elastic Compute Cloud (EC2)	EC2 server instances provide virtual versions of the servers you would run in your local data center. EC2 instances can be provisioned with the CPU, memory, storage, and network interface profile to meet any application need, from a simple web server to one part of a cluster of instances providing an integrated multi-tiered fleet architecture. Since EC2 instances are virtual, they're resource-efficient and deploy nearly instantly.
	Lambda	Serverless application architectures like the one provided by Amazon's Lambda service allow you to provide responsive public-facing services without the need for a server that's actually running 24/7. Instead, network events (like consumer requests) can trigger the execution of a predefined code-based operation. When the operation (which can currently run for as long as 15 minutes) is complete, the Lambda event ends, and all resources automatically shut down.
	Auto Scaling	Copies of running EC2 instances can be defined as image templates and automatically launched (or *scaled up*) when client demand can't be met by existing instances. As demand drops, unused instances can be terminated (or *scaled down*).
	Elastic Load Balancing	Incoming network traffic can be directed between multiple web servers to ensure that a single web server isn't overwhelmed while other servers are underused or that traffic isn't directed to failed servers.

TABLE 1.2 Core AWS services (by category) *(continued)*

Category	Service	Function
	Elastic Beanstalk	Beanstalk is a managed service that abstracts the provisioning of AWS compute and networking infrastructure. You are required to do nothing more than push your application code, and Beanstalk automatically launches and manages all the necessary services in the background.
Networking	Virtual Private Cloud (VPC)	VPCs are highly configurable networking environments designed to host your EC2 (and RDS) instances. You use VPC-based tools to secure and, if desired, isolate your instances by closely controlling inbound and outbound network access.
	Direct Connect	By purchasing fast and secure network connections to AWS through a third-party provider, you can use Direct Connect to establish an enhanced direct tunnel between your local data center or office and your AWS-based VPCs.
	Route 53	Route 53 is the AWS DNS service that lets you manage domain registration, record administration, routing protocols, and health checks, which are all fully integrated with the rest of your AWS resources
	CloudFront	CloudFront is Amazon's distributed global content delivery network (CDN). When properly configured, a CloudFront distribution can store cached versions of your site's content at edge locations around the world so that they can be delivered to customers on request with the greatest efficiency and lowest latency.
Storage	Simple Storage Service (S3)	S3 offers highly versatile, reliable, and inexpensive object storage that's great for data storage and backups. It's also commonly used as part of larger AWS production processes, including through the storage of script, template, and log files.
	S3 Glacier	A good choice for when you need large data archives stored cheaply over the long term and can live with retrieval delays measuring in the hours. Glacier's lifecycle management is closely integrated with S3.
	Elastic Block Store (EBS)	EBS provides the persistent virtual storage drives that host the operating systems and working data of an EC2 instance. They're meant to mimic the function of the storage drives and partitions attached to physical servers.

Category	Service	Function
	Storage Gateway	Storage Gateway is a hybrid storage system that exposes AWS cloud storage as a local, on-premises appliance. Storage Gateway can be a great tool for migration and data backup and as part of disaster recovery operations.
Database	Relational Database Service (RDS)	RDS is a managed service that builds you a stable, secure, and reliable database instance. You can run a variety of SQL database engines on RDS, including MySQL, Microsoft SQL Server, Oracle, and Amazon's own Aurora.
	DynamoDB	DynamoDB can be used for fast, flexible, highly scalable, and managed nonrelational (NoSQL) database workloads.
Application management	CloudWatch	CloudWatch can be set to monitor process performance and resource utilization and, when preset thresholds are met, either send you a message or trigger an automated response.
	CloudFormation	This service enables you to use template files to define full and complex AWS deployments. The ability to script your use of any AWS resources makes it easier to automate, standardizing and speeding up the application launch process.
	CloudTrail	CloudTrail collects records of all your account's API events. This history is useful for account auditing and troubleshooting purposes.
	Config	The Config service is designed to help you with change management and compliance for your AWS account. You first define a desired configuration state, and Config evaluates any future states against that ideal. When a configuration change pushes too far from the ideal baseline, you'll be notified.
Security and identity	Identity and Access Management (IAM)	You use IAM to administrate user and programmatic access and authentication to your AWS account. Through the use of users, groups, roles, and policies, you can control exactly who and what can access and/or work with any of your AWS resources.
	Key Management Service (KMS)	KMS is a managed service that allows you to administrate the creation and use of encryption keys to secure data used by and for any of your AWS resources.

TABLE 1.2 Core AWS services (by category) *(continued)*

Category	Service	Function
	Directory Service	For AWS environments that need to manage identities and relationships, Directory Service can integrate AWS resources with identity providers like Amazon Cognito and Microsoft AD domains.
Application integration	Simple Notification Service (SNS)	SNS is a notification tool that can automate the publishing of alert *topics* to other services (to an SQS Queue or to trigger a Lambda function, for instance), to mobile devices, or to recipients using email or SMS.
	Simple Workflow (SWF)	SWF lets you coordinate a series of tasks that must be performed using a range of AWS services or even non-digital (meaning, human) events.
	Simple Queue Service (SQS)	SQS allows for event-driven messaging within distributed systems that can decouple while coordinating the discrete steps of a larger process. The data contained in your SQS messages will be reliably delivered, adding to the fault-tolerant qualities of an application.
	API Gateway	This service enables you to create and manage secure and reliable APIs for your AWS-based applications.

AWS Platform Architecture

AWS maintains data centers for its physical servers around the world. Because the centers are so widely distributed, you can reduce your own services' network transfer latency by hosting your workloads geographically close to your users. It can also help you manage compliance with regulations requiring you to keep data within a particular legal jurisdiction.

Data centers exist within AWS regions, of which there are currently 21—not including private U.S. government AWS GovCloud regions—although this number is constantly growing. It's important to always be conscious of the region you have selected when you launch new AWS resources; pricing and service availability can vary from one to the next. Table 1.3 shows a list of all 21 (nongovernment) regions along with each region's name and core endpoint addresses. Note that accessing and authenticating to the two Chinese regions requires unique protocols.

TABLE 1.3 A list of publicly accessible AWS regions

Region name	Region	Endpoint
US East (Ohio)	us-east-2	us-east-2.amazonaws.com
US West (N. California)	us-west-1	us-west-1.amazonaws.com
US West (Oregon)	us-west-2	us-west-2.amazonaws.com
Asia Pacific (Hong Kong)	ap-east-1	ap-east-1.amazonaws.com
Asia Pacific (Mumbai)	ap-south-1	ap-south-1.amazonaws.com
Asia Pacific (Seoul)	ap-northeast-2	ap-northeast-2.amazonaws.com
Asia Pacific (Osaka-Local)	ap-northeast-3	ap-northeast-3.amazonaws.com
Asia Pacific (Singapore)	ap-southeast-1	ap-southeast-1.amazonaws.com
Asia Pacific (Sydney)	ap-southeast-2	ap-southeast-2.amazonaws.com
Asia Pacific (Tokyo)	ap-northeast-1	ap-northeast-1.amazonaws.com
Canada (Central)	ca-central-1	ca-central-1.amazonaws.com
China (Beijing)	cn-north-1	cn-north-1.amazonaws.com.cn
China (Ningxia)	cn-northwest-1	cn-northwest-1.amazonaws.com.cn
EU (Frankfurt)	eu-central-1	eu-central-1.amazonaws.com
EU (Ireland)	eu-west-1	eu-west-1.amazonaws.com
EU (London)	eu-west-2	eu-west-2.amazonaws.com
EU (Paris)	eu-west-3	eu-west-3.amazonaws.com
EU (Stockholm)	eu-north-1	eu-north-1.amazonaws.com
Middle East (Bahrain)	me-south-1	me-south-1.amazon.aws.com

> Endpoint addresses are used to access your AWS resources remotely from within application code or scripts. Prefixes like ec2, apigateway, or cloudformation are often added to the endpoints to specify a particular AWS service. Such an address might look like this: cloudformation.us-east-2.amazonaws.com. You can see a complete list of endpoint addresses and their prefixes at docs.aws.amazon.com/general/latest/gr/rande.html.

Because low-latency access is so important, certain AWS services are offered from designated edge network locations. These services include Amazon CloudFront, Amazon Route 53, AWS Firewall Manager, AWS Shield, and AWS WAF. For a complete and up-to-date list of available locations, see aws.amazon.com/about-aws/global-infrastructure/regional-product-services.

Physical AWS data centers are exposed within your AWS account as availability zones. There might be half a dozen availability zones within a region, like us-east-1a and us-east-1b, each consisting of one or more data centers.

You organize your resources from a region within one or more virtual private clouds (VPCs). A VPC is effectively a network address space within which you can create network subnets and associate them with availability zones. When configured properly, this architecture can provide effective resource isolation and durable replication.

AWS Reliability and Compliance

AWS has a lot of the basic regulatory, legal, and security groundwork covered before you even launch your first service.

AWS has invested significant planning and funds into resources and expertise relating to infrastructure administration. Its heavily protected and secretive data centers, layers of redundancy, and carefully developed best-practice protocols would be difficult or even impossible for a regular enterprise to replicate.

Where applicable, resources on the AWS platform are compliant with dozens of national and international standards, frameworks, and certifications, including ISO 9001, FedRAMP, NIST, and GDPR. (See aws.amazon.com/compliance/programs for more information.)

The AWS Shared Responsibility Model

Of course, those guarantees cover only the underlying AWS platform. The way you decide to use AWS resources is your business—and therefore your responsibility. So, it's important to be familiar with the AWS Shared Responsibility Model.

AWS guarantees the secure and uninterrupted operation of its "cloud." That means its physical servers, storage devices, networking infrastructure, and managed services. AWS customers, as illustrated in Figure 1.3, are responsible for whatever happens *within* that cloud. This covers the security and operation of installed operating systems, client-side data, the movement of data across networks, end-user authentication and access, and customer data.

FIGURE 1.3 The AWS Shared Responsibility Model

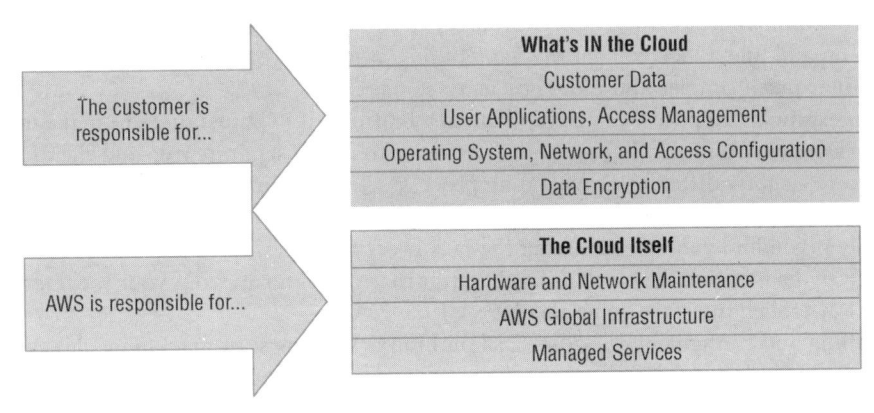

The customer is responsible for...	**What's IN the Cloud**
	Customer Data
	User Applications, Access Management
	Operating System, Network, and Access Configuration
	Data Encryption

AWS is responsible for...	**The Cloud Itself**
	Hardware and Network Maintenance
	AWS Global Infrastructure
	Managed Services

The AWS Service Level Agreement

By "guarantee," AWS doesn't mean that service disruptions or security breaches will *never* occur. Drives may stop spinning, major electricity systems may fail, and natural disasters may happen. But when something does go wrong, AWS will provide service credits to reimburse customers for their *direct* losses whenever uptimes fall below a defined threshold. Of course, that won't help you recover customer confidence or lost business.

The exact percentage of the guarantee will differ according to service. The service level agreement (SLA) rate for AWS EC2, for instance, is set to 99.99 percent—meaning that you can expect your EC2 instances, ECS containers, and EBS storage devices to be available for all but around four minutes of each month.

The important thing to remember is that it's not *if* things will fail but *when*. Build your applications to be geographically dispersed and fault tolerant so that when things do break, your users will barely notice.

Working with AWS

Whatever AWS services you choose to run, you'll need a way to manage them all. The browser-based management console is an excellent way to introduce yourself to a service's features and learn how it will perform in the real world. There are few AWS administration tasks that you can't do from the console, which provides plenty of useful visualizations and helpful documentation. But as you become more familiar with the way things work, and especially as your AWS deployments become more complex, you'll probably find yourself getting more of your serious work done away from the console.

The AWS CLI

The AWS Command Line Interface (CLI) lets you run complex AWS operations from your local command line. Once you get the hang of how it works, you'll discover that it can make things much simpler and more efficient.

As an example, suppose you need to launch a half-dozen EC2 instances to make up a microservices environment. Each instance is meant to play a separate role and therefore will require a subtly different provisioning process. Clicking through window after window to launch the instances from the console can quickly become tedious and time-consuming, especially if you find yourself repeating the task every few days. But the whole process can alternatively be incorporated into a simple script that you can run from your local terminal shell or PowerShell interface using the AWS CLI.

Installing and configuring the AWS CLI on Linux, Windows, or macOS machines is not hard at all, but the details might change depending on your platform. For the most up-to-date instructions, see `docs.aws.amazon.com/cli/latest/userguide/installing.html`.

AWS SDKs

If you want to incorporate access to your AWS resources into your application code, you'll need to use an AWS software development kit (SDK) for the language you're working with. AWS currently offers SDKs for nine languages, including Java, .NET, and Python, and a number of mobile SDKs that include Android and iOS. There are also toolkits available for IntelliJ, Visual Studio, and Visual Studio Team Services (VSTS).

You can see a full overview of AWS developer tools at `aws.amazon.com/tools`.

Technical Support and Online Resources

Things won't always go smoothly for you—on AWS just as everywhere else in your life. Sooner or later, you'll need some kind of technical or account support. There's a variety of types of support, and you should understand what's available.

One of the first things you'll be asked to decide when you create a new AWS account is which support plan you'd like. Your business needs and budget will determine the way you answer this question.

Support Plans

The Basic plan is free with every account and gives you access to customer service, along with documentation, white papers, and the support forum. Customer service covers billing and account support issues.

The Developer plan starts at $29/month and adds access for one account holder to a Cloud Support associate along with limited general guidance and "system impaired" response.

For $100/month (and up), the Business plan will deliver faster guaranteed response times to unlimited users for help with "impaired" systems, personal guidance and troubleshooting, and a support API.

Finally, Enterprise support plans cover all of the other features, plus direct access to AWS solutions architects for operational and design reviews, your own technical account manager, and something called a support concierge. For complex, mission-critical deployments, those benefits can make a big difference. But they'll cost you at least $15,000 each month.

You can read more about AWS support plans at aws.amazon.com/premiumsupport/compare-plans.

Other Support Resources

There's plenty of self-serve support available outside of the official support plans:

- AWS community help forums are open to anyone with a valid AWS account (forums.aws.amazon.com).
- Extensive and well-maintained AWS documentation is available at aws.amazon.com/documentation.
- The AWS Well-Architected page (aws.amazon.com/architecture/well-architected) is a hub that links to some valuable white papers and documentation addressing best practices for cloud deployment design.
- There are also plenty of third-party companies offering commercial support for your AWS deployments.

Summary

Cloud computing is built on the ability to efficiently divide physical resources into smaller but flexible virtual units. Those units can be "rented" by businesses on a pay-as-you-go basis and used to satisfy just about any networked application and/or workflow's needs in an affordable, scalable, and elastic way.

Amazon Web Services provides reliable and secure resources that are replicated and globally distributed across a growing number of regions and availability zones. AWS infrastructure is designed to be compliant with best-practice and regulatory standards—although the Shared Responsibility Model leaves you in charge of what you place *within* the cloud.

The growing family of AWS services covers just about any digital needs you can imagine, with core services addressing compute, networking, database, storage, security, and application management and integration needs.

You manage your AWS resources from the management console, with the AWS CLI, or through code generated using an AWS SDK.

Technical and account support is available through support plans and through documentation and help forums. AWS also makes white papers and user and developer guides available for free as Kindle books. You can access them at this address:
www.amazon.com/default/e/B007R6MVQ6/ref=dp_byline_cont_ebooks_1?redirectedFromKindleDbs=true

Exam Essentials

Understand AWS platform architecture.   AWS divides its servers and storage devices into globally distributed regions and, within regions, into availability zones. These divisions permit replication to enhance availability but also permit process and resource isolation for security and compliance purposes. You should design your own deployments in ways that take advantage of those features.

Understand how to use AWS administration tools.   Though you will use your browser to access the AWS administration console at least from time to time, most of your serious work will probably happen through the AWS CLI and, from within your application code, through an AWS SDK.

Understand how to choose a support plan.   Understanding which support plan level is right for a particular customer is an important element in building a successful deployment. Become familiar with the various options.

EXERCISE 1.1

Use the AWS CLI

Install (if necessary) and configure the AWS CLI on your local system and demonstrate it's running properly by listing the buckets that currently exist in your account. For extra practice, create an S3 bucket and then copy a simple file or document from your machine to your new bucket. From the browser console, confirm that the file reached its target.

To get you started, here are some basic CLI commands:

```
aws s3 ls
aws s3 mb <bucketname>
aws s3 cp /path/to/file.txt s3://bucketname
```

Review Questions

1. Your developers want to run fully provisioned EC2 instances to support their application code deployments but prefer not to have to worry about manually configuring and launching the necessary infrastructure. Which of the following should they use?
    A. AWS Lambda
    B. AWS Elastic Beanstalk
    C. Amazon EC2 Auto Scaling
    D. Amazon Route 53

2. Some of your application's end users are complaining of delays when accessing your resources from remote geographic locations. Which of these services would be the most likely to help reduce the delays?
    A. Amazon CloudFront
    B. Amazon Route 53
    C. Elastic Load Balancing
    D. Amazon Glacier

3. Which of the following is the best use-case scenario for Elastic Block Store?
    A. You need a cheap and reliable place to store files your application can access.
    B. You need a safe place to store backup archives from your local servers.
    C. You need a source for on-demand compute cycles to meet fluctuating demand for your application.
    D. You need persistent storage for the filesystem run by your EC2 instance.

4. You need to integrate your company's local user access controls with some of your AWS resources. Which of the following can help you control the way your local users access your AWS services and administration console? (Choose two.)
    A. AWS Identity and Access Management (IAM)
    B. Key Management Service (KMS)
    C. AWS Directory Service
    D. Simple WorkFlow (SWF)
    E. Amazon Cognito

5. The data consumed by the application you're planning will require more speed and flexibility than you can get from a closely defined relational database structure. Which AWS database service should you choose?
    A. Relational Database Service (RDS)
    B. Amazon Aurora
    C. Amazon DynamoDB
    D. Key Management Service (KMS)

6. You've launched an EC2 application server instance in the AWS Ireland region and you need to access it from the web. Which of the following is the correct endpoint address that you should use?
   A. compute.eu-central-1.amazonaws.com
   B. ec2.eu-central-1.amazonaws.com
   C. elasticcomputecloud.eu-west-2.amazonaws.com
   D. ec2.eu-west-1.amazonaws.com

7. When working to set up your first AWS deployment, you keep coming across the term *availability zone*. What exactly is an availability zone?
   A. An isolated physical data center within an AWS region
   B. A region containing multiple data centers
   C. A single network subnet used by resources within a single region
   D. A single isolated server room within a data center

8. As you plan your multi-tiered, multi-instance AWS application, you need a way to effectively organize your instances and configure their network connectivity and access control. Which tool will let you do that?
   A. Load Balancing
   B. Amazon Virtual Private Cloud (VPC)
   C. Amazon CloudFront
   D. AWS endpoints

9. You want to be sure that the application you're building using EC2 and S3 resources will be reliable enough to meet the regulatory standards required within your industry. What should you check?
   A. Historical uptime log records
   B. The AWS Program Compliance Tool
   C. The AWS service level agreement (SLA)
   D. The AWS Compliance Programs documentation page
   E. The AWS Shared Responsibility Model

10. Your organization's operations team members need a way to access and administer your AWS infrastructure via your local command line or shell scripts. Which of the following tools will let them do that?
    A. AWS Config
    B. AWS CLI
    C. AWS SDK
    D. The AWS Console

11. While building a large AWS-based application, your company has been facing configuration problems they can't solve on their own. As a result, they need direct access to AWS support for both development and IT team leaders. Which support plan should you purchase?

  A. Business
  B. Developer
  C. Basic
  D. Enterprise

Chapter 2

Amazon Elastic Compute Cloud and Amazon Elastic Block Store

THE AWS CERTIFIED SOLUTIONS ARCHITECT ASSOCIATE EXAM OBJECTIVES COVERED IN THIS CHAPTER MAY INCLUDE, BUT ARE NOT LIMITED TO, THE FOLLOWING:

✓ **Domain 1: Design Resilient Architectures**
- 1.1 Design a multi-tier architecture solution
- 1.2 Design highly available and/or fault-tolerant architectures
- 1.4 Choose appropriate resilient storage

✓ **Domain 2: Design High-Performing Architectures**
- 2.1 Identify elastic and scalable compute solutions for a workload
- 2.2 Select high-performing and scalable storage solutions for a workload

✓ **Domain 3: Design Secure Applications and Architectures**
- 3.1 Design secure access to AWS resources

✓ **Domain 4: Design Cost-Optimized Architectures**
- 4.1 Identify cost-effective storage solutions
- 4.2 Identify cost-effective compute and database services
- 4.3 Design cost-optimized network architectures

Introduction

The ultimate focus of a traditional data center/server room is its precious servers. But, to make those servers useful, you'll need to add racks, power supplies, cabling, switches, firewalls, and cooling.

AWS's Elastic Compute Cloud (EC2) is designed to replicate the data center/server room experience as closely as possible. At the center of it all is the EC2 virtual server, known as an *instance*. But, like the local server room I just described, EC2 provides a range of tools meant to support and enhance your instance's operations.

This chapter will explore the tools and practices used to fully leverage the power of the EC2 ecosystem, including the following:

- Provisioning an EC2 instance with the right hardware resources for your project
- Configuring the right base operating system for your application needs
- Building a secure and effective network environment for your instance
- Adding scripts to run as the instance boots to support (or start) your application
- Choosing the best EC2 pricing model for your needs
- Understanding how to manage and leverage the EC2 instance lifecycle
- Choosing the right storage drive type for your needs
- Securing your EC2 resources using key pairs, security groups, network access lists, and Identity and Access Management (IAM) roles
- Scaling the number of instances up and down to meet changing demand using Auto Scaling
- Accessing your instance as an administrator or end-user client

EC2 Instances

An EC2 instance may only be a virtualized and abstracted subset of a physical server, but it behaves just like the real thing. It will have access to storage, memory, and a network interface, and its primary storage drive will come with a fresh and clean operating system running.

It's up to you to decide what kind of hardware resources you want your instance to have, what operating system and software stack you'd like it to run, and, ultimately, how much you'll pay for it. Let's see how all that works.

Provisioning Your Instance

You configure your instance's operating system and software stack, hardware specs (the CPU power, memory, primary storage, and network performance), and environment before launching it. The OS is defined by the Amazon Machine Image (AMI) you choose, and the hardware follows the instance type.

EC2 Amazon Machine Images

An AMI is really just a template document that contains information telling EC2 what OS and application software to include on the root data volume of the instance it's about to launch. There are four kinds of AMIs:

Amazon Quick Start AMIs   Amazon Quick Start images appear at the top of the list in the console when you start the process of launching a new instance. The Quick Start AMIs are popular choices and include various releases of Linux or Windows Server OSs and some specialty images for performing common operations (like deep learning and database). These AMIs are up-to-date and officially supported.

AWS Marketplace AMIs   AMIs from the AWS Marketplace are official, production-ready images provided and supported by industry vendors like SAP and Cisco.

Community AMIs   More than 100,000 images are available as community AMIs. Many of these images are AMIs created and maintained by independent vendors and are usually built to meet a specific need. This is a good catalog to search if you're planning an application built on a custom combination of software resources.

Private AMIs   You can also store images created from your own instance deployments as private AMIs. Why would you want to do that? You might, for instance, want the ability to scale up the number of instances you've got running to meet growing demand. Having a reliable, tested, and patched instance image as an AMI makes incorporating autoscaling easy. You can also share images as AMIs or import VMs from your local infrastructure (by way of AWS S3) using the AWS VM Import/Export tool.

A particular AMI will be available in only a single region—although there will often be images with identical functionality in all regions. Keep this in mind as you plan your deployments: invoking the ID of an AMI in one region while working from within a different region will fail.

An Important Note About Billing

Besides the normal charges for running an EC2 instance, your AWS account might also be billed hourly amounts or license fees for the use of the AMI software itself. Although vendors make every effort to clearly display the charges for their AMIs, it's your responsibility to accept and honor those charges.

Instance Types

AWS allocates hardware resources to your instances according to the instance type—or hardware profile—you select. The particular workload you're planning for your instance will determine the type you choose. The idea is to balance cost against your need for compute power, memory, and storage space. Ideally, you'll find a type that offers exactly the amount of each to satisfy both your application and budget.

Should your needs change over time, you can easily move to a different instance type by stopping your instance, editing its instance type, and starting it back up again.

As listed in Table 2.1, there are currently more than 75 instance types organized into five instance families, although AWS frequently updates their selection. You can view the most recent collection at aws.amazon.com/ec2/instance-types.

TABLE 2.1 EC2 instance type family and their top-level designations

Instance type family	Types
General purpose	A1, T3, T3a, T2, M6g, M5, M5a, M5n, M4
Compute optimized	C5, C5n, C4
Memory optimized	R5, R5a, R5n, X1e, X1, High Memory, z1d
Accelerated computing	P3, P2, Inf1, G4, G3, F1
Storage optimized	I3, I3en, D2, H1

General Purpose The General Purpose family includes T3, T2, M5, and M4 types, which all aim to provide a balance of compute, memory, and network resources. T2 types, for instance, range from the t2.nano with one virtual CPU (vCPU0) and half a gigabyte of memory all the way up to the t2.2xlarge with its eight vCPUs and 32 GB of memory. Because it's eligible as part of the Free Tier, the t2.micro is often a good choice for experimenting. But there's nothing stopping you from using it for light-use websites and various development-related services.

> **NOTE** T2s are burstable, which means you can accumulate CPU credits when your instance is underutilized that can be applied during high-demand periods in the form of higher CPU performance.

M5 and M4 instances are recommended for many small and midsized data-centric operations. Unlike T2, which requires EBS virtual volumes for storage, some M* instances come with their own instance storage drives that are actually physically attached to the host server. M5 types range from m5.large (2 vCPUs and 8 GB of memory) to the monstrous m5d.metal (96 vCPUs and 384 GB of memory).

Compute Optimized   For more demanding web servers and high-end machine learning workloads, you'll choose from the Compute Optimized family that includes C5 and C4 types. C5 machines—currently available from the c5.large to the c5d.24xlarge—give you as much as 3.5 GHz of processor speed and strong network bandwidth.

Memory Optimized   Memory Optimized instances work well for intensive database, data analysis, and caching operations. The X1e, X1, and R4 types are available with as much as 3.9 terabytes of dynamic random-access memory (DRAM)-based memory and low-latency solid-state drive (SSD) storage volumes attached.

Accelerated Computing   You can achieve higher-performing general-purpose graphics processing unit (GPGPU) performance from the P3, P2, G3, and F1 types within the Accelerated Computing group. These instances make use of various generations of high-end NVIDIA GPUs or, in the case of the F1 instances, an Xilinx Virtex Ultra-Scale+ field-programmable gate array (FPGA—if you don't know what that is, then you probably don't need it). Accelerated Computing instances are recommended for demanding workloads such as 3D visualizations and rendering, financial analysis, and computational fluid dynamics.

Storage Optimized   The H1, I3, and D2 types currently make up the Storage Optimized family that have large, low-latency instance storage volumes (in the case of I3en, up to 60 TB of slower **hard disk drive** [HDD] storage). These instances work well with distributed filesystems and heavyweight data processing applications.

The specification details and instance type names will frequently change as AWS continues to leverage new technologies to support its customers' growing computing demands. But it's important to be at least familiar with the instance type families and the naming conventions AWS uses to identify them.

Configuring an Environment for Your Instance

Deciding where your EC2 instance will live is as important as choosing a performance configuration. Here, there are three primary details to get right: geographic region, virtual private cloud (VPC), and tenancy model.

AWS Regions

As you learned earlier, AWS servers are housed in data centers around the world and organized by geographical region. You'll generally want to launch an EC2 instance in the region that's physically closest to the majority of your customers or, if you're working with data that's subject to legal restrictions, within a jurisdiction that meets your compliance needs.

EC2 resources can be managed only when you're "located within" their region. You set the active region in the console through the drop-down menu at the top of the page and through default configuration values in the AWS CLI or your SDK. You can update your CLI configuration by running `aws configure`.

Bear in mind that the costs and even functionality of services and features might vary between regions. It's always a good idea to consult the most up-to-date official documentation.

VPCs

Virtual private clouds (VPCs) are easy-to-use AWS network organizers and great tools for organizing your infrastructure. Because it's so easy to isolate the instances in one VPC from whatever else you have running, you might want to create a new VPC for each one of your projects or project stages. For example, you might have one VPC for early application development, another for beta testing, and a third for production (see Figure 2.1).

FIGURE 2.1 A multi-VPC infrastructure for a development environment

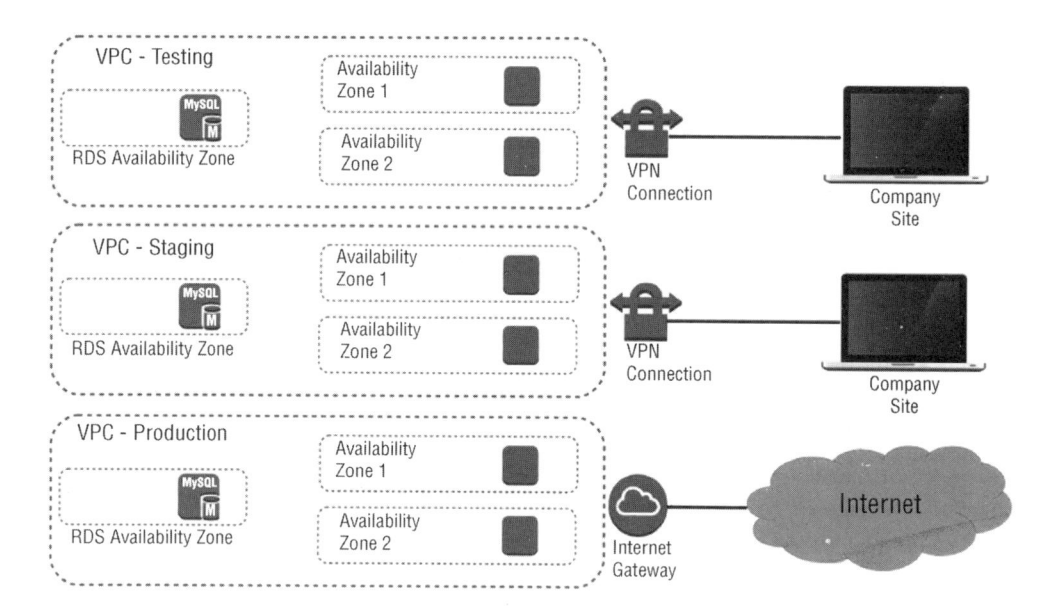

Adding a simple VPC that doesn't incorporate a network address translation (NAT) gateway (docs.aws.amazon.com/AmazonVPC/latest/UserGuide/vpc-nat-gateway.html) or VPN access (docs.aws.amazon.com/vpn/latest/s2svpn/VPC_VPN.html) won't cost you anything. You'll learn much more about all this in Chapter 4, "Amazon Virtual Private Cloud."

Tenancy

When launching an EC2 instance, you'll have the opportunity to choose a tenancy model. The default setting is *shared tenancy*, where your instance will run as a virtual machine on a physical server that's concurrently hosting other instances. Those other instances might well be owned and operated by other AWS customers, although the possibility of any kind of insecure interaction between instances is remote.

To meet special regulatory requirements, your organization's instances might need an extra level of isolation. The Dedicated Instance option ensures that your instance will run on a dedicated physical server. This means that it won't be sharing the server with resources owned by a different customer account. The Dedicated Host option allows you to actually identify and control the physical server you've been assigned to meet more restrictive licensing or regulatory requirements.

Naturally, dedicated instances and dedicated hosts will cost you more than instances using shared tenancy.

Exercise 2.1 will guide you through the launch of a simple EC2 Linux instance.

EXERCISE 2.1

Launch an EC2 Linux Instance and Log in Using SSH

1. From the EC2 Dashboard, click to launch a new instance and select a Linux AMI and instance type. Remember, the t2.micro is Free Tier–eligible if your AWS account is still within its first year.

2. Explore the Configure Instance Details, Add Storage, and Add Tags pages—although the default settings should work fine.

3. On the Configure Security Group page, make sure there's a rule permitting incoming SSH (port 22) traffic. It should be there by default.

4. Before letting you launch the instance, AWS will require you to select—or create—a key pair. Follow the instructions.

5. Once the instance is launched, you can return to the Instances Dashboard to wait a minute or two until everything is running properly.

6. Click the Actions pull-down and then the Connect item for instructions on how to connect to the instance from your local machine. Then connect and take a look at your virtual cloud server.

In Exercise 2.2, you'll see how changing an instance's type works.

EXERCISE 2.2

Assess the Free Capacity of a Running Instance and Change Its Instance Type

1. With an instance running, open the Instances Dashboard in the EC2 console. Select the instance you're interested in and click the Monitoring tab in the bottom half of the screen. That's a good place to see what percentage of compute and network resources you've been using over the past hours or weeks.

   Now pretend that your instance is nearly maxed out and change the instance type as follows.

EXERCISE 2.2 *(continued)*

2. Stop the instance. (Remember, your public IP address might be different when you start up again.)
3. From the Actions drop-down, click Instance Settings and then Change Instance Type. Select a new type.
4. Restart the instance and confirm that it's running properly.

Configuring Instance Behavior

You can optionally tell EC2 to execute commands on your instance as it boots by pointing to user data in your instance configuration (this is sometimes known as *bootstrapping*). Whether you specify the data during the console configuration process or by using the --user-data value with the AWS CLI, you can have script files bring your instance to any desired state.

User data can consist of a few simple commands to install a web server and populate its web root, or it can be a sophisticated script setting up the instance as a working node within a Puppet Enterprise–driven platform.

Placement Groups

By default AWS will attempt to spread your instances across their infrastructure to create a profile that will be optimal for most use cases. But the specific demands of your operation might require a different setup. EC2 placement groups give you the power to define nonstandard profiles to better meet your needs. There are, at this time, three kinds of placement groups:

- *Cluster* groups launch each associated instance into a single availability zone within close physical proximity to each other. This provides low-latency network interconnectivity and can be useful for high-performance computing (HPC) applications, for instance.
- *Spread* groups separate instances physically across distinct hardware racks and even availability zones to reduce the risk of failure-related data or service loss. Such a setup can be valuable when you're running hosts that can't tolerate multiple concurrent failures. If you're familiar with VMware's Distributed Resource Scheduler (DRS), this is similar to that.
- *Partition* groups let you associate some instances with each other, placing them in a single "partition." But the instances within that single partition can be kept physically separated from instances within other partitions. This differs from spread groups where no two instances will ever share physical hardware.

Instance Pricing

You can purchase the use of EC2 instances through one of three models.

For always-on deployments that you expect to run for less than 12 months, you'll normally pay for each hour your instance is running through the *on-demand model*. On-demand is the most flexible way to consume EC2 resources since you're able to closely control how much you pay by stopping and starting your instances according to your need. But, per hour, it's also the most expensive.

If you're planning to keep the lights burning 24/7 for more than a year, then you'll enjoy a significant discount by purchasing a *reserve instance*—generally over a term commitment of between one and three years. You can pay up front for the entire term of a reserve instance or, for incrementally higher rates, either partially up front and the rest in monthly charges or entirely through monthly charges. Table 2.2 gives you a sense of how costs can change between models. These estimates assume a Linux platform, all up-front payments, and default tenancy. Actual costs may vary over time and between regions.

TABLE 2.2 Pricing estimates comparing on-demand with reserve costs

Instance type	Pricing model	Cost/hour	Cost/year
t2.micro	On-demand	$0.0116	$102.00
t2.micro	Reserve (three-year term)		$38.33
g3.4xlarge	On-demand	$1.14	$9986.40
g3.4xlarge	Reserve (three-year term)		$4429.66

For workloads that can withstand unexpected disruption (like computation-intensive genome research applications), purchasing instances on Amazon's spot market can save you a lot of money. The idea is that you enter a maximum dollar-value bid for an instance type running in a particular region. The next time an instance in that region becomes available at a per-hour rate that's equal to or below your bid, it'll be launched using the AMI and launch template you specified. Once up, the instance will keep running either until you stop it—when your workload completes, for example—or until the instance's per-hour rate rises above your maximum bid. You'll learn more about the spot market and reserve instances in Chapter 13, "The Cost Optimization Pillar."

It will often make sense to combine multiple models within a single application infrastructure. An online store might, for instance, purchase one or two reserve instances to cover its normal customer demand but also allow autoscaling to automatically launch on-demand instances during periods of unusually high demand.

Use Exercise 2.3 to dive deeper into EC2 pricing.

> **EXERCISE 2.3**
>
> **Assess Which Pricing Model Will Best Meet the Needs of a Deployment**
>
> Imagine that your application will need to run two always-on f1.2xlarge instances (which come with instance storage and won't require any EBS volumes). To meet seasonal demand, you can expect to require as many as four more instances for a total of 100 hours through the course of a single year. How should you pay for this deployment?
>
> Bonus: Calculate your total estimated monthly and annual costs.

Instance Lifecycle

The state of a running EC2 instance can be managed in a number of ways. Terminating the instance will shut it down and cause its resources to be reallocated to the general AWS pool.

> **WARNING**
> Terminating an instance will, in most cases, destroy all data kept on the primary storage. The exception to this would be an Elastic Block Store (EBS) volume that has been set to persist after its instance is terminated.

If your instance won't be needed for some time but you don't want to terminate it, you can save money by simply stopping it and then restarting it when it's needed again. The data on an EBS volume will in this case not be lost, although that would not be true for an instance volume.

Later in this chapter, you'll learn about both EBS and instance store volumes and the ways they work with EC2 instances.

You should be aware that a stopped instance that had been using a nonpersistent public IP address will most likely be assigned a different address when it's restarted. If you need a predictable IP address that can survive restarts, allocate an elastic IP address and associate it with your instance.

You can edit or change an instance's security group (which we'll discuss a bit later in this chapter) to update access policies at any time—even while an instance is running. You can also change its instance type to increase or decrease its compute, memory, and storage capacity (just try doing *that* on a physical server). You will need to stop the instance, change the type, and then restart it.

Resource Tags

The more resources you deploy on your AWS account, the harder it can be to properly keep track of things. Having constantly changing numbers of EC2 instances—along with accompanying storage volumes, security groups, and elastic IP addresses—all spread across two or three VPCs can get complicated.

The best way to keep a lid on the chaos is to find a way to quickly identify each resource you've got running by its purpose and its relationships to other resources. The best way to do that is by establishing a consistent naming convention and applying it to tags.

AWS resource tags can be used to label everything you'll ever touch across your AWS account—they're certainly not restricted to just EC2. Tags have a key and, optionally, an associated value. So, for example, you could assign a tag with the key production-server to each element of a production deployment. Server instances could, in addition, have a value of server1, server2, and so on. A related security group could have the same production-server key but security-group1 for its value. Table 2.3 illustrates how that convention might play out over a larger deployment group.

TABLE 2.3 A sample key/value tagging convention

Key	Value
production-server	server1
production-server	server2
production-server	security-group1
staging-server	server1
staging-server	server2
staging-server	security-group1
test-server	server1
test-server	security-group1

Applied properly, tags can improve the visibility of your resources, making it much easier to manage them effectively, audit and control costs and billing trends, and avoid costly errors.

Service Limits

By default, each AWS account has limits to the number of instances of a particular service you're able to launch. Sometimes those limits apply to a single region within an account, and others are global. As examples, you're allowed only five VPCs per region and 5,000 Secure Shell (SSH) key pairs across your account. If necessary, you can ask AWS to raise your ceiling for a particular service.

You can find up-to-date details regarding the limits of all AWS services at docs.aws.amazon.com/general/latest/gr/aws_service_limits.html.

EC2 Storage Volumes

Storage drives (or *volumes* as they're described in AWS documentation) are for the most part virtualized spaces carved out of larger physical drives. To the OS running on your instance, though, all AWS volumes will present themselves exactly as though they were normal physical drives. But there's actually more than one kind of AWS volume, and it's important to understand how each type works.

Elastic Block Store Volumes

You can attach as many Elastic Block Store (EBS) volumes to your instance as you like (although one volume can be attached to no more than a single instance at a time) and use them just as you would hard drives, flash drives, or USB drives with your physical server. And as with physical drives, the type of EBS volume you choose will have an impact on both performance and cost.

The AWS SLA guarantees the reliability of the data you store on its EBS volumes (promising at least 99.99 percent availability), so you don't have to worry about failure. When an EBS drive does fail, its data has already been duplicated and will probably be brought back online before anyone notices a problem. So, practically, the only thing that should concern you is how quickly and efficiently you can access your data.

There are currently four EBS volume types, two using SSD technologies and two using the older spinning hard drives. The performance of each volume type is measured in maximum IOPS/volume (where IOPS means input/output operations per second).

EBS-Provisioned IOPS SSD

If your applications will require intense rates of I/O operations, then you should consider provisioned IOPS, which provides a maximum IOPS/volume of 64,000 and a maximum throughput/volume of 1,000 MB/s. Provisioned IOPS—which in some contexts is referred to as EBS Optimized—can cost $0.125/GB/month in addition to $0.065/provisioned IOPS.

EBS General-Purpose SSD

For most regular server workloads that, ideally, deliver low-latency performance, general-purpose SSDs will work well. You'll get a maximum of 16,000 IOPS/volume, and it will cost you $0.10/GB/month. For reference, a general-purpose SSD used as a typical 8 GB boot drive for a Linux instance would, at current rates, cost you $9.60/year.

Throughput-Optimized HDD

Throughput-optimized HDD volumes can provide reduced costs with acceptable performance where you're looking for throughput-intensive workloads, including log processing and big data operations. These volumes can deliver only 500 IOPS/volume but with a 500 MB/s maximum throughput/volume, and they'll cost you only $0.045/GB/month.

Cold HDD

When you're working with larger volumes of data that require only infrequent access, a 250 IOPS/volume type might meet your needs for only $0.025/GB/month.

Table 2.4 lets you compare the basic specifications and estimated costs of those types.

TABLE 2.4 Sample costs for each of the four EBS storage volume types

	EBS-provisioned IOPS SSD	EBS general-purpose SSD	Throughput-optimized HDD	Cold HDD
Volume size	4 GB–16 TB	1 GB–16 TB	500 GB–16 TB	500 GB–16 TB
Max IOPS/volume	64,000	16,000	500	250
Max throughput/volume (MB/s)	1,000	250	500	250
Price (/month)	$0.125/GB + $0.065/prov IOPS	$0.10/GB	$0.045/GB	$0.025/GB

EBS Volume Features

All EBS volumes can be copied by creating a snapshot. Existing snapshots can be used to generate other volumes that can be shared and/or attached to other instances or converted to images from which AMIs can be made. You can also generate an AMI image directly from a running instance-attached EBS volume—although, to be sure no data is lost, you should shut down the instance first.

EBS volumes can be encrypted to protect their data while at rest or as it's sent back and forth to the EC2 host instance. EBS can manage the encryption keys automatically behind the scenes or use keys that you provide through the AWS Key Management Service (KMS).

Exercise 2.4 will walk you through launching a new instance based on an existing snapshot image.

EXERCISE 2.4

Create and Launch an AMI Based on an Existing Instance Storage Volume

1. If necessary, launch an instance and make at least some token change to the root volume. This could be something as simple as typing **touch test.txt** on a Linux instance to create an empty file.
2. Create an image from the instance's volume (you'll access the dialog through the Actions pull-down menu in the Instance's Dashboard).
3. Launch an instance from the console and select the new AMI from the My AMIs tab.
4. Log into the instance and confirm that your previous change has persisted.

Instance Store Volumes

Unlike EBS volumes, instance store volumes are ephemeral. This means that when the instances they're attached to are shut down, their data is permanently lost. So, why would you want to keep your data on an instance store volume more than on EBS?

- Instance store volumes are SSDs that are physically attached to the server hosting your instance and are connected via a fast NVMe (Non-Volatile Memory Express) interface.
- The use of instance store volumes is included in the price of the instance itself.
- Instance store volumes work especially well for deployment models where instances are launched to fill short-term roles (as part of autoscaling groups, for instance), import data from external sources, and are, effectively, disposable.

Whether one or more instance store volumes are available for your instance will depend on the instance type you choose. This is an important consideration to take into account when planning your deployment.

Even with all the benefits of EBS and instance storage, it's worth noting that there will be cases where you're much better off keeping large data sets outside of EC2 altogether. For many use cases, Amazon's S3 service can be a dramatically less expensive way to store files or even databases that are nevertheless instantly available for compute operations.

You'll learn more about this in Chapter 3, "AWS Storage."

The bottom line is that EBS volumes are likely to be the right choice for instances whose data needs to persist beyond a reboot and for working with custom or off-the-shelf AMIs. Instance store volumes are, where available, useful for operations requiring low-latency access to large amounts of data that needn't survive a system failure or reboot. And non-EC2 storage can work well when you don't need fantastic read/write speeds, but you wish to enjoy the flexibility and cost savings of S3.

Accessing Your EC2 Instance

Like all networked devices, EC2 instances are identified by unique IP addresses. All instances are assigned at least one private IPv4 address that, by default, will fall within one of the blocks shown in Table 2.5.

TABLE 2.5 The three IP address ranges used by private networks

From	To
10.0.0.0	10.255.255.255
172.16.0.0	172.31.255.255
192.168.0.0	192.168.255.255

Out of the box, you'll only be able to connect to your instance from within its subnet, and the instance will have no direct connection to the Internet.

If your instance configuration calls for multiple network interfaces (to connect to otherwise unreachable resources), you can create and then attach one or more virtual elastic network interfaces to your instance. Each of these interfaces must be connected to an existing subnet and security group. You can optionally assign a static IP address within the subnet range.

Of course, an instance can also be assigned a public IP through which full Internet access is possible. As noted in the instance lifecycle discussion, the default public IP assigned to your instance is ephemeral and probably won't survive a reboot. Therefore, you'll usually want to allocate a permanent elastic IP for long-term deployments. As long as it's attached to a running instance, there's no charge for elastic IPs.

I'll talk about accessing an instance as an administrator a bit later within the context of security. But there's a lot you can learn about a running EC2 instance—including the IP addresses you'll need to connect—through the instance metadata system. Running the following curl command from the command line while logged into the instance will return a list of the kinds of data that are available:

```
$ curl http://169.254.169.254/latest/meta-data/
ami-id
ami-launch-index
ami-manifest-path
block-device-mapping/
```

```
hostname
instance-action
instance-id
instance-type
local-hostname
local-ipv4
mac
metrics/
network/
placement/
profile
public-hostname
public-ipv4
public-keys/
reservation-id
security-groups
```

> **NOTE** You'll use the 169.254.169.254 IP for the command no matter what your instance public or private IPs happen to be.

Entries ending with a trailing slash (/) contain further sublevels of information that can also be displayed by `curl`. Adding a data type to that `curl` command will then return the information you're after. This example displays the name of the security groups used by the instance:

```
$ curl http://169.254.169.254/latest/meta-data/security-groups launch-wizard-1
```

Securing Your EC2 Instance

You are responsible for configuring appropriate and effective access controls to protect your EC2 instances from unauthorized use. Broadly speaking, AWS provides four tools to help you with this task: security groups, Identity and Access Management (IAM) roles, network address translation (NAT) instances, and key pairs.

Security Groups

An EC2 security group plays the role of a firewall. By default, a security group will deny all incoming traffic while permitting all outgoing traffic. You define group behavior by setting policy rules that will either block or allow specified traffic types. From that point on, any data packet coming into or leaving the perimeter will be measured against those rules and processed accordingly.

Traffic is assessed by examining its source and destination, the network port it's targeting, and the protocol it's set to use. A TCP packet sent to the SSH port 22 could, for example, only be allowed access to a particular instance if its source IP address matches the local public IP used by computers in your office. This lets you open up SSH access on your instance without having to worry about anyone from outside your company getting in.

Using security groups, you can easily create sophisticated rule sets to finely manage access to your services. You could, say, open up a website to the whole world while blocking access to your backend servers for everyone besides members of your team.

If necessary, you can update your security group rules and/or apply them to multiple instances.

> Security groups control traffic at the instance level. However, AWS also provides you with network access control lists (NACLs) that are associated with entire subnets rather than individual instances. Chapter 4, "Amazon Virtual Private Cloud," discusses both security groups and NACLs.

IAM Roles

You can also control access to AWS resources—including EC2 instances—through the use of IAM roles. You define an IAM role by giving it permissions to perform actions on specified services or resources within your AWS account. When a particular role is assigned to a user or resource, they'll gain access to whichever resources were included in the role policies.

Using roles, you can give a limited number of entities (other resources or users) exclusive access to resources like your EC2 instances. But you can also assign an IAM role *to* an EC2 instance so that processes running within it can access the external tools—like an RDS database instance—it needs to do its work.

You'll learn more about IAM in Chapter 6, "Authentication and Authorization—AWS Identity and Access Management."

NAT Devices

Sometimes you'll need to configure an EC2 instance without a public IP address to limit its exposure to the network. Naturally, that means it won't have any Internet connectivity. But that can present a problem because you'll probably still need to give it Internet access so that it can receive security patches and software updates.

One solution is to use network address translation (NAT) to give your private instance access *to* the Internet without allowing access to it *from* the Internet. AWS gives you two ways to do that: a NAT instance and a NAT gateway (see Figure 2.2). They'll both do the job, but since a NAT gateway is a managed service, it doesn't require that you manually launch and maintain an instance. Both approaches will incur monthly charges.

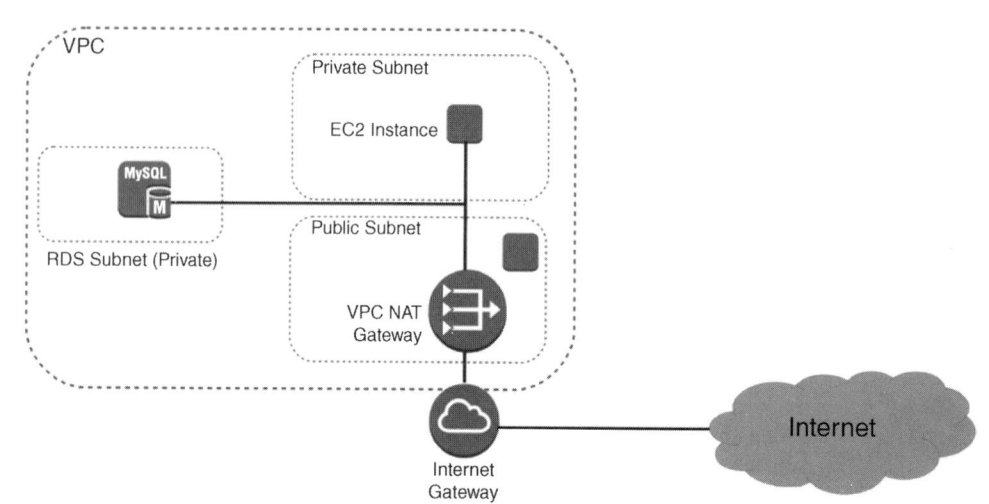

FIGURE 2.2 A NAT gateway providing network access to resources in private subnets

NAT will be discussed at greater length in Chapter 4.

Key Pairs

As any professional administrator will know, remote login sessions on your running instances should never be initiated over unencrypted plain-text connections. To ensure properly secured sessions, you'll need to generate a key pair, save the public key to your EC2 server, and save its private half to your local machine. If you're working with a Windows AMI, you'll use the private key file to retrieve the password you'll need to authenticate into your instance. For a Linux AMI, the private key will allow you to open an SSH session.

Each key pair that AWS generates for you will remain installed within its original region and available for use with newly launched instances until you delete it. You *should* delete the AWS copy in the event your public key is lost or exposed. Just be careful before you mess with your keys—your access to an instance might depend on it.

EC2 Auto Scaling

The *EC2 Auto Scaling* service offers a way to both avoid application failure and recover from it when it happens. Auto Scaling works by provisioning and starting on your behalf a specified number of EC2 instances. It can dynamically add more instances to keep up with increased demand. And when an instance fails or gets terminated, Auto Scaling will automatically replace it.

EC2 Auto Scaling uses either a *launch configuration* or a *launch template* to automatically configure the instances that it launches. Both perform the same basic function of defining the basic configuration parameters of the instance as well as what scripts (if any) run on it at launch time. Launch configurations have been around longer and are more familiar to you if you've been using AWS for a while. You're also more likely to encounter them if you're going into an existing AWS environment. Launch templates are newer and are what AWS now recommends. You'll learn about both, but which you use is up to you.

Launch Configurations

When you create an instance manually, you have to specify many configuration parameters, including an AMI, instance type, SSH key pair, security group, instance profile, block device mapping, whether it's EBS optimized, placement tenancy, and user data, such as custom scripts to install and configure your application. A launch configuration is essentially a named document that contains the same information you'd provide when manually provisioning an instance.

You can create a launch configuration from an existing EC2 instance. Auto Scaling will copy the settings from the instance for you, but you can customize them as needed. You can also create a launch configuration from scratch.

Launch configurations are for use only with EC2 Auto Scaling, meaning you can't manually launch an instance using a launch configuration. Also, once you create a launch configuration, you can't modify it. If you want to change any of the settings, you have to create an entirely new launch configuration.

Launch Templates

Launch templates are similar to launch configurations in that you can specify the same settings. But the uses for launch templates are more versatile. You can use a launch template with Auto Scaling, of course, but you can also use it for spinning up one-off EC2 instances or even creating a spot fleet.

Launch templates are also versioned, allowing you to change them after creation. Any time you need to make changes to a launch template, you create a new version of it. AWS keeps all versions, and you can then flip back and forth between versions as needed. This makes it easier to track your launch template changes over time. Complete Exercise 2.5 to create your own launch template.

> If you have an existing launch configuration, you can copy it to a launch template using the AWS web console. There's no need to create launch templates from scratch!

EXERCISE 2.5

Create a Launch Template

In this exercise, you'll create a launch template that installs and configures a simple web server. You'll then use the launch template to manually create an instance.

1. In the EC2 Dashboard, click Launch Templates.
2. Click the Create Launch Template button.
3. Give the launch template a name such as **MyTemplate**.
4. Click the Search For AMI link to locate one of the Ubuntu Server LTS AMIs (make sure the AMI you choose uses the 64-bit x86 architecture and not 64-bit ARM).
5. For Instance Type, select t2.micro.
6. Under Security Groups, select a security group that allows inbound HTTP access. Create a new security group if necessary.
7. Expand the Advanced Details section and enter the following in the User Data field:

   ```
   #!/bin/bash
   apt-get update
   apt-get install -y apache2
   echo "Welcome to my website" > index.html
   cp index.html /var/www/html
   ```
8. Click the Create Launch Template button.
9. Click the Launch Instance From This Template link.
10. Under Source Template Version, select 1 (Default).
11. Click the Launch Instance From Template button.
12. After the instance boots, browse to its public IP address. You should see a web page that says "Welcome to my website."
13. Terminate the instance when you're done with it.

Auto Scaling Groups

An *Auto Scaling group* is a group of EC2 instances that Auto Scaling manages. When creating an Auto Scaling group, you must first specify either the launch configuration or launch template you created. When you create an Auto Scaling group, you must specify how many running instances you want Auto Scaling to provision and maintain using the launch configuration or template you created. You must specify the minimum and maximum size of the Auto Scaling group. You may also optionally set the desired number of instances you want Auto Scaling to provision and maintain.

Minimum   Auto Scaling will ensure the number of healthy instances never goes below the minimum. If you set this to 0, Auto Scaling will not spawn any instances and will terminate any running instances in the group.

Maximum   Auto Scaling will make sure the number of healthy instances never exceeds this amount. This might seem strange but remember that you might have budget limitations and need to be protected from unexpected (and unaffordable) usage demands.

Desired Capacity   The desired capacity is an optional setting that must lie within the minimum and maximum values. If you don't specify a desired capacity, Auto Scaling will launch the number of instances as the minimum value. If you specify a desired capacity, Auto Scaling will add or terminate instances to stay at the desired capacity. For example, if you set the minimum to 1, the maximum to 10, and the desired capacity to 4, then Auto Scaling will create four instances. If one of those instances gets terminated—for example, because of human action or a host crash—Auto Scaling will replace it to maintain the desired capacity setting of 4. In the web console, desired capacity is also called the *group size*.

Specifying an Application Load Balancer Target Group

If you want to use an application load balancer (ALB) to distribute traffic to instances in your Auto Scaling group, just plug in the name of the ALB target group when creating the Auto Scaling group. Whenever Auto Scaling creates a new instance, it will automatically add it to the ALB target group.

Health Checks Against Application Instances

When you create an Auto Scaling group, Auto Scaling will strive to maintain the minimum number of instances, or the desired number if you've specified it. If an instance becomes unhealthy, Auto Scaling will terminate and replace it.

By default, Auto Scaling determines an instance's health based on EC2 health checks. Chapter 7, "CloudTrail, CloudWatch, and AWS Config," covers how EC2 automatically performs system and instance status checks. These checks monitor for instance problems such as memory exhaustion, filesystem corruption, or an incorrect network or startup configuration, as well as for system problems that require AWS involvement to repair. Although these checks can catch a variety of instance and host-related problems, they won't necessarily catch application-specific problems.

If you're using an application load balancer to route traffic to your instances, you can configure health checks for the load balancer's target group. Target group health checks can check for HTTP response codes from 200 to 499. You can then configure your Auto Scaling group to use the results of these health checks to determine whether an instance is healthy.

If an instance fails the ALB health check, it will route traffic away from the failed instance, ensuring that users don't reach it. At the same time, Auto Scaling will remove the instance, create a replacement, and add the new instance to the load balancer's target group. The load balancer will then route traffic to the new instance.

> **NOTE** A good design practice is to have a few recovery actions that work for a variety of circumstances. An instance may crash due to an out-of-memory condition, a bug, a deleted file, or an isolated network failure, but simply terminating and replacing the instance using Auto Scaling resolves all these cases. There's no need to come up with a separate recovery action for each cause when simply re-creating the instance solves them all.

Auto Scaling Options

Once you create an Auto Scaling group, you can leave it be and it will continue to maintain the minimum or desired number of instances indefinitely. However, maintaining the current number of instances is just one option. Auto Scaling provides several other options to scale out the number of instances to meet demand.

Manual Scaling

If you change the minimum, desired, or maximum values at any time after creating the group, Auto Scaling will immediately adjust. For example, if you have the desired capacity value set to 2 and change it to 4, Auto Scaling will launch two more instances. If you have four instances and set the desired capacity value to 2, Auto Scaling will terminate two instances. Think of the desired capacity as a thermostat.

Dynamic Scaling Policies

Most AWS-managed resources are elastic—that is, they automatically scale to accommodate increased load. Some examples include S3, load balancers, Internet gateways, and NAT gateways. Regardless of how much traffic you throw at them, AWS is responsible for ensuring that they remain available while continuing to perform well. But when it comes to your EC2 instances, you're responsible for ensuring that they're powerful and plentiful enough to meet demand.

Running out of instance resources—be it CPU utilization, memory, or disk space—will almost always result in the failure of whatever you're running on it. To ensure that your instances never become overburdened, dynamic scaling policies automatically provision more instances *before* they hit that point. Auto Scaling generates the following aggregate metrics for all instances within the group:

- Aggregate CPU utilization
- Average request count per target

- Average network bytes in
- Average network bytes out

You're not limited to using just these native metrics. You can also use metric filters to extract metrics from CloudWatch logs and use those. As an example, your application may generate logs that indicate how long it takes to complete a process. If the process takes too long, you could have Auto Scaling spin up new instances.

Dynamic scaling policies work by monitoring a CloudWatch alarm and scaling out—by increasing the desired capacity—when the alarm is breaching. You can choose from three dynamic scaling policies: simple, step, and target tracking.

Simple Scaling Policies

With a *simple scaling policy*, whenever the metric rises above the threshold, Auto Scaling simply increases the desired capacity. How much it increases the desired capacity, however, depends on which of the following *adjustment types* you choose:

ChangeInCapacity   Increases the capacity by a specified amount. For instance, you could start with a desired capacity value of 4 and then have Auto Scaling increase the value by 2 when the load increases.

ExactCapacity   Sets the capacity to a specific value, regardless of the current value. For example, suppose the desired capacity value is 4. You create a policy to change the value to 6 when the load increases.

PercentChangeInCapacity   Increases the capacity by a percentage of the current amount. If the current desired capacity value is 4 and you specify the percent change in capacity as 50 percent, then Auto Scaling will bump the desired capacity value to 6.

For example, suppose you have four instances and create a simple scaling policy that specifies a PercentChangeInCapacity adjustment of 50 percent. When the monitored alarm triggers, Auto Scaling will increment the desired capacity by 2, which will in turn add two instances to the Auto Scaling group, for a total of six.

After Auto Scaling completes the adjustment, it waits a *cooldown period* before executing the policy again, even if the alarm is still breaching. The default cooldown period is 300 seconds, but you can set it as high as you want or as low as 0—effectively disabling it. Note that if an instance is unhealthy, Auto Scaling will not wait for the cooldown period before replacing the unhealthy instance.

Referring to the preceding example, suppose that after the scaling adjustment completes and the cooldown period expires, the monitored alarm drops below the threshold. At this point, the desired capacity value is 6. If the alarm triggers again, the simple scaling action will execute again and add three more instances. Keep in mind that Auto Scaling will never increase the desired capacity beyond the group's maximum setting.

Step Scaling Policies

If the demand on your application is rapidly increasing, a simple scaling policy may not add enough instances quickly enough. Using a *step scaling policy*, you can instead add instances based on how much the aggregate metric exceeds the threshold.

To illustrate, suppose your group starts out with four instances. You want to add more instances to the group as the average CPU utilization of the group increases. When the utilization hits 50 percent, you want to add two more instances. When it goes above 60 percent, you want to add four more instances.

You'd first create a CloudWatch Alarm to monitor the average CPU utilization and set the alarm threshold to 50 percent, since this is the utilization level at which you want to start increasing the desired capacity.

You must then specify at least one step adjustment. Each step adjustment consists of the following:

- A lower bound
- An upper bound
- The adjustment type
- The amount by which to increase the desired capacity

The upper and lower bounds define a range that the metric has to fall within for the step adjustment to execute. Suppose that for the first step you set a lower bound of 50 and an upper bound of 60, with a ChangeInCapacity adjustment of 2. When the alarm triggers, Auto Scaling will consider the metric value of the group's average CPU utilization. Suppose it's 55 percent. Because 55 is between 50 and 60, Auto Scaling will execute the action specified in this step, which is to add two instances to the desired capacity.

Suppose now that you create another step with a lower bound of 60 and an upper bound of infinity. You also set a ChangeInCapacity adjustment of 4. If the average CPU utilization increases to 62 percent, Auto Scaling will note that 60 <= 62 < infinity and will execute the action for this step, adding four instances to the desired capacity.

You might be wondering what would happen if the utilization were 60 percent. Step ranges can't overlap. A metric of 60 percent would fall within the lower bound of the second step.

With a step scaling policy, you can optionally specify a *warm-up time*, which is how long Auto Scaling will wait until considering the metrics of newly added instances. The default warm-up time is 300 seconds. Note that there are no cooldown periods in step scaling policies.

Target Tracking Policies

If step scaling policies are too involved for your taste, you can instead use *target tracking policies*. All you do is select a metric and target value, and Auto Scaling will create a CloudWatch Alarm and a scaling policy to adjust the number of instances to keep the metric near that target.

The metric you choose must change proportionally to the instance load. Metrics like this include average CPU utilization for the group and request count per target. Aggregate metrics like the total request count for the ALB don't change proportionally to the load on an individual instance and aren't appropriate for use in a target tracking policy.

In addition to scaling out, target tracking will scale in by deleting instances to maintain the target metric value. If you don't want this behavior, you can disable scaling in. Also, just as with a step scaling policy, you can optionally specify a warm-up time.

Scheduled Actions

Scheduled actions are useful if you have a predictable load pattern and want to adjust your capacity proactively, ensuring you have enough instances *before* demand hits.

When you create a scheduled action, you must specify the following:

- A minimum, maximum, or desired capacity value
- A start date and time

You may optionally set the policy to recur at regular intervals, which is useful if you have a repeating load pattern. You can also set an end time, after which the scheduled policy gets deleted.

To illustrate how you might use a scheduled action, suppose you normally run only two instances in your Auto Scaling group during the week. But on Friday, things get busy, and you know you'll need four instances to keep up. You'd start by creating a scheduled action that sets the desired capacity to 2 and recurs every Saturday, as shown in Figure 2.3.

FIGURE 2.3 Scheduled action setting the desired capacity to 2 every Saturday

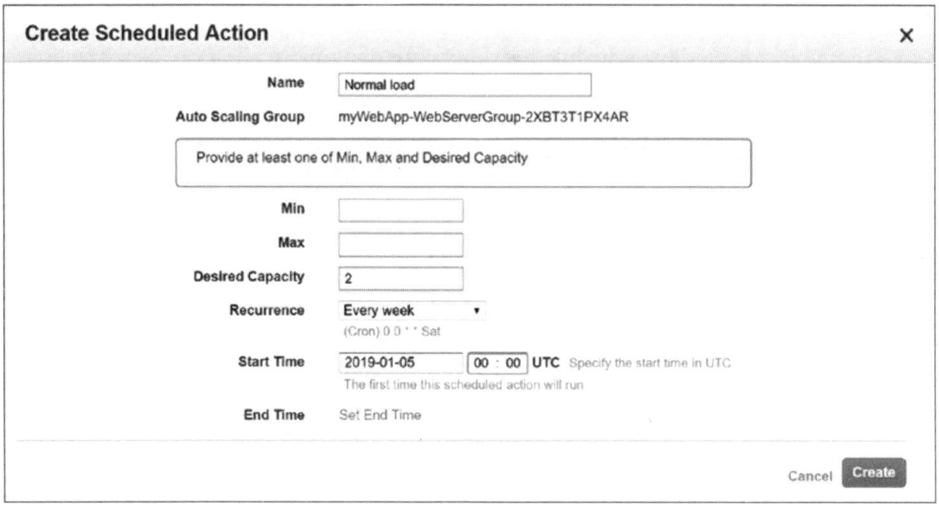

The start date is January 5, 2019, which is a Saturday. To handle the expected Friday spike, you'd create another weekly recurring policy to set the desired capacity to 4, as shown in Figure 2.4.

FIGURE 2.4 Scheduled action setting the desired capacity to 4 every Friday

Create Scheduled Action	✕
Name	Friday spike
Auto Scaling Group	myWebApp-WebServerGroup-2XBT3T1PX4AR
	Provide at least one of Min, Max and Desired Capacity
Min	
Max	
Desired Capacity	4
Recurrence	Every week ▼ (Cron) 0 10 * * Fri
Start Time	2019-01-04 10 : 00 UTC Specify the start time in UTC The first time this scheduled action will run
End Time	Set End Time
	Cancel **Create**

This action will run every Friday, setting the desired capacity to 4, prior to the anticipated increased load.

Note that you can combine scheduled actions with dynamic scaling policies. For example, if you're running an e-commerce site, you may use a scheduled action to increase the maximum group size during busy shopping seasons and then rely on dynamic scaling policies to increase the desired capacity as needed.

AWS Systems Manager

AWS Systems Manager, formerly known as EC2 Systems Manager and Simple Systems Manager (SSM), lets you automatically or manually perform *actions* against your AWS resources and on-premises servers.

From an operational perspective, Systems Manager can handle many of the maintenance tasks that often require manual intervention or writing scripts. For on-premises and EC2 instances, these tasks include upgrading installed packages, taking an inventory of installed software, and installing a new application. For your other AWS resources, such tasks may include creating an AMI golden image from an EBS snapshot, attaching IAM instance profiles, or disabling public read access to S3 buckets.

Systems Manager provides the following two capabilities:

- Actions
- Insights

Actions

Actions let you automatically or manually perform actions against your AWS resources, either individually or in bulk. These actions must be defined in *documents*, which are divided into three types:

- Automation—actions you can run against your AWS resources
- Command—actions you run against your Linux or Windows instances
- Policy—defined processes for collecting inventory data from managed instances

Automation

Automation enables you to perform actions against your AWS resources in bulk. For example, you can restart multiple EC2 instances, update CloudFormation stacks, and patch AMIs.

Automation provides granular control over how it carries out its individual actions. It can perform the entire automation task in one fell swoop, or it can perform one step at a time, enabling you to control precisely what happens and when. Automation also offers rate control so that you can specify as a number or a percentage how many resources to target at once.

Run Command

While automation lets you automate tasks against your AWS resources, Run commands let you execute tasks on your managed instances that would otherwise require logging in or using a third-party tool to execute a custom script.

Systems Manager accomplishes this via an agent installed on your EC2 and on-premises *managed instances*. The Systems Manager agent is installed by default on more recent Windows Server, Amazon Linux, and Ubuntu Server AMIs. You can manually install the agent on other AMIs and on-premises servers.

By default, Systems Manager doesn't have permissions to do anything on your instances. You first need to apply an instance profile role that contains the permissions in the AmazonEC2RoleforSSM policy.

AWS offers a variety of preconfigured command documents for Linux and Windows instances; for example, the AWS-InstallApplication document installs software on Windows, and the AWS-RunShellScript document allows you to execute arbitrary shell scripts against Linux instances. Other documents include tasks such as restarting a Windows service or installing the CodeDeploy agent.

You can target instances by tag or select them individually. As with automation, you may use rate limiting to control how many instances you target at once.

Session Manager

Session Manager lets you achieve interactive Bash and PowerShell access to your Linux and Windows instances, respectively, without having to open up inbound ports on a security group or network ACL or even having your instances in a public subnet. You don't need to set up a protective bastion host or worry about SSH keys. All Linux versions and Windows Server 2008 R2 through 2016 are supported.

You open a session using the web console or AWS CLI. You must first install the Session Manager plug-in on your local machine to use the AWS CLI to start a session. The Session Manager SDK has libraries for developers to create custom applications that connect to instances. This is useful if you want to integrate an existing configuration management system with your instances without opening ports in a security group or NACL.

Connections made via Session Manager are secured using TLS 1.2. Session Manager can keep a log of all logins in CloudTrail and store a record of commands run within a session in an S3 bucket.

Patch Manager

Patch Manager helps you automate the patching of your Linux and Windows instances. It will work for supporting versions of the following operating systems:

- Windows Server
- Ubuntu Server
- Red Hat Enterprise Linux (RHEL)
- SUSE Linux Enterprise Server (SLES)
- CentOS
- Amazon Linux
- Amazon Linux 2

You can individually choose instances to patch, patch according to tags, or create a *patch group*. A patch group is a collection of instances with the tag key Patch Group. For example, if you wanted to include some instances in the Webservers patch group, you'd assign tags to each instance with the tag key of Patch Group and the tag value of Webservers. Keep in mind that the tag key is case-sensitive.

Patch Manager uses *patch baselines* to define which available patches to install, as well as whether the patches will be installed automatically or require approval.

AWS offers default baselines that differ according to operating system but include patches that are classified as security related, critical, important, or required. The patch baselines for all operating systems except Ubuntu automatically approve these patches after seven days. This is called an *auto-approval delay*.

For more control over which patches get installed, you can create your own custom baselines. Each custom baseline contains one or more approval rules that define the operating system, the classification and severity level of patches to install, and an auto-approval delay.

You can also specify approved patches in a custom baseline configuration. For Windows baselines, you can specify knowledgebase and security bulletin IDs. For Linux baselines, you can specify Common Vulnerabilities and Exposures (CVE) IDs or full package names. If a patch is approved, it will be installed during a maintenance window that you specify. Alternatively, you can forego a maintenance window and patch your instances immediately. Patch Manager executes the AWS-RunPatchBaseline document to perform patching.

State Manager

While Patch Manager can help ensure your instances are all at the same patch level, State Manager is a configuration management tool that ensures your instances have the software you want them to have and are configured in the way you define. More generally, State Manager can automatically run command and policy documents against your instances, either one time only or on a schedule. For example, you may want to install antivirus software on your instances and then take a software inventory.

To use State Manager, you must create an *association* that defines the command document to run, any parameters you want to pass to it, the target instances, and the schedule. Once you create an association, State Manager will immediately execute it against the target instances that are online. Thereafter, it will follow the schedule.

There is currently only one policy document you can use with State Manager: AWS-GatherSoftwareInventory. This document defines what specific metadata to collect from your instances. Despite the name, in addition to collecting software inventory, you can have it collect network configurations, file information, CPU information, and for Windows, registry values.

Insights

Insights aggregate health, compliance, and operational details about your AWS resources into a single area of AWS Systems Manager. Some insights are categorized according to *AWS resource groups*, which are collections of resources in an AWS region. You define a resource group based on one or more tag keys and optionally tag values. For example, you can apply the same tag key to all resources related to a particular application—EC2 instances, S3 buckets, EBS volumes, security groups, and so on. Insight categories are covered next.

Built-in Insights

Built-in insights are monitoring views that Systems Manager makes available to you by default. Built-in insights include the following:

> **AWS Config Compliance**   This insight shows the total number of resources in a resource group that are compliant or noncompliant with AWS Config rules, as well as compliance by resource. It also shows a brief history of configuration changes tracked by AWS Config.

CloudTrail Events   This insight displays each resource in the group, the resource type, and the last event that CloudTrail recorded against the resource.

Personal Health Dashboard   The Personal Health Dashboard contains alerts when AWS experiences an issue that may impact your resources. For example, some service APIs occasionally experience increased latency. It also shows you the number of events that AWS resolved within the last 24 hours.

Trusted Advisor Recommendations   The AWS Trusted Advisor tool can check your AWS environment for optimizations and recommendations related to cost optimization, performance, security, and fault tolerance. It will also show you when you've exceeded 80 percent of your limit for a service.

Business and Enterprise support customers get access to all Trusted Advisor checks. All AWS customers get the following security checks for free:

- Public access to an S3 bucket, particularly upload and delete access
- Security groups with unrestricted access to ports that normally should be restricted, such as TCP port 1433 (MySQL) and 3389 (Remote Desktop Protocol)
- Whether you've created an IAM user
- Whether multifactor authentication is enabled for the root user
- Public access to an EBS or RDS snapshot

Inventory Manager

The Inventory Manager collects data from your instances, including operating system and application versions. Inventory Manager can collect data for the following:

- Operating system name and version
- Applications and filenames, versions, and sizes
- Network configuration, including IP and media access control (MAC) addresses
- Windows updates, roles, services, and registry values
- CPU model, cores, and speed

You choose which instances to collect data from by creating a regionwide *inventory association* by executing the AWS-GatherSoftwareInventory policy document. You can choose all instances in your account or select instances manually or by tag. When you choose all instances in your account, it's called a *global inventory association*, and new instances you create in the region are automatically added to it. Inventory collection occurs at least every 30 minutes.

When you configure the Systems Manager agent on an on-premises server, you specify a region for inventory purposes. To aggregate metadata for instances from different regions and accounts, you may configure Resource Data Sync in each region to store all inventory data in a single S3 bucket.

Compliance

Compliance insights show how the patch and association status of your instances stacks up against the rules you've configured. Patch compliance shows the number of instances that have the patches in their configured baseline, as well as details of the specific patches installed. Association compliance shows the number of instances that have had an association successfully executed against them.

AWS CLI Example

The following example code shows how you can use an AWS CLI command to deploy an EC2 instance that includes many of the features you learned about in this chapter. Naturally, the image-id, security-group-ids, and subnet-id values are not real. Those you would replace with actual IDs that fit your account and region.

```
aws ec2 run-instances --image-id ami-xxxxxxxx --count 1 \
--instance-type t2.micro --key-name MyKeyPair \
--security-group-ids sg-xxxxxxxx --subnet-id subnet-xxxxxxxx \
--user-data file://my_script.sh \
--tag-specifications \
'ResourceType=instance,Tags=[{Key=webserver,Value=production}]' \
'ResourceType=volume,Tags=[{Key=cost-center,Value=cc123}]'
```

This example launches a single (--count 1) instance that's based on the specified AMI. The desired instance type, key name, security group, and subnet are all identified. A script file (that must exist locally so it can be read) is added using the user-data argument, and two tags are associated with the instance (webserver:production and cost-center:cc123).

If you need to install the AWS CLI, perform Exercise 2.6.

EXERCISE 2.6

Install the AWS CLI and Use It to Launch an EC2 Instance

Need help? Learn how to install the AWS CLI for your OS here:

docs.aws.amazon.com/cli/latest/userguide/cli-chap-install.html

Refer to the previous AWS CLI example for help launching your instance. (Hint: You will need to fill in some xxxxx placeholders with actual resource IDs.)

Never leave any resources running after you've finished using them. Exercise 2.7 can help.

EXERCISE 2.7

Clean Up Unused EC2 Resources

Since you've probably been launching resources while experimenting with AWS, you'll want to make sure you haven't accidentally left anything running that could cost you money. So, take a good look through the console and kill off what shouldn't still be alive. Here are some things to consider:

- Remember to check any other AWS regions where you might have done some work—only a single region's resources will show up in the console at a time.

- Some resources can't be deleted because they're in use by other resources. A snapshot used by a private AMI is one example. You'll need to shut those down in the right order.

- When you're logged into your AWS account, you can check out your Billing and Cost Management dashboard in the console (`console.aws.amazon.com/billing`). This dashboard will show you whether, taking into account your current resource usage, you stand to run up a bill in the current month.

Summary

The base software stack that runs on an EC2 instance is defined by your choice of Amazon Machine Image and any scripts or user data you add at launch time, and the hardware profile is the product of an instance type. A tenancy setting determines whether your instance will share a physical host with other instances.

As with all your AWS resources, it's important to give your EC2 instances easily identifiable tags that conform to a systemwide naming convention. There are limits to the number of resources you'll be allowed to launch within a single region and account wide. Should you hit your limit, you can always request access to additional resources.

If you plan to run an instance for a year or longer, you can save a significant amount of money compared to paying for on-demand by purchasing a reserve instance. If your workload can withstand unexpected shutdowns, then a spot instance could also make sense.

There are four kinds of Elastic Block Store volumes: two high IOPS and low-latency SSD types and two traditional hard drives. Your workload and budget will inform your choice. In addition, some EC2 instance types come with ephemeral instance store volumes that offer fast data access but whose data is lost when the instance is shut down.

All EC2 instances are given at least one private IP address, and should they require Internet access, they can also be given a nonpermanent public IP. If you require a permanent public IP, you can assign an elastic IP to the instance.

You secure access to your EC2 instances using software firewalls known as security groups and can open up secure and limited access through IAM roles, NAT instances or NAT gateways, and key pairs.

EC2 Auto Scaling can help you avoid application failures caused by overloaded instances. By implementing dynamic scaling policies, you can ensure that you always have enough instances to handle increased demand. In the event of some failure, a well-designed Auto Scaling group will ensure that you always have a minimum number of healthy instances. When an instance becomes unhealthy, Auto Scaling will terminate and replace it.

Exam Essentials

Understand how to provision and launch an EC2 instance. You'll need to select the right AMI and instance type, configure a security group, add any extra storage volumes that might be needed, point to any necessary user data and scripts, and, ideally, tag all the elements using descriptive key values.

Understand how to choose the right hardware/software profile for your workload. Consider the benefits of building your own image against the ease and simplicity of using a marketplace, community, or official AMI. Calculate the user demand you expect your application to generate so that you can select an appropriate instance type. Remember that you can always change your instance type later if necessary.

Understand EC2 pricing models and how to choose one to fit your needs. Know how to calculate whether you'll be best off on the spot market, with on-demand, or with reserve—or some combination of the three.

Understand how to configure a security group to balance access with security to match your deployment profile. Security groups act as firewalls, applying policy rules to determine which network traffic is allowed through. You can control traffic based on a packet's protocol and network port and its source and intended destination.

Know how to access a running instance. Instance data, including private and public IP addresses, can be retrieved from the AWS Console, through the AWS CLI, and from metadata queries on the instance itself. You'll need this information so that you can log in to administer the instance or access its web-facing applications.

Understand the features and behavior of storage volume types. SSD volumes can achieve higher IOPS and, therefore, lower latency, but they come at a cost that's higher than traditional hard drives.

Know how to create a snapshot from a storage volume and how to attach the snapshot to a different instance. Any EBS drive can be copied and either attached to a different instance or used to generate an image that, in turn, can be made into an AMI and shared or used to launch any number of new instances.

Be able to configure EC2 Auto Scaling. Auto Scaling can help you avoid application failures by automatically provisioning new instances when you need them, avoiding instance failures caused by resource exhaustion. When an instance failure does occur, Auto Scaling steps in and creates a replacement.

Review Questions

1. You need to deploy multiple EC2 Linux instances that will provide your company with virtual private networks (VPNs) using software called OpenVPN. Which of the following will be the most efficient solutions? (Choose two.)
   A. Select a regular Linux AMI and bootstrap it using user data that will install and configure the OpenVPN package on the instance and use it for your VPN instances.
   B. Search the community AMIs for an official AMI provided and supported by the OpenVPN company.
   C. Search the AWS Marketplace to see whether there's an official AMI provided and supported by the OpenVPN company.
   D. Select a regular Linux AMI and SSH to manually install and configure the OpenVPN package.
   E. Create a Site-to-Site VPN Connection from the wizard in the AWS VPC dashboard.

2. As part of your company's long-term cloud migration strategy, you have a VMware virtual machine in your local infrastructure that you'd like to copy to your AWS account and run as an EC2 instance. Which of the following will be necessary steps? (Choose two.)
   A. Import the virtual machine to your AWS region using a secure SSH tunnel.
   B. Import the virtual machine using VM Import/Export.
   C. Select the imported VM from among your private AMIs and launch an instance.
   D. Select the imported VM from the AWS Marketplace AMIs and launch an instance.
   E. Use the AWS CLI to securely copy your virtual machine image to an S3 bucket within the AWS region you'll be using.

3. Your AWS CLI command to launch an AMI as an EC2 instance has failed, giving you an error message that includes `InvalidAMIID.NotFound`. What of the following is the most likely cause?
   A. You haven't properly configured the ~/.aws/config file.
   B. The AMI is being updated and is temporarily unavailable.
   C. Your key pair file has been given the wrong (overly permissive) permissions.
   D. The AMI you specified exists in a different region than the one you've currently specified.

4. The sensitivity of the data your company works with means that the instances you run must be secured through *complete* physical isolation. What should you specify as you configure a new instance?
   A. Dedicated Host tenancy
   B. Shared tenancy
   C. Dedicated Instance tenancy
   D. Isolated tenancy

5. Normally, two instances running m5.large instance types can handle the traffic accessing your online e-commerce site, but you know that you will face short, unpredictable periods of high demand. Which of the following choices should you implement? (Choose two.)
   A. Configure autoscaling.
   B. Configure load balancing.
   C. Purchase two m5.large instances on the spot market and as many on-demand instances as necessary.
   D. Shut down your m5.large instances and purchase instances using a more robust instance type to replace them.
   E. Purchase two m5.large reserve instances and as many on-demand instances as necessary.

6. Which of the following use cases would be most cost effective if run using spot market instances?
   A. Your e-commerce website is built using a publicly available AMI.
   B. You provide high-end video rendering services using a fault-tolerant process that can easily manage a job that was unexpectedly interrupted.
   C. You're running a backend database that must be reliably updated to keep track of critical transactions.
   D. Your deployment runs as a static website on S3.

7. In the course of a routine infrastructure audit, your organization discovers that some of your running EC2 instances are not configured properly and must be updated. Which of the following configuration details *cannot* be changed on an existing EC2 instance?
   A. AMI
   B. Instance type
   C. Security group
   D. Public IP address

8. For an account with multiple resources running as part of multiple projects, which of the following key/value combination examples would make for the most effective identification convention for resource tags?
   A. servers:server1
   B. project1:server1
   C. EC2:project1:server1
   D. server1:project1

9. Which of the following EBS options will you need to keep your data-hungry application that requires up to 20,000 IOPS happy?
   A. Cold HDD
   B. General-purpose SSD
   C. Throughput-optimized HDD
   D. Provisioned-IOPS SSD

10. Your organization needs to introduce Auto Scaling to its infrastructure and needs to generate a "golden image" AMI from an existing EBS volume. This image will need to be shared among multiple AWS accounts belonging to your organization. Which of the following steps will get you there? (Choose three.)
    A. Create an image from a detached EBS volume, use it to create a snapshot, select your new AMI from your private collection, and use it for your launch configuration.
    B. Create a snapshot of the EBS root volume you need, use it to create an image, select your new AMI from your private collection, and use it for your launch configuration.
    C. Create an image from the EBS volume attached to the instance, select your new AMI from your private collection, and use it for your launch configuration.
    D. Search the AWS Marketplace for the appropriate image and use it for your launch configuration.
    E. Import the snapshot of an EBS root volume from a different AWS account, use it to create an image, select your new AMI from your private collection, and use it for your launch configuration.

11. Which of the following are benefits of instance store volumes? (Choose two.)
    A. Instance volumes are physically attached to the server that's hosting your instance, allowing faster data access.
    B. Instance volumes can be used to store data even after the instance is shut down.
    C. The use of instance volumes does not incur costs (beyond those for the instance itself).
    D. You can set termination protection so that an instance volume can't be accidentally shut down.
    E. Instance volumes are commonly used as a base for the creation of AMIs.

12. According to default behavior (and AWS recommendations), which of the following IP addresses could be assigned as the private IP for an EC2 instance? (Choose two.)
    A. 54.61.211.98
    B. 23.176.92.3
    C. 172.17.23.43
    D. 10.0.32.176
    E. 192.140.2.118

13. You need to restrict access to your EC2 instance-based application to only certain clients and only certain targets. Which three attributes of an incoming data packet are used by a security group to determine whether it should be allowed through? (Choose three.)
    A. Network port
    B. Source address
    C. Datagram header size
    D. Network protocol
    E. Destination address

14. How are IAM roles commonly used to ensure secure resource access in relation to EC2 instances? (Choose two.)
    A. A role can assign processes running on the EC2 instance itself permission to access other AWS resources.
    B. A user can be given permission to authenticate as a role and access all associated resources.
    C. A role can be associated with individual instance-based processes (Linux instances only), giving them permission to access other AWS resources.
    D. A role can give users and resources permission to access the EC2 instance.

15. You have an instance running within a private subnet that needs external network access to receive software updates and patches. Which of the following can securely provide that access from a public subnet within the same VPC? (Choose two.)
    A. Internet gateway
    B. NAT instance
    C. Virtual private gateway
    D. NAT gateway
    E. VPN

16. What do you have to do to securely authenticate to the GUI console of a Windows EC2 session?
    A. Use the private key of your key pair to initiate an SSH tunnel session.
    B. Use the public key of your key pair to initiate an SSH tunnel session.
    C. Use the public key of your key pair to retrieve the password you'll use to log in.
    D. Use the private key of your key pair to retrieve the password you'll use to log in.

17. Your application deployment includes multiple EC2 instances that need low-latency connections to each other. Which of the following AWS tools will allow you to locate EC2 instances closer to each other to reduce network latency?
    A. Load balancing
    B. Placement groups
    C. AWS Systems Manager
    D. AWS Fargate

18. To save configuration time and money, you want your application to run only when network events trigger it but shut down immediately after. Which of the following will do that for you?
    A. AWS Lambda
    B. AWS Elastic Beanstalk
    C. Amazon Elastic Container Service (ECS)
    D. Auto Scaling

19. Which of the following will allow you to quickly copy a virtual machine image from your local infrastructure to your AWS VPC?
    A. AWS Simple Storage Service (S3)
    B. AWS Snowball
    C. VM Import/Export
    D. AWS Direct Connect

20. You've configured an EC2 Auto Scaling group to use a launch configuration to provision and install an application on several instances. You now need to reconfigure Auto Scaling to install an additional application on new instances. Which of the following should you do?
    A. Modify the launch configuration.
    B. Create a launch template and configure the Auto Scaling group to use it.
    C. Modify the launch template.
    D. Modify the CloudFormation template.

21. You create an Auto Scaling group with a minimum group size of 3, a maximum group size of 10, and a desired capacity of 5. You then manually terminate two instances in the group. Which of the following will Auto Scaling do?
    A. Create two new instances
    B. Reduce the desired capacity to 3
    C. Nothing
    D. Increment the minimum group size to 5

22. You're running an application that receives a spike in traffic on the first day of every month. You want to configure Auto Scaling to add more instances before the spike begins and then add additional instances in proportion to the CPU utilization of each instance. Which of the following should you implement? (Choose all that apply.)
    A. Target tracking policies
    B. Scheduled actions
    C. Step scaling policies
    D. Simple scaling policies
    E. Load balancing

23. As part of your new data backup protocols, you need to manually take EBS snapshots of several hundred volumes. Which type of Systems Manager document enables you to do this?
    A. Command
    B. Automation
    C. Policy
    D. Manual

Chapter 3

AWS Storage

THE AWS CERTIFIED SOLUTIONS ARCHITECT ASSOCIATE EXAM OBJECTIVES COVERED IN THIS CHAPTER MAY INCLUDE, BUT ARE NOT LIMITED TO, THE FOLLOWING:

✓ **Domain 1: Design Resilient Architectures**
- 1.1 Design a multi-tier architecture solution
- 1.2 Design highly available and/or fault-tolerant architectures
- 1.4 Choose appropriate resilient storage

✓ **Domain 2: Design High-Performing Architectures**
- 2.1 Identify elastic and scalable compute solutions for a workload
- 2.2 Select high-performing and scalable storage solutions for a workload

✓ **Domain 3: Design Secure Applications and Architectures**
- 3.1 Design secure access to AWS resources
- 3.2 Design secure application tiers
- 3.3 Select appropriate data security options

✓ **Domain 4: Design Cost-Optimized Architectures**
- 4.1 Identify cost-effective storage solutions

Introduction

Amazon Simple Storage Service (S3) is where individuals, applications, and a long list of AWS services keep their data. It's an excellent platform for the following:

- Maintaining backup archives, log files, and disaster recovery images
- Running analytics on big data at rest
- Hosting static websites

S3 provides inexpensive and reliable storage that can, if necessary, be closely integrated with operations running within or external to AWS.

This isn't the same as the operating system volumes you learned about in the previous chapter; those are kept on the *block storage* volumes driving your EC2 instances. S3, by contrast, provides a space for effectively unlimited *object storage*.

What's the difference between object and block storage? With block-level storage, data on a raw physical storage device is divided into individual blocks whose use is managed by a file system. NTFS is a common filesystem used by Windows, and Linux might use Btrfs or ext4. The filesystem, on behalf of the installed OS, is responsible for allocating space for the files and data that are saved to the underlying device and for providing access whenever the OS needs to read some data.

An object storage system like S3, on the other hand, provides what you can think of as a flat surface on which to store your data. This simple design avoids some of the OS-related complications of block storage and allows anyone easy access to any amount of professionally designed and maintained storage capacity.

When you write files to S3, they're stored along with up to 2 KB of metadata. The metadata is made up of keys that establish system details like data permissions and the appearance of a filesystem location within nested buckets.

In this chapter, you're going to learn the following:

- How S3 objects are saved, managed, and accessed
- How to choose from among the various classes of storage to get the right balance of durability, availability, and cost
- How to manage long-term data storage lifecycles by incorporating Amazon Glacier into your design
- What other AWS services exist to help you with your data storage and access operations

S3 Service Architecture

You organize your S3 files into buckets. By default, you're allowed to create as many as 100 buckets for each of your AWS accounts. As with other AWS services, you can ask AWS to raise that limit.

Although an S3 bucket and its contents exist only within a single AWS region, the name you choose for your bucket must be globally unique within the entire S3 system. There's some logic to this; you'll often want your data located in a particular geographical region to satisfy operational or regulatory needs. But at the same time, being able to reference a bucket without having to specify its region simplifies the process.

Here is the URL you would use to access a file called `filename` that's in a bucket called `bucketname` over HTTP:

`s3.amazonaws.com/bucketname/filename`

Naturally, this assumes you'll be able to satisfy the object's permissions requirements. This is how that same file would be addressed using the AWS CLI:

`s3://bucketname/filename`

Prefixes and Delimiters

As you've seen, S3 stores objects within a bucket on a flat surface without subfolder hierarchies. However, you can use prefixes and delimiters to give your buckets the *appearance* of a more structured organization.

A prefix is a common text string that indicates an organization level. For example, the word `contracts` when followed by the delimiter `/` would tell S3 to treat a file with a name like `contracts/acme.pdf` as an object that should be grouped together with a second file named `contracts/dynamic.pdf`.

S3 recognizes folder/directory structures as they're uploaded and emulates their hierarchical design within the bucket, automatically converting slashes to delimiters. That's why you'll see the correct folders whenever you view your S3-based objects through the console or the API.

Working with Large Objects

Although there's no theoretical limit to the total amount of data you can store within a bucket, a single object may be no larger than 5 TB. Individual *uploads* can be no larger than 5 GB. To reduce the risk of data loss or aborted uploads, AWS recommends that you use a feature called Multipart Upload for any object larger than 100 MB.

As the name suggests, Multipart Upload breaks a large object into multiple smaller parts and transmits them individually to their S3 target. If one transmission should fail, it can be repeated without impacting the others.

Multipart Upload will be used automatically when the upload is initiated by the AWS CLI or a high-level API, but you'll need to manually break up your object if you're working with a low-level API.

An application programming interface (API) is a programmatic interface through which operations can be run through code or from the command line. AWS maintains APIs as the primary method of administration for each of its services. AWS provides low-level APIs for cases when your S3 uploads require hands-on customization, and it provides high-level APIs for operations that can be more readily automated. This page contains specifics:

docs.aws.amazon.com/AmazonS3/latest/dev/uploadobjusingmpu.html

If you need to transfer large files to an S3 bucket, the Amazon S3 Transfer Acceleration configuration can speed things up. When a bucket is configured to use Transfer Acceleration, uploads are routed through geographically nearby AWS edge locations and, from there, routed using Amazon's internal network.

You can find out whether Transfer Acceleration would actually improve transfer speeds between your location and a particular AWS region by using the Amazon S3 Transfer Acceleration Speed Comparison tool (s3-accelerate-speedtest.s3-accelerate .amazonaws.com/en/accelerate-speed-comparsion.html). If your transfers are, in fact, good candidates for Transfer Acceleration, you should enable the setting in your bucket. You can then use special endpoint domain names (like bucketname.s3-accelerate .amazonaws.com) for your transfers.

Work through Exercise 3.1 and create your own bucket.

EXERCISE 3.1

Create a New S3 Bucket and Upload a File

1. From the AWS S3 dashboard, create a new bucket and use either the dashboard or the AWS CLI (aws s3 cp mylocalfile.txt s3://mybucketname/) to upload one or more files, giving public read access to the object.

2. From the file's Overview page in the S3 dashboard (displayed when the bucket name is clicked), copy its public link, paste it into the URL field of a browser that's not logged into your AWS account, and confirm that you can open the file.

Encryption

Unless it's intended to be publicly available—perhaps as part of a website—data stored on S3 should always be encrypted. You can use encryption keys to protect your data while it's at rest within S3 and—by using only Amazon's encrypted API endpoints for data transfers—protect data during its journeys between S3 and other locations.

Data at rest can be protected using either server-side or client-side encryption.

Server-Side Encryption

The "server-side" here is the S3 platform, and it involves having AWS encrypt your data objects as they're saved to disk and decrypt them when you send properly authenticated requests for retrieval.

You can use one of three encryption options:

- Server-Side Encryption with Amazon S3-Managed Keys (SSE-S3), where AWS uses its own enterprise-standard keys to manage every step of the encryption and decryption process
- Server-Side Encryption with AWS KMS-Managed Keys (SSE-KMS), where, beyond the SSE-S3 features, the use of an *envelope key* is added along with a full audit trail for tracking key usage. You can optionally import your own keys through the AWS KMS service.
- Server-Side Encryption with Customer-Provided Keys (SSE-C), which lets you provide your own keys for S3 to apply to its encryption

Client-Side Encryption

It's also possible to encrypt data before it's transferred to S3. This can be done using an AWS KMS–Managed Customer Master Key (CMK), which produces a unique key for each object before it's uploaded. You can also use a Client-Side Master Key, which you provide through the Amazon S3 encryption client.

Server-side encryption can greatly reduce the complexity of the process and is often preferred. Nevertheless, in some cases, your company (or regulatory oversight body) might require that you maintain full control over your encryption keys, leaving client-side as the only option.

Logging

Tracking S3 events to log files is disabled by default—S3 buckets can see a lot of activity, and not every use case justifies the log data that S3 can generate.

When you enable logging, you'll need to specify both a source bucket (the bucket whose activity you're tracking) and a target bucket (the bucket to which you'd like the logs saved). Optionally, you can also specify delimiters and prefixes (such as the creation date or time) to make it easier to identify and organize logs from multiple source buckets that are saved to a single target bucket.

S3-generated logs, which sometimes appear only after a short delay, will contain basic operation details, including the following:

- The account and IP address of the requestor
- The source bucket name

- The action that was requested (GET, PUT, POST, DELETE, etc.)
- The time the request was issued
- The response status (including error code)

S3 buckets are also used by other AWS services—including CloudWatch and CloudTrail—to store their logs or other objects (like EBS Snapshots).

S3 Durability and Availability

S3 offers more than one class of storage for your objects. The class you choose will depend on how critical it is that the data survives no matter what (durability), how quickly you might need to retrieve it (availability), and how much money you have to spend.

Durability

S3 measures durability as a percentage. For instance, the 99.999999999 percent durability guarantee for most S3 classes and Amazon Glacier is as follows:

> . . . corresponds to an average annual expected loss of 0.000000001% of objects. For example, if you store 10,000,000 objects with Amazon S3, you can on average expect to incur a loss of a single object once every 10,000 years.
>
> *Source:* aws.amazon.com/s3/faqs

In other words, realistically, there's pretty much no way that you can possibly lose data stored on one of the standard S3/Glacier platforms because of infrastructure failure. However, it would be irresponsible to rely on your S3 buckets as the only copies of important data. After all, there's a real chance that a misconfiguration, account lockout, or unanticipated external attack could permanently block access to your data. And, as crazy as it might sound right now, it's not unthinkable to suggest that AWS could one day go out of business. Kodak and Blockbuster Video once dominated their industries, right? You should always back up your data to multiple locations, using different services and media types. You'll learn how to do that in Chapter 10, "The Reliability Pillar." The high durability rates delivered by S3 are largely because they automatically replicate your data across at least three availability zones. This means that even if an entire AWS facility was suddenly wiped off the map, copies of your data would be restored from a different zone.

There is, however, one storage class that isn't quite so resilient. Amazon S3 Reduced Redundancy Storage (RRS) is rated at only 99.99 percent durability (because it's replicated across fewer servers than other classes). The RRS class is still available for historical reasons, but it's officially not recommended that you actually use it.

You can balance increased/decreased durability against other features like availability and cost to get the balance that's right for you. While all currently recommended classes are designed to withstand data loss 99.999999999% (11 nines) of the time, most will be maintained in at least three availability zones. The exception is S3 One Zone-IA that, as the name suggests, stores its data in only a single zone. The difference shows up in its slightly lower availability, which we'll discuss next.

Availability

Object availability is also measured as a percentage; this time, though, it's the percentage you can expect a given object to be instantly available on request through the course of a full year. The Amazon S3 Standard class, for example, guarantees that your data will be ready whenever you need it (meaning it will be available) for 99.99% of the year. That means there will be less than nine hours each year of downtime. If you feel downtime has exceeded that limit within a single year, you can apply for a service credit. Amazon's durability guarantee, by contrast, is designed to provide 99.999999999% data protection. This means there's practically no chance your data will be lost, even if you might sometimes not have instant access to it.

S3 Intelligent-Tiering is a relatively new storage class that saves you money while optimizing availability. For a monthly automation fee, Intelligent-Tiering will monitor the way you access data within the class over time. It will automatically move an object to the lower-cost infrequent access tier after it hasn't been accessed for 30 consecutive days.

Table 3.1 illustrates the availability guarantees for all S3 classes.

TABLE 3.1 Guaranteed availability standards for S3 storage

	S3 Standard	S3 Standard-IA	S3 One Zone-IA	S3 Intelligent-Tiering
Availability guarantee	99.99%	99.9%	99.5%	99.9%

Eventually Consistent Data

It's important to bear in mind that S3 replicates data across multiple locations. As a result, there might be brief delays while updates to existing objects propagate across the system. Uploading a new version of a file or, alternatively, deleting an old file altogether can result in one site reflecting the new state with another still unaware of any changes.

To ensure that there's never a conflict between versions of a single object—which could lead to serious data and application corruption—you should treat your data according to an eventually consistent standard. That is, you should expect a delay (usually just two seconds or less) and design your operations accordingly.

Because there isn't the risk of corruption when creating *new* objects, S3 provides read-after-write consistency for creation (PUT) operations.

S3 Object Lifecycle

Many of the S3 workloads you'll launch will probably involve backup archives. But the thing about backup archives is that, when properly designed, they're usually followed regularly by more backup archives. Maintaining some previous archive versions is critical, but you'll also want to retire and delete older versions to keep a lid on your storage costs.

S3 lets you automate all this with its versioning and lifecycle features.

Versioning

Within many file system environments, saving a file using the same name and location as a preexisting file will overwrite the original object. That ensures you'll always have the most recent version available to you, but you will lose access to older versions—including versions that were overwritten by mistake.

By default, objects on S3 work the same way. But if you enable versioning at the bucket level, then older overwritten copies of an object will be saved and remain accessible indefinitely. This solves the problem of accidentally losing old data, but it replaces it with the potential for archive bloat. Here's where lifecycle management can help.

Lifecycle Management

In addition to the S3 Intelligent-Tiering storage class we discussed earlier, you can manually configure lifecycle rules for a bucket that will automatically transition an object's storage class after a set number of days. You might, for instance, have new objects remain in the S3 Standard class for their first 30 days, after which they're moved to the cheaper One Zone IA for another 30 days. If regulatory compliance requires that you maintain older versions, your files could then be moved to the low-cost, long-term storage service Glacier for 365 more days before being permanently deleted.

> You can optionally use prefixes to apply lifecycle rules to only certain objects within a bucket. You should also be aware that there are minimum times (30 days, for instance) an object must remain within one class before it can be moved. In addition, you can't transition directly from S3 Standard to Reduced Redundancy.

Try it yourself with Exercise 3.2.

> **EXERCISE 3.2**
>
> **Enable Versioning and Lifecycle Management for an S3 Bucket**
>
> 1. Select your bucket and edit its properties to enable versioning.
> 2. Upload a file to that bucket, edit the copy on your local computer, and upload the new copy (keeping the filename the same). If you're working with a file from your static website, make sure you give the new file permissions allowing public access.
> 3. With the contents of the bucket displayed in the dashboard, select Show Versions. You should now see two versions of your file.
> 4. Add a couple of directories with files to your bucket.
> 5. From the bucket's Management tab, select Lifecycle and specify a prefix/tag filter that matches the directory name of one of the directories you uploaded.
> 6. Add a lifecycle rule by adding transitions and configuring the transition timing (in days) and target for each one.
>
> You'll need to be patient to test this configuration because the minimum lag between transitions is 30 days.

Accessing S3 Objects

If you didn't think you'd ever need your data, you wouldn't go to the trouble of saving it to S3. So, you'll need to understand how to access your S3-hosted objects and, just as important, how to restrict access to only those requests that match your business and security needs.

Access Control

Out of the box, new S3 buckets and objects will be fully accessible to your account but to no other AWS accounts or external visitors. You can strategically open up access at the bucket and object levels using access control list (ACL) rules, finer-grained S3 bucket policies, or Identity and Access Management (IAM) policies.

There is more than a little overlap between those three approaches. In fact, ACLs are really leftovers from before AWS created IAM. As a rule, Amazon recommends applying S3 bucket policies or IAM policies instead of ACLs.

S3 bucket policies—which are formatted as JavaScript Object Notation (JSON) text and attached to your S3 bucket—will make sense for cases where you want to control access to a single S3 bucket for multiple external accounts and users. On the other hand, IAM policies—because they exist at the account level within IAM—will probably make sense when you're trying to control the way individual users and roles access multiple resources, including S3.

The following code is an example of an S3 bucket policy that allows both the root user and the user Steve from the specified AWS account to access the S3 `MyBucket` bucket and its contents. Both users are considered *principals* within this rule.

```
{
  "Version": "2012-10-17",
  "Statement": [
    {
      "Effect": "Allow",
      "Principal": {
        "AWS": ["arn:aws:iam::xxxxxxxxxxxx:root",
        "arn:aws:iam::xxxxxxxxxxxx:user/Steve"]
      },
      "Action": "s3:*",
      "Resource": ["arn:aws:s3:::MyBucket",
                   "arn:aws:s3:::MyBucket/*"]
    }
  ]
}
```

> **NOTE** You can also limit access within bucket policies by specifying time of day or source Classless Inter-Domain Routing (CIDR) IP address blocks.

When it's attached to an IAM entity (a user, group, or role), the following IAM policy will accomplish the same thing as the previous S3 bucket policy:

```
{
  "Version": "2012-10-17",
  "Statement":[{
    "Effect": "Allow",
    "Action": "s3:*",
    "Resource": ["arn:aws:s3:::MyBucket",
                 "arn:aws:s3:::MyBucket/*"]
    }
  ]
}
```

IAM roles and policies will be discussed in greater detail in Chapter 6, "Authentication and Authorization—AWS Identity and Access Management."

You can also closely control the way users and services access objects within your buckets by using Amazon S3 Access Points. An access point is a hostname that can point to a carefully defined subset of objects in a bucket. Depending on how you configure your access points, clients invoking the hostname will be able to read or write only the data you allow, and only as long as you allow it.

A simple AWS CLI command to request an access point might look something like this:

```
aws s3control create-access-point --name my-vpc-ap \
    --account-id 123456789012 --bucket my-bucket \
    --vpc-configuration VpcId=vpc-2b9d3c
```

Presigned URLs

If you want to provide temporary access to an object that's otherwise private, you can generate a presigned URL. The URL will be usable for a specified period of time, after which it will become invalid. You can build presigned URL generation into your code to provide object access programmatically.

The following AWS CLI command will return a URL that includes the required authentication string. The authentication will become invalid after 10 minutes (600 seconds). The default expiration value is 3,600 seconds (one hour).

```
aws s3 presign s3://MyBucketName/PrivateObject --expires-in 600
```

Try it yourself with Exercise 3.3.

EXERCISE 3.3

Generate and Use a Presigned URL

1. Use the complete URL of a private object in an S3 bucket belonging to you to generate a presigned URL using a variation of this command:

    ```
    aws s3 presign s3://MyBucketName/PrivateObject --expires-in 600
    ```

2. Copy the full URL from the command output and, from a browser that's not logged into your AWS account, try to open the file.

3. Wait for the URL to expire and try again. This time, it should not work.

Static Website Hosting

S3 buckets can be used to host the HTML files for entire static websites. A website is static when the system services used to render web pages and scripts are all client- rather than server-based. This architecture permits simple and lean HTML code that's designed to be executed by the client browser.

S3, because it's such an inexpensive yet reliable platform, is an excellent hosting environment for such sites. When an S3 bucket is configured for static hosting, traffic directed at the bucket's URL can be automatically made to load a specified root document, usually named index.html. Users can click links within HTML pages to be sent to the target page or media resource. Error handling and redirects can also be incorporated into the profile.

If you want requests for a DNS domain name (like mysite.com) routed to your static site, you can use Amazon Route 53 to associate your bucket's endpoint with any registered name. This will work only if your domain name is also the name of the S3 bucket. You'll learn more about domain name records in Chapter 8, "The Domain Name System and Network Routing: Amazon Route 53 and Amazon CloudFront."

You can also get a free SSL/TLS certificate to encrypt your site by requesting a certificate from AWS Certificate Manager (ACM) and importing it into a CloudFront distribution that specifies your S3 bucket as its origin.

Build your own static website using Exercise 3.4.

EXERCISE 3.4

Enable Static Website Hosting for an S3 Bucket

1. From the S3 dashboard, select (or create) an S3 bucket that contains at least one file with some simple text named index.html. Any files you want to be accessible should be readable by the public.

2. From the bucket Properties tab, enable static website hosting and specify your index.html file as your index document.

3. Paste the static website endpoint into the URL field of a browser that's not logged into your AWS account and confirm you can access the website.

Note that you can also enable static website hosting from the AWS CLI using a variation of these two commands:

```
aws s3api put-bucket-acl --bucket my-bucket --acl public-read
aws s3 website s3://my-bucket/ --index-document index.html --error-document error.html
```

AWS provides a different way to access data stored on either S3 Select or Glacier Select. The feature lets you apply SQL-like queries to stored objects so that only relevant data from within objects is retrieved, permitting significantly more efficient and cost-effective operations.

One possible use case would involve large comma-separated values (CSV) files containing sales and inventory data from multiple retail sites. Your company's marketing team might need to periodically analyze only sales data and only from certain stores. Using S3 Select, they'll be able to retrieve exactly the data they need—just a fraction of the full data set—while bypassing the bandwidth and cost overhead associated with downloading the whole thing.

Amazon S3 Glacier

At first glance, Glacier looks a bit like just another S3 storage class. After all, like most S3 classes, Glacier guarantees 99.999999999 percent durability and, as you've seen, can be incorporated into S3 lifecycle configurations.

Nevertheless, there are important differences. Glacier, for example, supports archives as large as 40 TB, whereas S3 buckets have no size limit. Its archives are encrypted by default, whereas encryption on S3 is an option you need to select. And unlike S3's "human-readable" key names, Glacier archives are given machine-generated IDs.

But the biggest difference is the time it takes to retrieve your data. Retrieving the objects from an existing Glacier archive can take a number of hours, compared to nearly instant access from S3. That last feature really defines the purpose of Glacier: to provide inexpensive long-term storage for data that will be needed only in unusual and infrequent circumstances.

> **NOTE** Glacier does offer Expedited retrievals, getting you your data in minutes rather than hours. Expedited will incur a premium charge.

In the context of Glacier, the term *archive* is used to describe an object like a document, video, or a TAR or ZIP file. Archives are stored in *vaults*—the Glacier equivalent of S3's buckets. Glacier vault names do not have to be globally unique.

There are currently two Glacier storage tiers: Standard and Deep Archive. As you can probably guess, Glacier Deep Archive will cost you less but will require longer waits for data retrieval. Storing one terabyte of data for a month on Glacier standard will, at current rates, cost you $4.10, while leaving the same terabyte for the same month on Glacier Deep Archive will cost only $1.02. Retrieval from Deep Archive, however, will take between 12 and 48 hours.

Table 3.2 lets you compare the costs of retrieving 100 GB of data from Glacier using each of its five retrieval tiers.

TABLE 3.2 Sample retrieval costs for Glacier data in the US East region

Tier	Amount retrieved	Cost
Glacier Standard	100 GB	$0.90
Glacier Expedited	100 GB	$3.00
Glacier Bulk	100 GB	$0.25
Deep Archive Standard	100 GB	$2.00

Storage Pricing

To give you a sense of what S3 and Glacier might cost you, here's a typical usage scenario. Imagine you make weekly backups of your company sales data that generate 5 GB archives. You decide to maintain each archive in the S3 Standard Storage and Requests class for its first 30 days and then convert it to S3 One Zone (S3 One Zone-IA), where it will remain for 90 more days. At the end of those 120 days, you will move your archives once again, this time to Glacier, where they will be kept for another 730 days (two years) and then deleted.

Once your archive rotation is in full swing, you'll have a steady total of (approximately) 20 GB in S3 Standard, 65 GB in One Zone-IA, and 520 GB in Glacier. Table 3.3 shows what that storage will cost in the US East region at rates current as of this writing.

TABLE 3.3 Sample storage costs for data in the US East region

Class	Storage amount	Rate/GB/month	Cost/month
Standard	20 GB	$0.023	$0.46
One Zone-IA	65 GB	$0.01	$0.65
Glacier	520 GB	$0.004	$2.08
Total			$3.19

Of course, storage is only one part of the mix. You'll also be charged for operations including data retrievals; PUT, COPY, POST, or LIST requests; and lifecycle transition requests. Full, up-to-date details are available at aws.amazon.com/s3/pricing.

Exercise 3.5 will introduce you to an important cost-estimating tool.

> **EXERCISE 3.5**
>
> **Calculate the Total Lifecycle Costs for Your Data**
>
> Use the AWS Simple Monthly Calculator (calculator.s3.amazonaws.com/index.html) to estimate the total monthly costs of the scenario described at the beginning of this section. Even better, use a scenario that fits your own business. Try to include a full usage scenario, including requests, scans, and data retrieval. Note that you access the S3 part of the calculator by clicking the Amazon S3 tab on the left, and you can keep track of your itemized estimate using the Estimate Of Your Monthly Bill tab along the top.

Other Storage-Related Services

It's worth being aware of some other storage-related AWS services that, while perhaps not as common as the others you've seen, can make a big difference for the right deployment.

Amazon Elastic File System

The Elastic File System (EFS) provides automatically scalable and shareable file storage to be accessed from Linux instances. EFS-based files are designed to be accessed from within a virtual private cloud (VPC) via Network File System (NFS) mounts on EC2 Linux instances or from your on-premises servers through AWS Direct Connect connections. The goal is to make it easy to enable secure, low-latency, and durable file sharing among multiple instances.

Amazon FSx

Amazon FSx comes in two flavors: FSx for Lustre and Amazon FSx for Windows File Server. Lustre is an open source distributed filesystem built to give Linux clusters access to high-performance filesystems for use in compute-intensive operations. Amazon's FSx service brings Lustre capabilities to your AWS infrastructure.

FSx for Windows File Server, as you can tell from the name, offers the kind of file-sharing service EFS provides but for Windows servers rather than Linux. FSx for Windows File Server integrates operations with Server Message Block (SMB), NTFS, and Microsoft Active Directory.

AWS Storage Gateway

Integrating the backup and archiving needs of your local operations with cloud storage services can be complicated. AWS Storage Gateway provides software gateway appliances (based on an on-premises hardware appliance or virtual machines built on VMware ESXi, Microsoft Hyper-V, Linux KVM, VMware Cloud on AWS, or EC2 images) with multiple virtual connectivity interfaces. Local devices can connect to the appliance as though it's a physical backup device like a tape drive, and the data itself is saved to AWS platforms like S3 and EBS. Data can be maintained in a local cache to make it locally available.

AWS Snowball

Migrating large data sets to the cloud over a normal Internet connection can sometimes require far too much time and bandwidth to be practical. If you're looking to move terabyte- or even petabyte-scaled data for backup or active use within AWS, ordering a Snowball device might be the best option.

When requested, AWS will ship you a physical, 256-bit, encrypted Snowball storage device onto which you'll copy your data. You then ship the device back to Amazon, where its data will be uploaded to your S3 bucket(s).

Choosing the best method for transferring data to your AWS account will require a bit of arithmetic. You'll have to know the real-world upload speeds you get from your Internet connection and how much of that bandwidth wouldn't be used by other operations. In this chapter, you've learned about Multipart Upload and Transfer Acceleration for moving larger objects into an S3 bucket. But some objects are just so large that uploading them over your existing Internet connection isn't practical. Think about it; if your Internet service provider (ISP) gives you 10 MB/second, then, assuming no one else is using the connection, uploading a one-terabyte archive would take you around 10 days!

So, if you really need to move that data to the cloud, you're going to have to either invest in an expensive AWS Direct Connect connection or introduce yourself to an AWS Snowball device (or, for really massive volumes of data, AWS Snowmobile at `aws.amazon.com/snowmobile`).

AWS DataSync

DataSync specializes in moving on-premises data stores into your AWS account with a minimum of fuss. It works over your regular Internet connection, so it's not as useful as Snowball for really large data sets. But it is much more flexible, since you're not limited to S3 (or RDS as you are with AWS Database Migration Service). Using DataSync, you can drop your data into any service within your AWS account. That means you can do the following:

- Quickly and securely move old data out of your expensive data center into cheaper S3 or Glacier storage.
- Transfer data sets directly into S3, EFS, or FSx, where it can be processed and analyzed by your EC2 instances.
- Apply the power of any AWS service to any class of data as part of an easy-to-configure automated system.

DataSync can handle transfer rates of up to 10 Gbps (assuming your infrastructure has that capacity) and offers both encryption and data validation.

AWS CLI Example

This example will use the AWS CLI to create a new bucket and recursively copy the `sales-docs` directory to it. Then, using the low-level s3api CLI (which should have been installed along with the regular AWS CLI package), you'll check for the current lifecycle configuration of your new bucket with the `get-bucket-lifecycle-configuration` subcommand, specifying your bucket name. This will return an error, of course, since there currently is no configuration.

Next, you'll run the `put-bucket-lifecycle-configuration` subcommand, specifying the bucket name. You'll also add some JSON code to the `--lifecycle-configuration` argument. The code (which could also be passed as a file) will transition all objects using the `sales-docs` prefix to the Standard-IA after 30 days and to Glacier after 60 days. The objects will be deleted (or "expire") after a full year (365 days).

Finally, you can run `get-bucket-lifecycle-configuration` once again to confirm that your configuration is active. Here are the commands you would need to run to make all this work:

```
$ aws s3 mb s3://bucket-name
$ aws s3 cp --recursive sales-docs/ s3://bucket-name
$ aws s3api get-bucket-lifecycle-configuration \
   --bucket bucket-name
$ aws s3api put-bucket-lifecycle-configuration \
   --bucket bucket-name \
   --lifecycle-configuration '{
    "Rules": [
        {
            "Filter": {
                "Prefix": "sales-docs/"
            },
            "Status": "Enabled",
            "Transitions": [
                {
                    "Days": 30,
                    "StorageClass": "STANDARD_IA"
                },
                {
                    "Days": 60,
```

```
                    "StorageClass": "GLACIER"
                }
            ],
            "Expiration": {
                "Days": 365
            },
            "ID": "Lifecycle for bucket objects."
        }
    ]
}'
$ aws s3api get-bucket-lifecycle-configuration \
    --bucket bucket-name
```

Summary

Amazon S3 provides reliable and highly available object-level storage for low-maintenance, high-volume archive and data storage. Objects are stored in buckets on a "flat" surface. However, through the use of prefixes, objects can be made to appear as though they're part of a normal filesystem.

You can—and usually should—encrypt your S3 data using either AWS-provided or self-serve encryption keys. Encryption can take place when your data is at rest using either server-side or client-side encryption.

There are multiple storage classes within S3 relying on varying degrees of data replication that allow you to balance durability, availability, and cost. Lifecycle management lets you automate the transition of your data between classes until it's no longer needed and can be deleted.

You can control who and what get access to your S3 buckets—and when—through legacy ACLs or through more powerful S3 bucket policies and IAM policies. Presigned URLs are also a safe way to allow temporary and limited access to your data.

You can reduce the size and cost of your requests against S3 and Glacier-based data by leveraging the SQL-like Select feature. You can also provide inexpensive and simple static websites through S3 buckets.

Amazon Glacier stores your data archives in vaults that might require hours to retrieve but that cost considerably less than the S3 storage classes.

Exam Essentials

Understand the way S3 resources are organized.   S3 objects are stored in buckets whose names must be globally unique. Buckets are associated with AWS regions. Objects are stored within buckets on a "flat" surface, but prefixes and delimiters can give data the appearance of a folder hierarchy.

Understand how to optimize your data transfers.   Although the individual objects you store within an S3 bucket can be as large as 5 TB, anything larger than 100 MB *should* be uploaded with Multipart Upload, and objects larger than 5 GB *must* use Multipart Upload.

Understand how to secure your S3 data.   You can use server-side encryption to protect data within S3 buckets using either AWS-generated or your own privately generated keys. Data can be encrypted even before being transferred to S3 using client-side encryption.

Understand how S3 object durability and availability are measured.   The various S3 classes (and Glacier) promise varying levels of infrastructure reliability, along with data availability.

Understand S3 object versioning and lifecycle management.   Older versions of S3 objects can be saved even after they've been overwritten. To manage older objects, you can automate transition of objects between more accessible storage classes to less expensive but less accessible classes. You can also schedule object deletion.

Understand how to secure your S3 objects.   You can control access through the legacy, bucket, and object-based ACL rules or by creating more flexible S3 bucket policies or, at the account level, IAM policies. You can also provide temporary access to an object using a presigned URL.

Understand how to create a static website.   S3-based HTML and media files can be exposed as a website that, with the help of Route 53 and CloudFront, can even live behind a DNS domain name and use encrypted HTTPS pages.

Understand the differences between S3 and Glacier.   Glacier is meant for inexpensive long-term storage for data archives that you're unlikely to need often.

Review Questions

1. Your organization runs Linux-based EC2 instances that all require low-latency read/write access to a single set of files. Which of the following AWS services are your best choices? (Choose two.)

    A. AWS Storage Gateway

    B. AWS S3

    C. Amazon Elastic File System

    D. AWS Elastic Block Store

2. Your organization expects to be storing and processing large volumes of data in many small increments. When considering S3 usability, you'll need to know whether you'll face any practical limitations in the use of AWS account resources. Which of the following will normally be available only in limited amounts?

    A. PUT requests/month against an S3 bucket

    B. The volume of data space available per S3 bucket

    C. Account-wide S3 storage space

    D. The number of S3 buckets within a single account

3. You have a publicly available file called filename stored in an S3 bucket named bucketname. Which of the following addresses will successfully retrieve the file using a web browser?

    A. s3.amazonaws.com/bucketname/filename

    B. filename/bucketname.s3.amazonaws.com

    C. s3://bucketname/filename

    D. s3://filename/bucketname

4. If you want the files stored in an S3 bucket to be accessible using a familiar directory hierarchy system, you'll need to specify prefixes and delimiters. What are prefixes and delimiters?

    A. A prefix is the name common to the objects you want to group, and a delimiter is the bar character (|).

    B. A prefix is the DNS name that precedes the amazonaws.com domain, and a delimiter is the name you want to give your file directory.

    C. A prefix is the name common to the objects you want to group, and a delimiter is a forward slash character (/).

    D. A prefix is the name common to the file type you want to identify, and a delimiter is a forward slash character (/).

5. Your web application relies on data objects stored in AWS S3 buckets. Compliance with industry regulations requires that those objects are encrypted and that related events can be closely tracked. Which combination of tools should you use? (Choose two.)

    A. Server-side encryption

    B. Amazon S3-Managed Keys

C. AWS KMS-Managed Keys
D. Client-side encryption
E. AWS End-to-End managed keys

6. You are engaged in a deep audit of the use of your AWS resources and you need to better understand the structure and content of your S3 server access logs. Which of the following operational details are likely to be included in S3 server access logs? (Choose three.)
    A. Source bucket name
    B. Action requested
    C. Current bucket size
    D. API bucket creation calls
    E. Response status

7. You're assessing the level of durability you'll need to sufficiently ensure the long-term viability of a new web application you're planning. Which of the following risks are covered by S3's data durability guaranties? (Choose two.)
    A. User misconfiguration
    B. Account security breach
    C. Infrastructure failure
    D. Temporary service outages
    E. Data center security breach

8. Which of the following explains the difference in durability between S3's One Zone-IA and Reduced Redundancy classes?
    A. One Zone-IA data is heavily replicated but only within a single availability zone, whereas Reduced Redundancy data is only lightly replicated.
    B. Reduced Redundancy data is heavily replicated but only within a single availability zone, whereas One Zone-IA data is only lightly replicated.
    C. One Zone-IA data is replicated across AWS regions, whereas Reduced Redundancy data is restricted to a single region.
    D. One Zone-IA data is automatically backed up to Amazon Glacier, whereas Reduced Redundancy data remains within S3.

9. Which of the following is the 12-month availability guarantee for the S3 Standard-IA class?
    A. 99.99 percent
    B. 99.9 percent
    C. 99.999999999 percent
    D. 99.5 percent

10. Your application regularly writes data to an S3 bucket, but you're worried about the potential for data corruption as a result of conflicting concurrent operations. Which of the following data operations would *not* be subject to concerns about eventual consistency?
    A. Operations immediately preceding the deletion of an existing object
    B. Operations subsequent to the updating of an existing object

- C. Operations subsequent to the deletion of an existing object
- D. Operations subsequent to the creation of a new object

11. You're worried that updates to the important data you store in S3 might incorrectly overwrite existing files. What must you do to protect objects in S3 buckets from being accidentally lost?
    - A. Nothing. S3 protects existing files by default.
    - B. Nothing. S3 saves older versions of your files by default.
    - C. Enable versioning.
    - D. Enable file overwrite protection.

12. Your S3 buckets contain many thousands of objects. Some of them could be moved to less expensive storage classes and others still require instant availability. How can you apply transitions between storage classes for only certain objects within an S3 bucket?
    - A. By specifying particular prefixes when you define your lifecycle rules
    - B. This isn't possible. Lifecycle rules must apply to all the objects in a bucket.
    - C. By specifying particular prefixes when you create the bucket
    - D. By importing a predefined lifecycle rule template

13. Which of the following classes will usually make the most sense for long-term storage when included within a sequence of lifecycle rules?
    - A. Glacier
    - B. Reduced Redundancy
    - C. S3 One Zone-IA
    - D. S3 Standard-IA

14. Which of the following are the recommended methods for providing secure and controlled access to your buckets? (Choose two.)
    - A. S3 access control lists (ACLs)
    - B. S3 bucket policies
    - C. IAM policies
    - D. Security groups
    - E. AWS Key Management Service

15. In the context of an S3 bucket policy, which of the following statements describes a principal?
    - A. The AWS service being defined (S3 in this case)
    - B. An origin resource that's given permission to alter an S3 bucket
    - C. The resource whose access is being defined
    - D. The user or entity to which access is assigned

Review Questions 81

16. You don't want to open up the contents of an S3 bucket to anyone on the Internet, but you need to share the data with specific clients. Generating and then sending them a presigned URL is a perfect solution. Assuming you didn't explicitly set a value, how long will the presigned URL remain valid?

    A. 24 hours
    B. 3,600 seconds
    C. 5 minutes
    D. 360 seconds

17. Which non-S3 AWS resources can improve the security and user experience of your S3-hosted static website? (Choose two.)

    A. AWS Certificate Manager
    B. Elastic Compute Cloud (EC2)
    C. Relational Database Service (RDS)
    D. Route 53
    E. AWS Key Management Service

18. What is the largest single archive supported by Amazon Glacier?

    A. 5 GB
    B. 40 TB
    C. 5 TB
    D. 40 GB

19. You need a quick way to transfer very large (peta-scale) data archives to the cloud. Assuming your Internet connection isn't up to the task, which of the following will be both (relatively) fast and cost-effective?

    A. Direct Connect
    B. Server Migration Service
    C. Snowball
    D. Storage Gateway

20. Your organization runs Windows-based EC2 instances that all require low-latency read/write access to a single set of files. Which of the following AWS services is your best choice?

    A. Amazon FSx for Windows File Server
    B. Amazon FSx for Lustre
    C. Amazon Elastic File System
    D. Amazon Elastic Block Store

Chapter 4

Amazon Virtual Private Cloud

THE AWS CERTIFIED SOLUTIONS ARCHITECT ASSOCIATE EXAM OBJECTIVES COVERED IN THIS CHAPTER MAY INCLUDE, BUT ARE NOT LIMITED TO, THE FOLLOWING:

✓ **Domain 1: Design Resilient Architectures**

- 1.3 Design decoupling mechanisms using AWS services

✓ **Domain 3: Design Secure Applications and Architectures**

- 3.1 Design secure access to AWS resources
- 3.3 Select appropriate data security options

Introduction

Amazon's Virtual Private Cloud service provides the networking layer of EC2. A VPC is a virtual network that can contain EC2 instances as well as network resources for other AWS services. By default, every virtual private cloud (VPC) is isolated from all other networks. You can, however, connect your VPC to other networks, including the Internet and other VPCs.

In addition to EC2, VPCs are foundational to many AWS services, so understanding how they work is fundamental to your success on the exam and as an AWS architect. Don't assume you can ignore VPCs just because you're not using EC2.

A VPC can exist only within an AWS region. When you create a VPC in a region, it won't show up in any other regions. You can have multiple VPCs in your account and create multiple VPCs in a single region. To keep things simple, we'll start by assuming only one VPC in one region. Later, we'll cover considerations for multiple VPCs.

If you're familiar with the components of a traditional network, you'll recognize many VPC components. But although VPCs function like a traditional TCP/IP network, they are scalable, allowing you to expand and extend your network without having to add physical hardware. To make this scalability possible, some components that you'd find in a traditional network—such as routers, switches, and VLANs—don't exist in VPCs. Instead, they're abstracted into software functions and called by different names.

VPC CIDR Blocks

Like a traditional network, a VPC consists of at least one range of contiguous IP addresses. This address range is represented as a *Classless Inter-Domain Routing* (CIDR) block. The CIDR block determines which IP addresses may be assigned to instances and other resources within the VPC. You must assign a primary CIDR block when creating a VPC. After creating a VPC, you divide the primary VPC CIDR block into subnets that hold your AWS resources.

There are different ways to represent a range of IP addresses. The shortest way is by CIDR notation, sometimes called *slash notation*. For example, the CIDR 172.16.0.0/16 includes all addresses from 172.16.0.0 to 172.16.255.255—a total of 65,536 addresses!

You may also hear the CIDR block referred to as an *IP prefix*. The /16 portion of the CIDR is the *prefix length*. The prefix length refers to the length of the subnet mask, which in the case of a VPC CIDR can range from /16 to /28. An inverse relationship exists between the prefix length and the number of IP addresses in the CIDR. The smaller the prefix length, the greater the number of IP addresses in the CIDR. A /28 prefix length gives you only 16 possible addresses, whereas a /16 prefix length gives you 65,536 possible addresses!

The acronym IP refers to Internet Protocol version 4 (IPv4). Valid IPv4 prefix lengths range from /0 to /32. Although you can specify any valid IP range for your VPC CIDR, it's best to use one in the following RFC 1918 (tools.ietf.org/rfc/rfc1918.txt) range to avoid conflicts with public Internet addresses.

- 10.0.0.0–10.255.255.255 (10.0.0.0/8)
- 172.16.0.0–172.31.255.255 (172.16.0.0/12)
- 192.168.0.0–192.168.255.255 (192.168.0.0/16)

If you plan to connect your VPC to another network—whether an on-premises network or another VPC—be sure the VPC CIDR you choose doesn't overlap with addresses already in use on the other network.

You can't change the primary CIDR block after you create your VPC, so think carefully about your address requirements before creating a VPC.

Secondary CIDR Blocks

You may optionally specify secondary CIDR blocks for a VPC after you've created it. These blocks must come from either the same address range as the primary or a publicly routable range, but they must not overlap with the primary or other secondary blocks. For example, if the VPC's primary CIDR is 172.16.0.0/16, you may specify a secondary CIDR of 172.17.0.0/16 because it resides in the 172.16.0.0/12 range (172.16.0.0–172.31.255.255). But you may not specify 192.168.0.0/16 as a secondary CIDR.

If you think you might ever need a secondary CIDR, be careful about your choice of primary CIDR. If you choose 192.168.0.0/16 as your primary CIDR, you won't be able to create a secondary CIDR using any of the other RFC 1918 ranges.

IPv6 CIDR Blocks

You may let AWS assign an IPv6 CIDR to your VPC. Unlike the primary CIDR, which is an IP prefix you choose, you can't choose your own IPv6 CIDR. Instead, AWS assigns one to your VPC at your request. The IPv6 CIDR will be a publicly routable prefix from the global unicast IPv6 address space, so all IPv6 addresses are reachable from the Internet. For example, AWS may assign you the CIDR 2600:1f18:2551:8900/56. Note that the prefix length of an IPv6 VPC CIDR is always /56.

Complete Exercise 4.1 to create your own VPC.

EXERCISE 4.1

Create a New VPC

Create a VPC with the primary CIDR 172.16.0.0/16.

To complete this exercise using the AWS Command Line Interface, enter the following command:

EXERCISE 4.1 (continued)

```
aws ec2 create-vpc
--cidr-block 172.16.0.0/16
```

If the command succeeds, you should see output similar to the following:

```
{
    "Vpc": {
        "CidrBlock": "172.16.0.0/16",
        "DhcpOptionsId": "dopt-21500a47",
        "State": "pending",
        "VpcId": "vpc-0d19e8153b4d142ed",
        "OwnerId": "158826777352",
        "InstanceTenancy": "default",
        "Ipv6CidrBlockAssociationSet": [],
        "CidrBlockAssociationSet": [
            {
                "AssociationId": "vpc-cidr-assoc-0a99f2470e29710b7",
                "CidrBlock": "172.16.0.0/16",
                "CidrBlockState": {
                    "State": "associated"
                }
            }
        ],
        "IsDefault": false,
        "Tags": []
    }
}
```

Note that the VPC is initially in the pending state as AWS creates it. To view its state, note the VPC identifier and issue the following command:

```
aws ec2 describe-vpcs --vpc-ids [vpc-id]
```

You should see output similar to the following:

```
{
    "Vpcs": [
        {
            "CidrBlock": "172.16.0.0/16",
            "DhcpOptionsId": "dopt-21500a47",
            "State": "available",
```

```
            "VpcId": "vpc-0d19e8153b4d142ed",
            "OwnerId": "158826777352",
            "InstanceTenancy": "default",
            "CidrBlockAssociationSet": [
                {
                    "AssociationId": "vpc-cidr-assoc-0a99f2470e29710b7",
                    "CidrBlock": "172.16.0.0/16",
                    "CidrBlockState": {
                        "State": "associated"
                    }
                }
            ],
            "IsDefault": false
        }
    ]
}
```

Your new VPC is now ready to use. Keep it handy, as you'll be using it in subsequent exercises.

Subnets

A subnet is a logical container within a VPC that holds VPC resources, including your EC2 instances. A subnet lets you isolate instances from each other, control how traffic flows to and from your instances, and organize them by function. For example, you can create one subnet for public web servers that need to be accessible from the Internet and create another subnet for database servers that only the web instances can access. In concept, subnets are similar to virtual LANs (VLANs) in a traditional network.

Every instance must exist within a subnet. You'll often hear the phrase "launch an instance into a subnet." After you create an instance in a subnet, you can't move the instance. By extension, this also means you can't move an instance from one VPC to another. You can, however, terminate it and create a different instance in another subnet. To preserve the data on the instance's EBS volume, you can snapshot the volume, create an AMI, and then use the AMI to create a new instance in another subnet.

Subnet CIDR Blocks

From the VPC CIDR block, you carve out a smaller CIDR block for each subnet. For example, if your VPC has a CIDR of 172.16.0.0/16, one of your subnets may have a CIDR of 172.16.100.0/24. This range includes all IP addresses from 172.16.100.0 to 172.16.100.255, which yields a total of 256 addresses.

AWS reserves the first four and last IP addresses in every subnet. You can't assign these addresses to any instances. Assuming a subnet CIDR of 172.16.100.0/24, the following addresses would be reserved:

- 172.16.100.0
- 172.16.100.1–Implied router
- 172.16.100.2–Amazon-provided DNS server
- 172.16.100.3–Reserved
- 172.16.100.255

The restrictions on prefix lengths for a subnet CIDR are the same as VPC CIDRs. Subnet CIDR blocks in a single VPC can't overlap with each other. Also, once you assign an IP prefix to a subnet, you can't change it.

It's possible for a subnet and VPC to share the same CIDR. For example, you might assign the CIDR 192.168.0.0/16 to both a VPC and a subnet within the VPC. This is uncommon and won't leave you room for additional subnets, which means your VPC will be effectively limited to one availability zone. We explain availability zones in the next section. More commonly, each subnet's prefix length will be longer than the VPC's CIDR block to allow for multiple subnets to exist in the same VPC. For example, if you assign the CIDR 192.168.0.0/16 to a VPC, you might assign the CIDR 192.168.3.0/24 to a subnet within that VPC, leaving room for additional subnets.

A subnet can't have multiple CIDRs. Unlike a VPC that can have secondary CIDRs, a subnet can have only one. However, if a VPC has a primary CIDR and a secondary CIDR, your subnet's CIDR can be derived from either. For example, if your VPC has the primary CIDR of 172.16.0.0/16 and a secondary CIDR of 172.17.0.0/16, a subnet in that VPC could be 172.17.12.0/24, as it's derived from the secondary VPC CIDR.

Availability Zones

A subnet can exist within only one *availability zone* (AZ, or *zone* for short), which is roughly analogous to a relatively small geographic location such as a data center. Although availability zones in an AWS region are connected, they are designed so that a failure in one zone doesn't cause a failure in another.

You can achieve resiliency for your applications by creating two subnets each in a different availability zone and then spreading your instances across those zones. Table 4.1 provides an example of two subnets in different availability zones.

TABLE 4.1 Subnets in different availability zones

Subnet	Availability zone	Instance
web-subnet1	us-east-1a	web1
web-subnet2	us-east-1b	web2

In this example, if the us-east-1a zone fails, the instance web1 will fail because it's in that zone. But web2, which is in the us-east-1b zone, will remain available.

Having subnets in different availability zones is not a requirement. You can place all of your subnets in the same zone, but keep in mind that if that zone fails, all instances in those subnets will fail as well. Refer to Figure 4.1 for an example of two instances in different availability zones.

FIGURE 4.1 VPC with subnets and instances

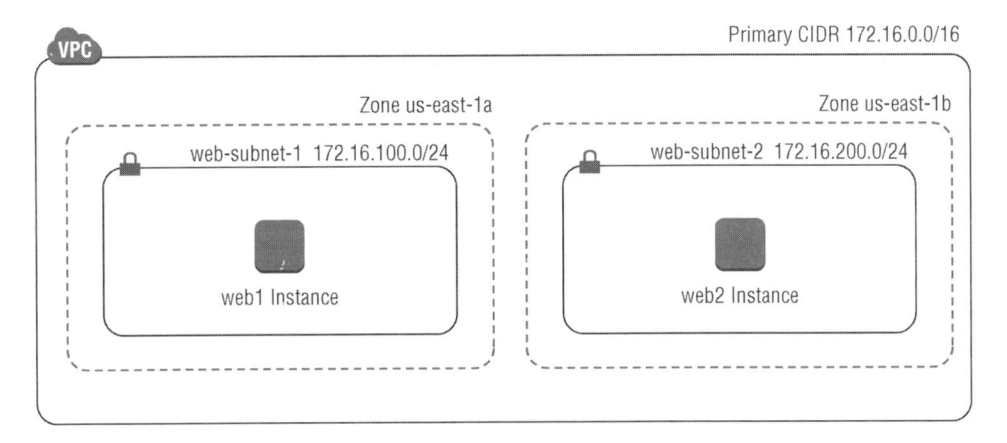

Complete Exercise 4.2 to practice creating a subnet.

EXERCISE 4.2

Create a New Subnet

Create a subnet in the VPC you created earlier. Choose an availability zone and assign the CIDR block 172.16.100.0/24.

To complete this using the following AWS Command Line Interface commands, you'll need the resource ID of the VPC:

```
aws ec2 create-subnet
--vpc-id [vpc-id]
--cidr-block 172.16.100.0/24
--availability-zone us-east-1a
```

You should see output similar to the following:

```
{
    "Subnet": {
        "AvailabilityZone": "us-east-1a",
```

EXERCISE 4.2 *(continued)*

```
            "AvailabilityZoneId": "use1-az2",
            "AvailableIpAddressCount": 251,
            "CidrBlock": "172.16.100.0/24",
            "DefaultForAz": false,
            "MapPublicIpOnLaunch": false,
            "State": "pending",
            "SubnetId": "subnet-0e398e93c154e8757",
            "VpcId": "vpc-0d19e8153b4d142ed",
            "OwnerId": "158826777352",
            "AssignIpv6AddressOnCreation": false,
            "Ipv6CidrBlockAssociationSet": [],
            "SubnetArn": "arn:aws:ec2:us-east-1:158826777352:subnet/subnet-
                0e398e93c154e8757"
    }
```

Wait a few moments, and then check the status of the subnet as follows:

```
aws ec2 describe-subnets --subnet-ids [subnet-id]
```

You should see the subnet in an `available` state, like so:

```
{
    "Subnets": [
        {
            "AvailabilityZone": "us-east-1a",
            "AvailabilityZoneId": "use1-az2",
            "AvailableIpAddressCount": 251,
            "CidrBlock": "172.16.100.0/24",
            "DefaultForAz": false,
            "MapPublicIpOnLaunch": false,
            "State": "available",
            "SubnetId": "subnet-0e398e93c154e8757",
            "VpcId": "vpc-0d19e8153b4d142ed",
            "OwnerId": "158826777352",
            "AssignIpv6AddressOnCreation": false,
            "Ipv6CidrBlockAssociationSet": [],
            "SubnetArn": "arn:aws:ec2:us-east-1:158826777352:subnet/subnet-
                0e398e93c154e8757"
        }
    ]
}
```

IPv6 CIDR Blocks

If you've allocated an IPv6 CIDR to your VPC, you can assign IPv6 CIDRs to subnets within that VPC. The prefix length for an IPv6 subnet is fixed at /64. If your VPC's IPv6 CIDR is 2600:1f18:2551:8900/56, you can assign a subnet the CIDR 2600:1f18:2551:8900/64 through 2600:1f18:2551:89FF/64.

> **NOTE**
> You must always assign an IPv4 CIDR block to a subnet, even if you intend to use only IPv6.

Elastic Network Interfaces

An *elastic network interface* (ENI) allows an instance to communicate with other network resources, including AWS services, other instances, on-premises servers, and the Internet. It also makes it possible for you to use Secure Shell (SSH) or Remote Desktop Protocol (RDP) to connect to the operating system running on your instance to manage it. As the name suggests, an ENI performs the same basic function as a network interface on a physical server, although ENIs have more restrictions on how you can configure them.

Every instance must have a *primary network interface* (also known as the *primary ENI*), which is connected to only one subnet. This is the reason you have to specify a subnet when launching an instance. You can't remove the primary ENI from an instance, and you can't change its subnet.

Primary and Secondary Private IP Addresses

Each instance must have a *primary private IP address* from the range specified by the subnet. The primary private IP address is bound to the primary ENI of the instance. You can't change or remove this address, but you can assign secondary private IP addresses to the primary ENI. Any secondary addresses must come from the same subnet that the ENI is attached to.

It's possible to attach additional ENIs to an instance. Those ENIs may be in a different subnet, but they must be in the same availability zone as the instance. As always, any addresses assigned to the ENI must come from the subnet to which it is attached.

Attaching Elastic Network Interfaces

An ENI can exist independently of an instance. You can create an ENI first and then attach it to an instance later. For example, you can create an ENI in one subnet and then attach it to an instance in the same subnet as the primary ENI when you launch the instance. If you disable the Delete On Termination attribute of the ENI, you can terminate the instance without deleting the ENI. You can then associate the ENI with another instance.

You can also take an existing ENI that's not attached to an instance and attach it to an existing instance as a secondary ENI. This lets you redirect traffic from a failed instance to a working instance by detaching the ENI from the failed instance and reattaching it to the working instance. Complete Exercise 4.3 to practice creating an ENI and attaching it to an instance.

EXERCISE 4.3

Create and Attach a Primary ENI

Create an ENI in the subnet of your choice.

Using the subnet ID from the previous exercise, issue the following AWS CLI command:

```
aws ec2 create-network-interface
--private-ip-address 172.16.100.99
--subnet-id [subnet-id]
```

You should see output similar to the following:

```
{
    "NetworkInterface": {
        "AvailabilityZone": "us-east-1a",
        "Description": "",
        "Groups": [
            {
                "GroupName": "default",
                "GroupId": "sg-011c3fe63d256f5d9"
            }
        ],
        "InterfaceType": "interface",
        "Ipv6Addresses": [],
        "MacAddress": "12:6a:74:e1:95:a9",
        "NetworkInterfaceId": "eni-0863f88f670e8ea06",
        "OwnerId": "158826777352",
        "PrivateIpAddress": "172.16.100.99",
        "PrivateIpAddresses": [
            {
                "Primary": true,
                "PrivateIpAddress": "172.16.100.99"
            }
        ],
        "RequesterId": "AIDAJULMQVBYWWAB6IQQS",
```

```
            "RequesterManaged": false,
            "SourceDestCheck": true,
            "Status": "pending",
            "SubnetId": "subnet-0e398e93c154e8757",
            "TagSet": [],
            "VpcId": "vpc-0d19e8153b4d142ed"
    }
```

After a few moments, verify the status of the network interface as follows, using the NetworkInterfaceId from the preceding output:

```
aws ec2 describe-network-interfaces
--network-interface-ids [network-interface-id]
```

Enhanced Networking

Enhanced networking offers higher network throughput speeds and lower latency than ENIs. Enhanced networking uses single-root input/output virtualization (SR-IOV). SR-IOV allows multiple instances on the same physical host to bypass the hypervisor, resulting in lower CPU utilization and better network performance. You can take advantage of enhanced networking in two ways:

Elastic Network Adapter   The elastic network adapter (ENA) supports throughput speeds up to 100 Gbps. Most instance types support it.

Intel 82599 Virtual Function Interface   The Intel 82599 virtual function (VF) interface supports speeds up to 10 Gbps. It's available for only a few instance types that don't support ENAs.

Your instance's operating system must include the appropriate drivers to support enhanced networking. Amazon Linux and Ubuntu hardware virtual machine (HVM) AMIs have ENA support enabled by default.

Internet Gateways

An *Internet gateway* gives instances the ability to receive a public IP address, connect to the Internet, and receive requests from the Internet. The default VPC has an Internet gateway associated with it by default. But when you create a custom VPC, it does not have an Internet gateway associated with it. You must create an Internet gateway and associate it with a VPC manually. You can associate only one Internet gateway with a VPC, but you may create multiple Internet gateways and associate each one with a different VPC.

An Internet gateway is somewhat analogous to an Internet router an Internet service provider may install on-premises. But in AWS, an Internet gateway doesn't behave exactly like a router.

In a traditional network, to give your servers access to the Internet, you might configure your core router with a default route pointing to the Internet router's internal IP address. An Internet gateway, however, doesn't have an IP address or network interface. Instead, AWS identifies an Internet gateway by its resource ID, which begins with igw- followed by an alphanumeric string. To use an Internet gateway, you must create a *default route* in a *route table* that points to the Internet gateway as a target.

Route Tables

To control how traffic ingresses, egresses, and moves within your VPC, you need to use routes stored in route tables. Rather than using physical or virtual routers that you configure, the VPC architecture implements IP routing as a software function that AWS calls an *implied router* (also sometimes called an *implicit router*). This means there's no virtual router on which to configure interface IP addresses or dynamic routing protocols such as BGP. Rather, you only have to manage the route table that the implied router uses.

Each route table consists of one or more routes and at least one subnet association. Think of a route table as being connected to multiple subnets in much the same way a traditional router would be. When you create a VPC, AWS automatically creates a default route table called the *main route table* and associates it with every subnet in that VPC. You can use the main route table or create a custom one that you can manually associate with one or more subnets.

A subnet cannot exist without a route table association. If you do not explicitly associate a subnet with a custom route table you've created, AWS will implicitly associate it with the main route table.

Routes

Routes determine how to forward traffic to or from resources within the subnets associated with the route table. IP routing is destination-based, meaning that routing decisions are based only on the destination IP prefix, not the source IP address. When you create a route, you must provide the following elements:

- Destination IP prefix
- Target resource

The destination must be an IPv4 or IPv6 prefix in CIDR notation. The target must be an AWS network resource such as an Internet gateway or an ENI. It cannot be an IP prefix.

Every route table contains a *local route* that allows instances in different subnets to communicate with each other. Table 4.2 shows what this route would look like in a VPC with the CIDR 172.31.0.0/16.

TABLE 4.2 The local route

Destination	Target
172.31.0.0/16	Local

The local route is the only mandatory route that exists in every route table. It's what allows communication between instances in the same VPC. Because there are no routes for any other IP prefixes, any traffic destined for an address outside of the VPC CIDR range will get dropped.

The Default Route

To enable Internet access for your instances, you must create a default route pointing to the Internet gateway. The default route is what allows Internet traffic to ingress and egress the subnet. After adding a default route, you would end up with Table 4.3.

TABLE 4.3 Route table with default route

Destination	Target
172.31.0.0/16	Local
0.0.0.0/0	igw-0e538022a0fddc318

The 0.0.0.0/0 prefix encompasses all IP addresses, including those of hosts on the Internet. This is why it's always listed as the destination in a default route. Any subnet that is associated with a route table containing a route pointing to an Internet gateway is called a *public subnet*. Contrast this with a *private subnet*, which does not have a route with an Internet gateway as a target.

Notice that the 0.0.0.0/0 and 172.31.0.0/16 prefixes overlap. When deciding where to route traffic, the implied router will route based on the closest match. Put another way, the order of routes doesn't matter. Suppose an instance sends a packet to the Internet address 198.51.100.50. Because 198.51.100.50 does not match the 172.31.0.0/16 prefix but does match the 0.0.0.0/0 prefix, the implied router will use the default route and send the packet to the Internet gateway.

AWS documentation speaks of one implied router per VPC. It's important to understand that the implied router doesn't actually exist as a discrete resource. It's an abstraction of an IP routing function. Nevertheless, you may find it helpful to think of each route table as a separate virtual router with a connection to one or more subnets. Follow the steps in Exercise 4.4 to create an Internet gateway and a default route.

EXERCISE 4.4

Create an Internet Gateway and Default Route

In this exercise, you'll create an Internet gateway and attach it to the VPC you used in the previous exercises.

1. Create an Internet gateway using the following command:

    ```
    aws ec2 create-internet-gateway
    ```

    You should see the following output:

    ```
    {
        "InternetGateway": {
            "Attachments": [],
            "InternetGatewayId": "igw-0312f81aa1ef24715",
            "Tags": []
        }
    }
    ```

2. Attach the Internet gateway to the VPC you used in the previous exercises using the following command:

    ```
    aws ec2 attach-internet-gateway
    --internet-gateway-id [internet-gateway-id]
    --vpc-id [vpc-id]
    ```

3. Retrieve the route table ID of the main route table for the VPC:

    ```
    aws ec2 describe-route-tables
    --filters Name=vpc-id,Values=[vpc-id]
    ```

    You should see something like the following:

    ```
    {
        "RouteTables": [
            {
                "Associations": [
                    {
    ```

Route Tables

```
                "Main": true,
                "RouteTableAssociationId": "rtbassoc-00f60edb255be0332",
                "RouteTableId": "rtb-097d8be97649e0584",
                "AssociationState": {
                    "State": "associated"
                }
            }
        ],
        "PropagatingVgws": [],
        "RouteTableId": "rtb-097d8be97649e0584",
        "Routes": [
            {
                "DestinationCidrBlock": "172.16.0.0/16",
                "GatewayId": "local",
                "Origin": "CreateRouteTable",
                "State": "active"
            }
        ],
        "Tags": [],
        "VpcId": "vpc-0d19e8153b4d142ed",
        "OwnerId": "158826777352"
    }
  ]
}
```

4. To create a default route in the main route table, issue the following command:

   ```
   aws ec2 create-route
   --route-table-id rtb-097d8be97649e0584
   --destination-cidr-block "0.0.0.0/0"
   --gateway-id igw-0312f81aa1ef24715
   ```

   If the command succeeds, you should see the following output:

   ```
   {
       "Return": true
   }
   ```

5. Confirm the presence of the route by rerunning the command from step 3.

Security Groups

A *security group* functions as a firewall that controls traffic to and from an instance by permitting traffic to ingress or egress that instance's ENI. Every ENI must have at least one security group associated with it. One ENI can have multiple security groups attached, and the same security group can be attached to multiple ENIs.

In practice, because most instances have only one ENI, people often think of a security group as being attached to an instance. When an instance has multiple ENIs, take care to note whether those ENIs use different security groups.

When you create a security group, you must specify a group name, description, and VPC for the group to reside in. Once you create the group, you create inbound and outbound rules to specify what traffic the security group allows. If you don't explicitly allow traffic using a rule, the security group will block it.

Inbound Rules

Inbound rules specify what traffic is allowed into the attached ENI. An inbound rule consists of three required elements:

- Source
- Protocol
- Port range

When you create a security group, it doesn't contain any inbound rules. Security groups use a default-deny approach, also called *whitelisting*, which denies all traffic that is not explicitly allowed by a rule. When you create a new security group and attach it to an instance, all inbound traffic to that instance will be blocked. You must create inbound rules to allow traffic to your instance. The order of rules in a security group doesn't matter.

Suppose you have an instance running an HTTPS-based web application. You want to allow anyone on the Internet to connect to this instance, so you'd need an inbound rule to allow all TCP traffic coming in on port 443 (the default port and protocol for HTTPS). To manage this instance using SSH, you'd need another inbound rule for TCP port 22. However, you don't want to allow SSH access from just anyone. You need to allow SSH access only from the IP address 198.51.100.10. To achieve this, you would use a security group containing the inbound rules listed in Table 4.4.

TABLE 4.4 Inbound rules allowing SSH and HTTPS access from any IP address

Source	Protocol	Port range
198.51.100.10/32	TCP	22
0.0.0.0/0	TCP	443

The prefix 0.0.0.0/0 covers all valid IP addresses, so using the preceding rule would allow HTTPS access not only from the Internet but from all instances in the VPC as well.

Outbound Rules

Outbound rules specify what traffic the instance may send out of the attached ENI. The format of an outbound rule mirrors an inbound rule and contains three elements:

- Destination
- Protocol
- Port range

In many cases, the outbound rules of a security group will be less restrictive than the inbound rules. When you create a security group, AWS automatically creates the outbound rule listed in Table 4.5.

TABLE 4.5 Outbound rule allowing Internet access

Destination	Protocol	Port range
0.0.0.0/0	All	All

The main purpose of this rule is to allow the instance to access the Internet and other AWS resources. You may delete this rule, but if you do, the security group won't permit your instance to access the Internet or anything else!

Sources and Destinations

The source or destination in a rule can be any CIDR. The source can also be the resource ID of a security group. If you specify a security group as the source, the rule will permit the traffic to any instance with that security group attached. This makes it easy to allow instances to communicate with each other by simply assigning the same security group to all of them.

The source security group you use can exist in a different AWS account. You'll need to specify the AWS account owner ID in order to specify a source security group in a different account.

Stateful Firewall

A security group acts as a stateful firewall. Stateful means that when a security group allows traffic to pass in one direction, it intelligently allows reply traffic in the opposite direction. For instance, when you allow an instance outbound access to download updates from a software repository on the Internet, the security group automatically allows reply traffic back into the instance.

Security groups use connection tracking to determine whether to allow response traffic to pass. For TCP and UDP traffic, connection tracking looks at the flow information for each packet that is allowed by a security group rule. By tracking flows, the security group can identify reply traffic that belongs to the same flow and distinguish it from unsolicited traffic. Flow information includes:

- Protocol
- Source and destination IP address
- Source and destination port number

Default Security Group

Each VPC contains a default security group that you can't delete. You don't have to use it, but if you do, you can modify its rules to suit your needs. Alternatively, you may opt to create your own custom group and use it instead. Complete Exercise 4.5 to practice creating a custom security group.

EXERCISE 4.5

Create a Custom Security Group

In this exercise, you'll create a new security group that allows SSH, HTTP, and HTTPS access from any IP address.

1. Create a security group named **web-ssh** in the VPC you created previously:

   ```
   aws ec2 create-security-group
   --group-name "web-ssh"
   --description "Web and SSH traffic"
   --vpc-id [vpc-id]
   ```

   Make a note of the security group ID in the output:

   ```
   {
       "GroupId": "sg-0f076306080ac2c91"
   }
   ```

2. Using the security group ID, create three rules to allow SSH, HTTP, and HTTPS access from any IP address.

   ```
   aws ec2 authorize-security-group-ingress
   --group-id [group-id]
   --protocol "tcp"
   --cidr "0.0.0.0/0"
   --port "22"
   ```

```
aws ec2 authorize-security-group-ingress
--group-id [group-id]
--protocol "tcp"
--cidr "0.0.0.0/0"
--port "80"
aws ec2 authorize-security-group-ingress
--group-id [group-id]
--protocol "tcp"
--cidr "0.0.0.0/0"
--port "443"
```

3. Verify the rules by viewing the security group:

   ```
   aws ec2 describe-security-groups
   --group-id [group-id]
   ```

Network Access Control Lists

Like a security group, a *network access control list* (NACL) functions as a firewall in that it contains inbound and outbound rules to allow traffic based on a source or destination CIDR, protocol, and port. Also, each VPC has a default NACL that can't be deleted. But the similarities end there.

An NACL differs from a security group in many respects. Instead of being attached to an ENI, an NACL is attached to a subnet. The NACL associated with a subnet controls what traffic may enter and exit that subnet. This means that NACLs can't be used to control traffic between instances in the same subnet. If you want to do that, you have to use security groups.

A subnet can have only one NACL associated with it. When you create a new subnet in a VPC, the VPC's default NACL is associated with the subnet by default. You can modify the default NACL, or you can create a new one and associate it with the subnet. You can also associate the same NACL with multiple subnets, provided those subnets are all in the same VPC as the NACL.

Unlike a security group, which is stateful, an NACL is stateless, meaning that it doesn't use connection tracking and doesn't automatically allow reply traffic. This is much like an access control list (ACL) on a traditional switch or router. The stateless nature of the NACL is why each one is preconfigured with rules to allow all inbound and outbound traffic, as discussed in the following sections.

Inbound Rules

Inbound rules determine what traffic is allowed to ingress the subnet. Each rule contains the following elements:

- Rule number
- Protocol
- Port range
- Source CIDR
- Action—Allow or deny

The default NACL for a VPC that has no IPv6 CIDR assigned comes prepopulated with the two inbound rules listed in Table 4.6.

TABLE 4.6 Default NACL inbound rules

Rule number	Protocol	Port range	Source	Action
100	All	All	0.0.0.0/0	Allow
*	All	All	0.0.0.0/0	Deny

NACL rules are processed in ascending order of the rule number. Rule 100 is the lowest-numbered rule, so it gets processed first. This rule allows all traffic from any source. You can delete or modify this rule or create additional rules before or after it. For example, if you wanted to block only HTTP (TCP port 80), you could add the rule in Table 4.7.

TABLE 4.7 Blocking rule

Rule number	Protocol	Port range	Source	Action
90	TCP	80	0.0.0.0/0	Deny

This rule denies all TCP traffic with a destination port of 80. Because it's the lowest-numbered rule in the list, it gets processed first. Any traffic not matching this rule would be processed by rule 100, which allows all traffic. The net effect of rules 90 and 100 is that all inbound traffic except for that destined to TCP port 80 would be allowed.

Network Access Control Lists

> **NOTE**
>
> After traffic matches a rule, the action specified in the rule occurs. That means if traffic matches a rule with a deny action, that traffic will be denied, even if there's a subsequent (higher-numbered) rule that would allow the traffic.

The last rule in Table 4.6 is the default rule. It's designated by an asterisk (*) instead of a number and is always the last rule in the list. You can't delete or otherwise change the default rule. The default rule causes the NACL to deny any traffic that isn't explicitly allowed by any of the preceding rules. Complete Exercise 4.6 to create a custom NACL.

EXERCISE 4.6

Create an Inbound Rule to Allow Remote Access from Any IP Address

NACL rule order matters! Create a new NACL and attach it to the subnet you used in Exercise 4.3.

1. Create a new network ACL using the following command:

   ```
   aws ec2 create-network-acl
   --vpc-id [vpc-id]
   ```

   Note the network ACL ID in the output, which should look something like the following:

   ```
   {
       "NetworkAcl": {
           "Associations": [],
           "Entries": [
               {
                   "CidrBlock": "0.0.0.0/0",
                   "Egress": true,
                   "IcmpTypeCode": {},
                   "PortRange": {},
                   "Protocol": "-1",
                   "RuleAction": "deny",
                   "RuleNumber": 32767
               },
               {
                   "CidrBlock": "0.0.0.0/0",
                   "Egress": false,
                   "IcmpTypeCode": {},
   ```

EXERCISE 4.6 (continued)

```
            "PortRange": {},
            "Protocol": "-1",
            "RuleAction": "deny",
            "RuleNumber": 32767
        }
    ],
    "IsDefault": false,
    "NetworkAclId": "acl-052f05f358d96cfb3",
    "Tags": [],
    "VpcId": "vpc-0d19e8153b4d142ed",
    "OwnerId": "158826777352"
    }
}
```

Notice that the network ACL includes the default ingress and egress rules to deny all traffic. Even though the rule number is 32767, it shows up in the AWS Management Console as an asterisk.

2. Create an inbound rule entry to allow SSH (TCP port 22) access from any IP address. We'll use the rule number 70.

```
aws ec2 create-network-acl-entry
--ingress
--cidr-block "0.0.0.0/0"
--protocol "tcp"
--port-range "From=22,To=22"
--rule-action "allow"
--network-acl-id [network-acl-id]
--rule-number 70
```

The From= and To= values after the --port-range flag refer to the range of ports to allow, *not* to the source port number. In this rule, we just want to allow traffic to TCP port 22.

3. Modify the command from step 2 to create rule number 80 that allows RDP (TCP port 3389) access from your public IP address (Use https://ifconfig.co to find it if you don't already know it). Use a /32 prefix length.

4. Use the following command to verify that the rules made it into your NACL:

```
aws ec2 describe-network-acls
--network-acl-id [network-acl-id]
```

Outbound Rules

As you might expect, the outbound NACL rules follow an almost identical format as the inbound rules. Each rule contains the following elements:

- Rule number
- Protocol
- Port range
- Destination
- Action

Each default NACL comes with the outbound rules listed in Table 4.8. Notice that the rules are identical to the default inbound rules except for the Destination element.

TABLE 4.8 Default NACL outbound rules

Rule number	Protocol	Port range	Destination	Action
100	All	All	0.0.0.0/0	Allow
*	All	All	0.0.0.0/0	Deny

Because an NACL is stateless, it won't allow return traffic unless you add a rule to permit it. Therefore, if you permit HTTPS traffic with an inbound rule, you must also explicitly permit the return traffic using an outbound rule. In this case, rule 100 permits the return traffic.

If you do need to restrict access from the subnet—to block Internet access, for example—you will have to create an outbound rule to allow return traffic over *ephemeral ports*. Ephemeral ports are reserved TCP or UDP ports that clients listen for reply traffic on. As an example, when a client sends an HTTPS request to your instance over TCP port 80, that client may listen for a reply on TCP port 36034. Your NACL's outbound rules must allow traffic to TCP port 36034.

The range of ephemeral ports varies by client operating system. Many modern operating systems use ephemeral ports in the range of 49152–65535. But don't assume that allowing only this range will be sufficient. The range for TCP ports may differ from the range for UDP, and older or customized operating systems may use a different range altogether. To maintain compatibility, do not restrict outbound traffic using an NACL. Instead, use a security group to restrict outbound traffic.

> **NOTE** If your VPC includes an IPv6 CIDR, AWS will automatically add inbound and outbound rules to permit IPv6 traffic.

Using Network Access Control Lists and Security Groups Together

You may want to use an NACL in addition to a security group so that you aren't dependent on AWS administrators to specify the correct security group when they launch an instance. Because an NACL is applied to the subnet, the rules of the NACL apply to all traffic ingressing and egressing the subnet, regardless of how the security groups are configured.

When you make a change to an NACL or security group rule, that change takes effect immediately (practically, within several seconds). Avoid changing security groups and NACLs simultaneously. If your changes don't work as expected, it can become difficult to identify whether the problem is with a security group or an NACL. Complete your changes on one before moving to the other. Additionally, be cautious about making changes when there are active connections to an instance, because an incorrectly ordered NACL rule or missing security group rule can terminate those connections.

> **Note:** In an NACL rule, you can specify only a CIDR as the source or destination. This is unlike a security group rule, for which you can specify another security group for the source or destination.

Public IP Addresses

A public IP address is reachable over the public Internet. This is in contrast to RFC 1918 addresses (e.g., 192.168.1.1), which cannot be routed over the Internet but can be routed within private networks.

You need a public IP address for an instance if you want others to directly connect to it via the Internet. Naturally, this requires an Internet gateway attached to the VPC that the instance resides in. You *may* also give an instance a public IP address if you want it to have outbound-only Internet access. You *don't* need a public IP address for your instances to communicate with each other within the VPC infrastructure, as this instance-to-instance communication happens using private IP addresses.

When you launch an instance into a subnet, you can choose to automatically assign it a public IP. This is convenient, but there are a couple of potential downsides to this approach.

First, if you forget to choose this option when you launch the instance, you cannot go back and have AWS automatically assign a public IP later. Second, automatically assigned public IP addresses aren't persistent. When you stop or terminate the instance, you will lose the public IP address. If you stop and restart the instance, it will receive a different public IP address.

Even if you don't plan to stop your instance, keep in mind that AWS may perform maintenance events that cause your instance to restart. If this happens, its public IP address will change.

If your instance doesn't need to maintain the same IP address for a long period of time, this approach may be acceptable. If not, you may opt instead for an *elastic IP address*.

Elastic IP Addresses

An elastic IP (EIP) address is a type of public IP address that AWS allocates to your account when you request it. After AWS allocates an EIP to your account, you have exclusive use of that address until you manually release it. Outside of AWS, there's no noticeable difference between an EIP and an automatically assigned public IP.

When you initially allocate an EIP, it is not bound to any instance. Instead, you must associate it with an ENI. You can move an EIP around to different ENIs, although you can associate it with only one ENI at a time. When you associate an EIP with an ENI, it will remain associated for the life of the ENI or until you disassociate it.

If you associate an EIP to an ENI that already has an automatically assigned public IP address allocated, AWS will replace the public IP address with the EIP. Complete Exercise 4.7 to practice allocating and using an EIP.

An EIP is tied to an AWS region and can't be moved outside of that region. It is possible, however, to transfer any public IP addresses that you own into your AWS account as EIPs. This is called bring your own IP address (BYOIP). You can bring up to five address blocks per region.

EXERCISE 4.7

Allocate and Use an Elastic IP Address

Allocate an elastic IP address and associate it with the instance you created earlier.

1. Allocate an EIP using the following command:

   ```
   aws ec2 allocate-address
   ```

   You should see a public IP address and allocation ID in the output:

   ```
   {
       "PublicIp": "3.210.93.49",
       "AllocationId": "eipalloc-0e14547dac92e8a75",
       "PublicIpv4Pool": "amazon",
       "NetworkBorderGroup": "us-east-1",
       "Domain": "vpc"
   }
   ```

2. Associate the EIP to the elastic network interface you created earlier:

   ```
   aws ec2 associate-address
   --allocation-id eipalloc-0e14547dac92e8a75
   --network-interface-id eni-0863f88f670e8ea06
   ```

EXERCISE 4.7 *(continued)*

3. Verify that the EIP is associated with the elastic network interface:

   aws ec2 describe-network-interfaces
   --network-interface-ids eni-0863f88f670e8ea06

   You should see the public IP address in the output:

   ```
   {
       "NetworkInterfaces": [
           {
               "Association": {
                   "AllocationId": "eipalloc-0e14547dac92e8a75",
                   "AssociationId": "eipassoc-045cc2cab27d6338b",
                   "IpOwnerId": "158826777352",
                   "PublicDnsName": "",
                   "PublicIp": "3.210.93.49"
               },
               "AvailabilityZone": "us-east-1a",
               "Description": "",
               "Groups": [
                   {
                       "GroupName": "default",
                       "GroupId": "sg-011c3fe63d256f5d9"
                   }
               ],
               "InterfaceType": "interface",
               "Ipv6Addresses": [],
               "MacAddress": "12:6a:74:e1:95:a9",
               "NetworkInterfaceId": "eni-0863f88f670e8ea06",
               "OwnerId": "158826777352",
               "PrivateIpAddress": "172.16.100.99",
               "PrivateIpAddresses": [
                   {
                       "Association": {
                           "AllocationId": "eipalloc-0e14547dac92e8a75",
                           "AssociationId": "eipassoc-045cc2cab27d6338b",
                           "IpOwnerId": "158826777352",
                           "PublicDnsName": "",
                           "PublicIp": "3.210.93.49"
                       },
   ```

```
                "Primary": true,
                "PrivateIpAddress": "172.16.100.99"
            }
        ],
        "RequesterId": "AIDAJULMQVBYWWAB6IQQS",
        "RequesterManaged": false,
        "SourceDestCheck": true,
        "Status": "available",
        "SubnetId": "subnet-0e398e93c154e8757",
        "TagSet": [],
        "VpcId": "vpc-0d19e8153b4d142ed"
    }
]
}
```

AWS Global Accelerator

If you have AWS resources in multiple regions, having separate EIPs for each region can be cumbersome to manage. AWS Global Accelerator solves this problem by giving you two anycast static IPv4 addresses that you can use to route traffic to resources in any region. Unlike EIPs, which are tied to an AWS region, Global Accelerator static addresses are spread across different AWS points-of-presence (POPs) in over 30 countries. These static addresses are also called anycast addresses because they are simultaneously advertised from multiple POPs.

Users connecting to a static address are automatically routed to the nearest POP. The Global Accelerator listener receives the TCP or UDP connection, and then proxies it to the resources you've specified in an endpoint group. An endpoint group can contain elastic IP addresses, elastic load balancers, or EC2 instances.

Global Accelerator routes traffic to the fastest endpoint. Because the static addresses it uses are anycast, the failure of one POP doesn't cause an interruption of service. Users are automatically and transparently routed to another POP.

Network Address Translation

When you associate an ENI with a public IP address, the ENI maintains its private IP address. Associating a public IP with an ENI doesn't reconfigure the ENI with a new address. Instead, the Internet gateway maps the public IP address to the ENI's private IP address using a process called *network address translation* (NAT).

When an instance with a public IP connects to a host on the Internet, the host sees the traffic as originating from the instance's public IP. For example, assume an instance with a private IP address of 172.31.7.10 is associated with the EIP 35.168.241.48. When the instance attempts to send a packet to the Internet host 198.51.100.11, it will send the following packet to the Internet gateway:

Source IP address: 172.31.7.10

Destination IP address: 35.168.241.48

The Internet gateway will translate this packet to change the source IP address to the instance's public IP address. The translated packet, which the Internet gateway forwards to the host, looks like this:

Source IP address: 35.168.241.48

Destination IP address: 198.51.100.11

Likewise, when a host on the Internet sends a packet to the instance's EIP, the Internet gateway will perform network address translation on the incoming packet. The packet that reaches the Internet gateway from the Internet host will look like this:

Source IP address: 198.51.100.11

Destination IP address: 35.168.241.48

The Internet gateway will translate this packet, replacing the destination IP address with the instance's private IP address, as follows:

Source IP address: 198.51.100.11

Destination IP address: 172.31.7.10

Network address translation occurs automatically at the Internet gateway when an instance has a public IP address. You can't change this behavior.

> Network address translation as described here is also sometimes called one-to-one NAT because one private IP address gets mapped to one public IP address.

Network Address Translation Devices

Although network address translation occurs at the Internet gateway, there are two other resources that can also perform NAT:

- NAT gateway
- NAT instance

AWS calls these *NAT devices*. The purpose of a NAT device is to allow an instance to access the Internet while preventing hosts on the Internet from reaching the instance directly. This is useful when an instance needs to go out to the Internet to fetch updates or to upload data but does not need to service requests from clients.

When you use a NAT device, the instance needing Internet access does not have a public IP address allocated to it. Incidentally, this makes it impossible for hosts on the Internet to reach it directly. Instead, only the NAT device is configured with a public IP. Additionally, the NAT device has an interface in a public subnet. Refer to Table 4.9 for an example.

TABLE 4.9 IP address configuration when using a NAT device

Name	Subnet	Private IP	Public IP
db1	Private	172.31.7.11	None
db2	Private	172.31.7.12	None
NAT device	Public	172.31.8.10	18.209.220.180

When db1 sends a packet to a host on the Internet with the address 198.51.100.11, the packet must first go to the NAT device. The NAT device translates the packet as follows:

Original Packet's Source IP Address	Original Packet's Destination IP Address	Translated Packet's Source IP Address	Translated Packet's Destination IP Address
172.31.7.11 (db1)	198.51.100.11	172.31.8.10 (NAT device)	198.51.100.11

The NAT device then takes the translated packet and forwards it to the Internet gateway. The Internet gateway performs NAT translation on this packet as follows:

NAT Device Packet's Source IP Address	NAT Device Packet's Destination IP Address	Translated Packet's Source IP Address	Translated Packet's Destination IP Address
172.31.8.10 (NAT device)	198.51.100.11	18.209.220.180 (NAT device's EIP)	198.51.100.11

Multiple instances can use the same NAT device, thus sharing the same public IP address for outbound connections. The function that NAT devices perform is also called *port address translation* (PAT).

Configuring Route Tables to Use NAT Devices

Instances that use the NAT device must send Internet-bound traffic to it, and the NAT device must send Internet-bound traffic to an Internet gateway. Hence, the NAT device and the instances that use it must use different default routes. Furthermore, they must also use different route tables and hence must reside in separate subnets.

Refer to Table 4.9 again. Notice that the instances reside in the Private subnet, and the NAT device is in the Public subnet. The default routes for these subnets would follow the pattern in Table 4.10.

TABLE 4.10 Default routes for the Private and Public subnets

Subnet	Destination	Target
Private	0.0.0.0/0	NAT device
Public	0.0.0.0/0	igw-0e538022a0fddc318

Refer to the diagram in Figure 4.2 to see the relationship between both of the route tables. Recall that a route target must be a VPC resource such as an instance, Internet gateway, or ENI. The specific target you choose depends on the type of NAT device you use: a NAT gateway or a NAT instance.

FIGURE 4.2 Network address translation using a NAT device

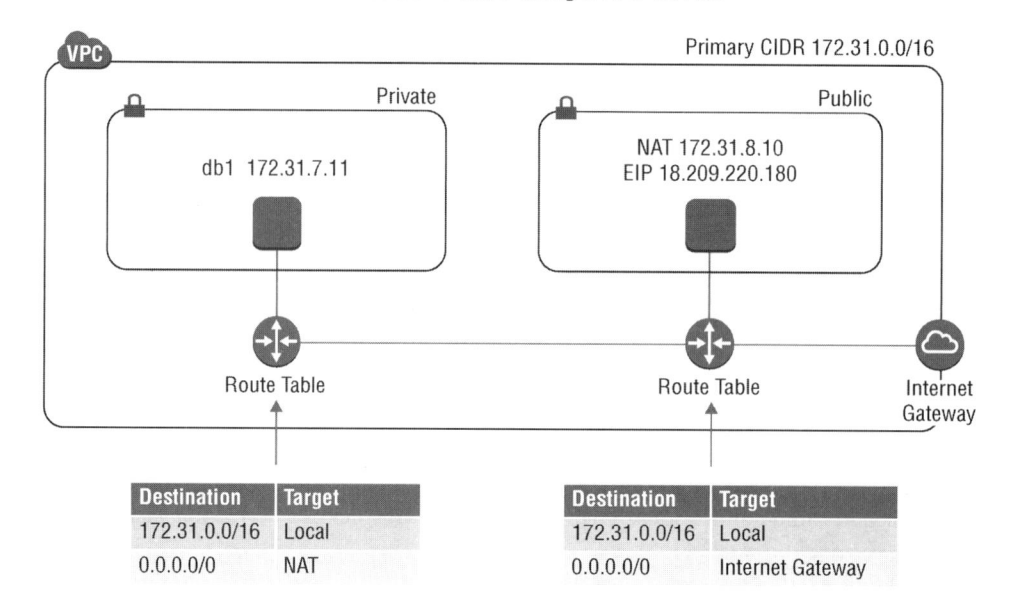

NAT Gateway

A NAT gateway is a NAT device managed by AWS. Like an Internet gateway, it's a one-size-fits-all resource. It doesn't come in a variety of flavors, and there's nothing to manage or access. It automatically scales to accommodate your bandwidth requirements. You set it and forget it.

When you create a NAT gateway, you must assign it an EIP. A NAT gateway can reside in only one subnet, which must be a public subnet for it to access the Internet. AWS selects a private IP address from the subnet and assigns it to the NAT gateway. For redundancy, you may create additional NAT gateways in different availability zones.

After creating a NAT gateway, you must create a default route to direct Internet-bound traffic from your instances to the NAT gateway. The target you specify will be the NAT gateway ID, which follows the format `nat-0750b9c8de7e75e9f`. If you use multiple NAT gateways, you can create multiple default routes, each pointing to a different NAT gateway as the target.

Because a NAT gateway doesn't use an ENI, you can't apply a security group to it. You can, however, apply an NACL to the subnet that it resides in.

NAT Instance

A NAT instance is a normal EC2 instance that uses a preconfigured Linux-based AMI. You have to perform the same steps to launch it as you would any other instance. It functions like a NAT gateway in many respects, but there are some key differences.

Unlike a NAT gateway, a NAT instance doesn't automatically scale to accommodate increased bandwidth requirements. Therefore, it's important that you select an appropriately robust instance type. If you choose an instance type that's too small, you must manually upgrade to a larger instance type.

Also, a NAT instance has an ENI, so you must apply a security group to it. You also must remember to assign it a public IP address. Lastly, you must disable the *source/destination check* on the NAT instance's ENI. This allows the NAT instance to receive traffic addressed to an IP other than its own, and it also allows the instance to send traffic using a source IP that it doesn't own.

One advantage of a NAT instance is that you can use it as a *bastion host*, sometimes called a *jump host*, to connect to instances that don't have a public IP. You can't do this with a NAT gateway.

You must create a default route to direct Internet-bound traffic to the NAT instance. The target of the default route will be the NAT instance's ID, which follows the format `i-0a1674fe5671dcb00`.

If you want to guard against instance or availability zone failures, it's not as simple as just spinning up another NAT instance. You cannot create multiple default routes pointing to different NAT instances. If you need this level of resiliency, you're better off using NAT gateways instead.

VPC Peering

You can configure VPC peering to allow instances in one VPC to communicate with VPCs in another over the private AWS network. You may want to do this if you have instances in different regions that need to communicate. You may also want to connect your instances to another AWS customer's instances.

To enable VPC peering, you must set up a *VPC peering connection* between two VPCs. A VPC peering connection is a point-to-point connection between two and only two VPCs. You can have at most one peering connection between a pair of VPCs. Also, peered VPCs must not have overlapping CIDR blocks.

With one exception, a VPC peering connection allows only instance-to-instance communication. This means an instance in one VPC can use the peering connection only to connect to another instance in the peered VPC. You can't use it to share Internet gateways or NAT devices. You can, however, use it to share a network load balancer (NLB).

If you have more than two VPCs you need to connect, you *cannot* daisy-chain VPC peering connections together and route through them. You must create a peering connection between each pair. This configuration is called *transitive routing*.

To use a peering connection, you must create new routes in both VPCs to allow traffic to travel in both directions. For each route, the destination prefix should exist within the destination VPC. The target of each route must be the peering connection's identifier, which begins with pcx-. Refer to Table 4.11 for examples of two routes you would create to enable peering between a pair of VPCs.

TABLE 4.11 Routes for VPC peering

Source VPC CIDR	Destination VPC CIDR	Target
172.31.0.0/16	10.0.0.0/16	pcx-076781cf11220b9dc
10.0.0.0/16	172.31.0.0/16	pcx-076781cf11220b9dc

Notice that the routes are mirrors of each other. Again, this is to allow traffic in both directions. The destination CIDR doesn't need to exactly match that of the destination VPC. If you want to enable peering only between specific subnets, you can specify the subnet CIDR instead.

Interregion VPC peering is not available for some AWS regions. Peering connections between regions have a maximum transmission unit (MTU) of 1,500 bytes and do not support IPv6.

Hybrid Cloud Networking

So far, we've covered connecting your VPC resources to the Internet. But in many cases, you want your resources to essentially be an extension of your data center, and to remain private and inaccessible from the Internet. AWS offers three options for private connectivity between your on-premises network and your VPCs:

- Virtual private networks (VPNs)
- AWS Transit Gateway
- AWS Direct Connect

Virtual Private Networks

Using a virtual private network (VPN), you can connect a VPC to an on-premises network such as a data center or office via a secure connection that goes over the public Internet. You create a VPN connection by configuring a VPC resource called a virtual private gateway and then configuring your on-premises router or firewall—what AWS calls your customer gateway—to build an encrypted VPN tunnel with the virtual private gateway. Virtual private gateways support AES 256-bit and AES 128-bit encryption.

If you want to connect your on-premises network to multiple VPCs, you'll need to create a separate virtual private gateway and VPN tunnel for each VPC. Keep in mind that VPCs don't support transitive routing, so you can't just peer VPCs together, create a VPN connection to one of them, and then route through that one VPC.

However, if you want to connect a large number of VPCs to your on-premises network, or if you need to connect many on-premises networks to a VPC, creating a separate VPN connection for each one can become cumbersome. In that case, you'll want to use AWS Transit Gateway.

> **NOTE** You can optionally have AWS create a Global Accelerator endpoint for your VPN connection. This approach can simplify configuration and improve performance if you have multiple VPN customer gateways in different geographical areas.

AWS Transit Gateway

AWS Transit Gateway is a highly available service that lets you connect multiple VPCs and on-premises networks via Direct Connect links or virtual private networks. In addition to simplifying connectivity, AWS Transit Gateway gives you granular control over how traffic is routed among your VPCs and on-premises networks.

After you create a transit gateway, you attach it to a VPC, VPN connection, Direct Connect gateway, or another transit gateway. You then associate these attachments with a special type of route table called a transit gateway route table.

Transit Gateway Route Table

The transit gateway route table controls how traffic flows among the attached network resources. A transit gateway route table is similar to a VPC route table, such as a main or custom route table that you associate with a VPC subnet. But some key differences exist.

The target of a route in a transit gateway route table can only be an attachment to a VPC, VPN connection, Direct Connect gateway, or other transit gateway. You can't specify an ENI or Internet gateway as a target.

> **Note:** You're billed hourly for each transit gateway attachment. As of this writing, the current rate is about $0.05 USD per hour.

There are many uses for AWS Transit Gateway, but the following five are the most common:

- Centralized router
- Isolated router
- Shared services
- Peering
- Multicast

Centralized Router

You can configure a transit gateway to function as a centralized router that controls how traffic is routed among all of your VPCs and on-premises networks. In the centralized router model, you have one transit gateway route table associated with all your attachments.

When you use a transit gateway to connect a VPC to an on-premises network, you don't use a virtual private gateway. Instead, the transit gateway terminates the VPN connection with your on-premises router or firewall. The transit gateway advertises and receives routes via BGP and holds those routes in the transit gateway route table. Likewise, routes to connected VPC networks are stored in the transit gateway route table. Dynamically learning routes and storing them in a routing table is called *route propagation*.

To better understand how transit gateways work, consider an example. Suppose that your on-premises network is 192.168.0.0/16 and your VPC subnet is 10.98.76.0/24. To enable communication between your VPC subnet and your on-premises network using a transit gateway, you first must attach the transit gateway to the VPC. Doing this will create in the transit gateway routing table a dynamic (propagated) route to the subnet, as shown in Table 4.12. Second, you create a static route in the subnet's associated route table pointing to the transit gateway as a target. Third, you must configure a VPN connection between the transit gateway and your on-premises network. Your on-premises device terminating the VPN connection will send routes via BGP, and those routes will dynamically get put into the transit gateway routing table.

TABLE 4.12 Route table entries for using a transit gateway

Route table	Destination	Target	Route type
VPC subnet	192.168.0.0/16	Transit gateway	Static
Transit gateway	192.168.0.0/16	VPN connection	Propagated
Transit gateway	10.98.76.0/24	VPC attachment	Propagated

Isolated VPCs

A transit gateway can have multiple transit gateway route tables, each associated with different attachments. This allows you to effectively create multiple isolated virtual routers within one transit gateway.

One instance where this can come in handy is when you have multiple VPCs that you want to connect to an on-premises network but want to keep the VPCs isolated from each other.

Isolated VPCs with Shared Services

If you're hosting shared services (such as Active Directory or Link Layer Discovery Protocol [LLDP]) in one VPC, you can use a transit gateway to securely share those resources with other networks, while maintaining isolation among them. Follow the steps in Exercise 4.8 to create a transit gateway that connects two VPCs.

EXERCISE 4.8

Create a Transit Gateway

In this exercise, you'll create another VPC and subnet. You'll then create a transit gateway and attach it to both VPCs. Finally, you'll configure the transit gateway to route between the VPC subnets.

1. Create a new VPC and subnet:

   ```
   aws ec2 create-vpc --cidr-block 172.17.0.0/16
   aws ec2 create-subnet
   --vpc-id vpc-08edadb7e52eedd37
   --cidr-block 172.17.100.0/24
   --availability-zone us-east-1b
   ```

EXERCISE 4.8 (continued)

2. Use the following AWS CLI command to create a new transit gateway:

   ```
   aws ec2 create-transit-gateway
   ```

   You should see in the output a transit gateway ID and the default transit gateway route table ID:

   ```
   {
       "TransitGateway": {
           "TransitGatewayId": "tgw-0fe8e470174ca0be8",
           "TransitGatewayArn": "arn:aws:ec2:us-east-1:158826777352:transit-
           gateway/tgw-0fe8e470174ca0be8",
           "State": "pending",
           "OwnerId": "158826777352",
           "CreationTime": "2020-05-11T17:26:35+00:00",
           "Options": {
               "AmazonSideAsn": 64512,
               "AutoAcceptSharedAttachments": "disable",
               "DefaultRouteTableAssociation": "enable",
               "AssociationDefaultRouteTableId": "tgw-rtb-01e158d45848e8522",
               "DefaultRouteTablePropagation": "enable",
               "PropagationDefaultRouteTableId": "tgw-rtb-01e158d45848e8522",
               "VpnEcmpSupport": "enable",
               "DnsSupport": "enable",
               "MulticastSupport": "disable"
           }
       }
   }
   ```

3. Attach the transit gateway to the VPC subnets:

   ```
   aws ec2 create-transit-gateway-vpc-attachment
   --transit-gateway-id "tgw-0fe8e470174ca0be8"
   --vpc-id "vpc-0d19e8153b4d142ed"
   --subnet-ids "subnet-0e398e93c154e8757"
   aws ec2 create-transit-gateway-vpc-attachment
   --transit-gateway-id "tgw-0fe8e470174ca0be8"
   --vpc-id "vpc-08edadb7e52eedd37"
   --subnet-ids "subnet-08cce691ef4bfae40"
   ```

4. After you attach the transit gateway to the VPC subnets, the default transit gateway route table is populated with routes to the subnets. You can view these routes with the following command:

```
aws ec2 search-transit-gateway-routes
--transit-gateway-route-table-id tgw-rtb-01e158d45848e8522
--filters "Name=type,Values=static,propagated"
```

You should see the CIDRs for both subnets as follows:

```
{
    "Routes": [
        {
            "DestinationCidrBlock": "172.16.0.0/16",
            "TransitGatewayAttachments": [
                {
                    "ResourceId": "vpc-0d19e8153b4d142ed",
                    "TransitGatewayAttachmentId": "tgw-attach-0421300408cf0a63b",
                    "ResourceType": "vpc"
                }
            ],
            "Type": "propagated",
            "State": "active"
        },
        {
            "DestinationCidrBlock": "172.17.0.0/16",
            "TransitGatewayAttachments": [
                {
                    "ResourceId": "vpc-08edadb7e52eedd37",
                    "TransitGatewayAttachmentId":"tgw-attach-
                        07cfbfa14a60cbce2",
                    "ResourceType": "vpc"
                }
            ],
            "Type": "propagated",
            "State": "active"
        }
    ],
    "AdditionalRoutesAvailable": false
}
```

The transit gateway is now configured to pass traffic between the subnets.

EXERCISE 4.8 *(continued)*

5. To use the transit gateway, you need to add to each subnet's route table a route for the other subnet. To do this, you'll need the route table IDs and CIDRs for each subnet.

   ```
   aws ec2 create-route
   --route-table-id rtb-097d8be97649e0584
   --destination-cidr-block "172.17.0.0/16"
   --transit-gateway-id tgw-0fe8e470174ca0be8
   aws ec2 create-route
   --route-table-id rtb-01bbf401d3b503764
   --destination-cidr-block "172.16.0.0/16"
   --transit-gateway-id tgw-0fe8e470174ca0be8
   ```

6. Verify the routes using the following command:

   ```
   aws ec2 describe-route-tables
   --filters Name=route.transit-gateway-id,
   Values=tgw-0fe8e470174ca0be8
   ```

   Your output should look as follows:

   ```
   {
       "RouteTables": [
           {
               "Associations": [
                   {
                       "Main": true,
                       "RouteTableAssociationId": "rtbassoc-00f60edb255be0332",
                       "RouteTableId": "rtb-097d8be97649e0584",
                       "AssociationState": {
                           "State": "associated"
                       }
                   }
               ],
               "PropagatingVgws": [],
               "RouteTableId": "rtb-097d8be97649e0584",
               "Routes": [
                   {
                       "DestinationCidrBlock": "172.16.0.0/16",
                       "GatewayId": "local",
                       "Origin": "CreateRouteTable",
                       "State": "active"
                   },
                   {
   ```

```
                "DestinationCidrBlock": "172.17.0.0/16",
                "TransitGatewayId": "tgw-0fe8e470174ca0be8",
                "Origin": "CreateRoute",
                "State": "active"
            },
            {
                "DestinationCidrBlock": "0.0.0.0/0",
                "GatewayId": "igw-0312f81aa1ef24715",
                "Origin": "CreateRoute",
                "State": "active"
            }
        ],
        "Tags": [],
        "VpcId": "vpc-0d19e8153b4d142ed",
        "OwnerId": "158826777352"
    },
    {
        "Associations": [
            {
                "Main": true,
                "RouteTableAssociationId": "rtbassoc-051c79bc6d208c008",
                "RouteTableId": "rtb-01bbf401d3b503764",
                "AssociationState": {
                    "State": "associated"
                }
            }
        ],
        "PropagatingVgws": [],
        "RouteTableId": "rtb-01bbf401d3b503764",
        "Routes": [
            {
                "DestinationCidrBlock": "172.16.0.0/16",
                "TransitGatewayId": "tgw-0fe8e470174ca0be8",
                "Origin": "CreateRoute",
                "State": "active"
            },
            {
                "DestinationCidrBlock": "172.17.0.0/16",
                "GatewayId": "local",
                "Origin": "CreateRouteTable",
```

EXERCISE 4.8 *(continued)*

```
                "State": "active"
            }
        ],
        "Tags": [],
        "VpcId": "vpc-08edadb7e52eedd37",
        "OwnerId": "158826777352"
    }
  ]
}
```

Transit Gateway Peering

You can peer transit gateways together, even between different regions. If you have resources spread across different regions, this can reduce the number of VPN connections and VPC peering connections.

Multicast

AWS Transit Gateway supports multicast traffic between VPCs. For each multicast domain, you specify the ENI of an instance that will serve as the multicast source (also called the multicast group). You also specify the multicast group address and the EC2 instances that should receive multicast traffic. Note that a multicast receiver can be any EC2 instance, but the sender must be a Nitro instance.

You can configure multiple multicast groups in a single multicast domain. Also, you can associate a VPC subnet with only one multicast domain. Keep in mind that multicast routes do not appear in any route tables.

Blackhole Routes

If you want to block a specific route, you can create a blackhole entry for that route in the transit gateway route table. A blackhole route causes the transit gateway to drop any traffic that matches the route. Blackhole routes are useful for permanently blocking specific networks such as known malicious IP addresses. Blackhole routes are also useful for temporarily stopping traffic from a VPC without detaching it. Complete Exercise 4.9 for an example.

EXERCISE 4.9

Create a Blackhole Route

In this exercise, you'll create a blackhole route for the range of addresses 172.16.100.64–172.16.100.71. You'll then delete your transit gateway.

1. Use the following command to create a blackhole route for the 172.16.100.64/29 subnet.

   ```
   aws ec2 create-transit-gateway-route
   --destination-cidr-block 172.16.100.64/29
   --transit-gateway-route-table-id tgw-rtb-01e158d45848e8522
   --blackhole
   ```

   When you're done practicing, you'll want to delete the transit gateway attachments to avoid charges. You'll need the transit gateway attachment IDs, which you can get using the same command you used in Exercise 4.8:

   ```
   aws ec2 search-transit-gateway-routes
   --transit-gateway-route-table-id tgw-rtb-01e158d45848e8522
   --filters "Name=type,Values=static,propagated"
   ```

2. Delete the attachments using the following commands:

   ```
   aws ec2 delete-transit-gateway-vpc-attachment
   --transit-gateway-attachment-id tgw-attach-0421300408cf0a63b
   aws ec2 delete-transit-gateway-vpc-attachment
   --transit-gateway-attachment-id tgw-attach-07cfbfa14a60cbce2
   ```

   It may take a few minutes for the attachments to be deleted.

3. Delete the transit gateway using the following command:

   ```
   aws ec2 delete-transit-gateway --transit-gateway-id tgw-0fe8e470174ca0be8
   ```

AWS Direct Connect

The AWS Direct Connect service offers private, low-latency connectivity to your AWS resources. One advantage of Direct Connect is that it lets you bypass the Internet altogether when accessing AWS resources, letting you avoid the unpredictable and high latency of a broadband Internet connection. This approach is useful when you need to transfer large data sets or real-time data, or you need to meet regulatory requirements that preclude transferring data over the Internet. Using Direct Connect, you can access any AWS services in a given region that you could otherwise access via the Internet, including your EC2 and RDS instances and Amazon S3. There are two types of Direct Connect connections: dedicated and hosted.

Dedicated

A dedicated connection is a single physical connection that terminates at an AWS Direct Connect location. To use a dedicated connection, you will need your own equipment in a Direct Connect location. For a list of locations, see aws.amazon.com/directconnect/features/#AWS_Direct_Connect_Locations. Each location is associated with an AWS region and gives you access to AWS resources in that region. For example, the Digital Realty ATL1 Direct Connect facility offers access to the us-east-1 region. For dedicated connections, you can choose from 1 Gbps or 10 Gbps connections.

Hosted

If you need less than 1 Gbps of bandwidth, or if placing your equipment in a Direct Connect location is infeasible, you can get a hosted Direct Connect link that supports speeds between 50 Mbps and 10 Gbps. A hosted connection extends the "last-mile" connection from a Direct Connect location to your data center or office.

Direct Connect Gateways

A Direct Connect gateway is a global resource that provides a single connection point to multiple VPCs in a region. On the AWS side, a Direct Connect gateway does this by connecting to an AWS Transit Gateway or virtual private gateway. On your end, a Direct Connect gateway maintains a BGP session with your on-premises equipment and advertises and receives IPv4 and IPv6 route prefixes via BGP.

Virtual Interfaces

Depending on how you plan to use your Direct Connect link, you must create one or more virtual interfaces. There are three types of virtual interfaces:

Private Virtual Interface Allows you to connect to the private IP addresses of resources in a single VPC, such as EC2 and RDS instances.

Public Virtual Interface Lets you use the public IP address of AWS services that have a public endpoint, such as S3 and DynamoDB. This is useful if you have on-premises applications already configured to access AWS services using public endpoints.

Transit Virtual Interface Provides connectivity to one or more AWS transit gateways. You typically use an AWS transit gateway when you need to connect resources in multiple VPCs. This type of virtual interface is available with connection speeds of 1 Gbps and greater.

A Direct Connect link with a speeds of 1 Gbps or greater can support multiple virtual interfaces. A lower-speed link can support only one virtual interface.

> **NOTE** Unlike VPN connections, Direct Connect links are not encrypted. However, because AWS endpoints are all secured by TLS, traffic between your on-premises network and AWS is secure.

High-Performance Computing

High-performance computing (HPC) is a computing paradigm that uses multiple instances to simultaneously process computationally intensive workloads in parallel. These instances compose an HPC cluster. HPC clusters fall into one of two categories, which are differentiated by the degree of interaction required between instances.

Loosely Coupled   Loosely coupled workloads can be broken up into smaller tasks that each instance can work on independently. Image processing is one example. Another is DNA sequencing, wherein a genome can be broken up into pieces and those pieces can be doled out to different nodes for independent analysis.

In a loosely coupled cluster, one instance doesn't depend on another to complete its work. Because instances don't need low-latency network connectivity to each other, they don't have to be in the same cluster placement group.

Tightly Coupled   Tightly coupled workloads require massive computing power and can't be broken up into pieces. Instead, multiple instances must work in concert as a single supercomputer. This requires instances to have low-latency, high-speed network connectivity to each other. Hence, you'll want to place tightly coupled instances in the same cluster placement group.

Tightly coupled workloads include simulations wherein one variable can affect multiple other variables. One example of this is machine learning. Another is weather forecasting. Each instance in a cluster begins with a set of weather conditions and works on a particular aspect of the weather forecast, such as location, temperature, precipitation, humidity, or wind speed. Because weather doesn't exist in a vacuum, all the instances in the cluster must continually share the results of their simulations with each other.

Typically, because a delay in one instance can slow down the entire cluster, instances in a tightly coupled cluster will have similar or identical specifications. Also, because the failure of one instance can impact the others, it's necessary for each instance to periodically save its simulation state so that the simulation can be restarted in the event of a failure.

> Generally, when you hear the term HPC, it refers to tightly coupled HPC clusters. Thus, a defining characteristic of HPC is incredibly high-speed, low-latency, and reliable network connectivity among nodes.

Elastic Fabric Adapter

The elastic fabric adapter (EFA) is a special type of ENA that supports traditional TCP/IP networking capabilities. What makes it unique is that it allows HPC applications to use the

Libfabric API to bypass the operating system's TCP/IP stack and access the EFA directly, resulting in more throughput and reduced latency.

When using an EFA for HPC applications, all instances must be in the same subnet. EFA traffic is not routable. Also, every EFA in a cluster must be attached to the same security group, and that group must allow all traffic to and from the group. And although it's not required, you should put your instances in the same cluster placement group to ensure minimal network latency.

Only about a dozen of the pricier instance types support EFA. You can attach only one EFA to an instance, and you can do so only at launch time or when the instance is stopped.

AWS ParallelCluster

AWS ParallelCluster can automatically manage your Linux-based HPC cluster so that you don't have to do it manually. ParallelCluster provisions your cluster instances and creates a 15 GB shared filesystem for them to use. The shared filesystem is stored on an EBS volume that's attached to a master instance and served up to the other instances via the Network File System (NFS). However, you can alternatively use a shared filesystem such as provided by Amazon Elastic File System (EFS) or Amazon FSx for Lustre. ParallelCluster also creates a batch scheduler using AWS Batch. You submit your HPC computing jobs to the scheduler, and ParallelCluster takes care of scaling your cluster in or out as necessary.

Summary

The Virtual Private Cloud service provides the networking foundation for EC2 and other AWS services. AWS abstracts some networking components in such a way that their configuration is easier than in a traditional network, but you still need to have a solid grasp of networking fundamentals to architect VPCs.

In each region, AWS automatically provides a default VPC with default subnets, a main route table, a default security group, and a default NACL. Many use a default VPC for a long time without ever having to configure a VPC from scratch. This makes it all the more important that you as an AWS architect understand how to configure a virtual network infrastructure from scratch. There's a good chance you won't be allowed to modify an infrastructure that was built on top of a default VPC. Instead, you may be tasked with replicating it from the ground up—troubleshooting various issues along the way. Practice what you've learned in this chapter until creating fully functional VPCs becomes second nature to you.

In a traditional network, you're free to reconfigure server IP addresses, move them to different subnets, and even move them to different physical locations. You have tremendous flexibility to change your plans midstream. When creating a VPC, you don't have that luxury. You must carefully plan your entire application infrastructure up front, not just your network. It's therefore crucial that you understand how all the VPC and EC2 components fit together.

You begin by defining a contiguous IP address range for the VPC and representing it as a CIDR. The primary CIDR should ideally be sufficiently large to accommodate all your instances but small enough to leave room for a secondary CIDR.

After that, you have to divvy up your VPC CIDR into subnets. Because a subnet is a container that resides in only one availability zone, you must make up-front decisions about where to place your instances. After you create an instance in a subnet, you can't move it.

Prior to launching instances, you need to configure security groups, since every instance's ENI must have at least one attached. One area where you do have some flexibility is with network access control lists. You can associate an NACL with a subnet at any time or forego NACLs altogether.

If you want your instances to be accessible from the Internet, you must provision an Internet gateway, create a default route, and assign public IP addresses. Those are the basics. If you choose to use a NAT gateway or instance or a VPC peering connection, you'll have to modify multiple route tables.

Exam Essentials

Be able to determine the correct prefix length for a CIDR block based on the number of IP addresses you need in a VPC or subnet. Allowed prefix lengths range from /16 to /28. The longer the prefix length, the fewer the number of IP addresses you'll have available.

Understand the significance of a subnet. A subnet is a logical container that holds EC2 instances. Each subnet has a CIDR block derived from the CIDR of the VPC that it resides in. An instance in the subnet must take its private IP address from the subnet's CIDR. But AWS reserves the first four and last IP addresses in every subnet.

Know the impact of an availability zone failure. If a zone fails, every subnet in that zone—and every instance in that subnet—will likewise fail. To tolerate a zone failure, build redundancy into your instance deployments by spreading them across different zones.

Understand the rules for creating and using elastic network interfaces (ENIs). Every instance must have a primary network interface with a primary private IP address. Any additional ENI you attach to an instance must be in the same subnet as the primary ENI.

Be able to create, modify, and use route tables. Know the purpose of the main route table in a VPC and its relationship to the VPC's subnets. Also understand how to create a public subnet using an Internet gateway and a default route.

Know the differences between security groups and network access control lists. Understand why a stateful security group requires different rules than a stateless NACL to achieve the same result.

Understand how network address translation works. Know the difference between NAT that occurs at the Internet gateway and NAT that occurs on a NAT device. NAT that occurs on a NAT device is also called port address translation (PAT). Multiple instances can share a single public IP address using a NAT device.

Be able to create and configure VPC peering among multiple VPCs. Know the limitations of VPC peering connections. VPC peering connections do not support transitive routing or IPv6. Inter-region peering is available for some regions.

Review Questions

1. What is the range of allowed IPv4 prefix lengths for a VPC CIDR block?
   A. /16 to /28
   B. /16 to /56
   C. /8 to /30
   D. /56 only

2. You've created a VPC with the CIDR 192.168.16.0/24. You want to assign a secondary CIDR to this VPC. Which CIDR can you use?
   A. 172.31.0.0/16
   B. 192.168.0.0/16
   C. 192.168.0.0/24
   D. 192.168.16.0/23

3. You need to create two subnets in a VPC that has a CIDR of 10.0.0.0/16. Which of the following CIDRs can you assign to one of the subnets while leaving room for an additional subnet? (Choose all that apply.)
   A. 10.0.0.0/24
   B. 10.0.0.0/8
   C. 10.0.0.0/16
   D. 10.0.0.0/23

4. What is the relationship between a subnet and an availability zone?
   A. A subnet can exist in multiple availability zones.
   B. An availability zone can have multiple subnets.
   C. An availability zone can have only one subnet.
   D. A subnet's CIDR is derived from its availability zone.

5. Which is true regarding an elastic network interface?
   A. It must have a private IP address from the subnet that it resides in.
   B. It cannot exist independently of an instance.
   C. It can be connected to multiple subnets.
   D. It can have multiple IP addresses from different subnets.

6. Which of the following statements is true of security groups?
   A. Only one security group can be attached to an ENI.
   B. A security group must always be attached to an ENI.
   C. A security group can be attached to a subnet.
   D. Every VPC contains a default security group.

7. How does an NACL differ from a security group?
   A. An NACL is stateless.
   B. An NACL is stateful.
   C. An NACL is attached to an ENI.
   D. An NACL can be associated with only one subnet.

8. What is an Internet gateway?
   A. A resource that grants instances in multiple VPCs' Internet access
   B. An implied router
   C. A physical router
   D. A VPC resource with no management IP address

9. What is the destination for a default IPv4 route?
   A. 0.0.0.0/0
   B. ::0/0
   C. An Internet gateway
   D. The IP address of the implied router

10. You create a new route table in a VPC but perform no other configuration on it. You then create a new subnet in the same VPC. Which route table will your new subnet be associated with?
    A. The main route table
    B. The route table you created
    C. The default route table
    D. None of these

11. You create a Linux instance and have AWS automatically assign a private IP address but not a public IP address. What will happen when you stop and restart the instance?
    A. You won't be able to establish an SSH session directly to the instance from the Internet.
    B. The instance won't be able to access the Internet.
    C. The instance will receive the same private IP address.
    D. The instance will be unable to reach other instances in its subnet.

12. How can you assign a public IP address to a running instance that doesn't have one?
    A. Allocate an ENI and associate it with the instance's primary EIP.
    B. Allocate an EIP and associate it with the instance's primary ENI.
    C. Configure the instance to use an automatically assigned public IP.
    D. Allocate an EIP and change the private IP address of the instance's ENI to match.

13. When an instance with an automatically assigned public IP sends a packet to another instance's EIP, what source address does the destination instance see?
    A. The public IP
    B. The EIP
    C. The private IP
    D. 0.0.0.0

14. Why must a NAT device reside in a different subnet than an instance that uses it?
    A. Both must use different default gateways.
    B. Both must use different NACLs.
    C. Both must use different security groups.
    D. The NAT device requires a public interface and a private interface.

15. Which of the following is a difference between a NAT instance and NAT gateway?
    A. There are different NAT gateway types.
    B. A NAT instance scales automatically.
    C. A NAT gateway can span multiple availability zones.
    D. A NAT gateway scales automatically.

16. Which VPC resource performs network address translation?
    A. Internet gateway
    B. Route table
    C. EIP
    D. ENI

17. What must you do to configure a NAT instance after creating it?
    A. Disable the source/destination check on its ENI.
    B. Enable the source/destination check on its ENI.
    C. Create a default route in its route table with a NAT gateway as the target.
    D. Assign a primary private IP address to the instance.

18. Which of the following is true regarding VPC peering?
    A. Transitive routing is not supported.
    B. A VPC peering connection requires a public IP address.
    C. You can peer up to three VPCs using a single peering connection.
    D. You can use a peering connection to share an Internet gateway among multiple VPCs.

19. You've created one VPC peering connection between two VPCs. What must you do to use this connection for bidirectional instance-to-instance communication? (Choose all that apply.)
    A. Create two routes with the peering connection as the target.
    B. Create only one default route with the peering connection as the target.
    C. Create another peering connection between the VPCs.
    D. Configure the instances' security groups correctly.

20. Which of the following is a *not* a limitation of interregion VPC peering?
    A. It's not supported in some regions.
    B. The maximum MTU is 1,500 bytes.
    C. You can't use IPv4.
    D. You can't use IPv6.

21. Which over which of the following connection types is always encrypted?
    A. Direct Connect
    B. VPN
    C. VPC peering
    D. Transit gateway

22. Which of the following allows EC2 instances in different regions to communicate using private IP addresses? (Choose three.)
    A. VPN
    B. Direct Connect
    C. VPC peering
    D. Transit gateway

23. Which of the following is true of a route in a transit gateway route table?
    A. It can be multicast.
    B. It can be a blackhole route.
    C. It can have an Internet gateway as a target.
    D. It can have an ENI as a target.

24. Which of the following is an example of a tightly coupled HPC workload?
    A. Image processing
    B. Audio processing
    C. DNA sequencing
    D. Hurricane track forecasting
    E. Video processing

Chapter 5

Database Services

THE AWS CERTIFIED SOLUTIONS ARCHITECT ASSOCIATE EXAM OBJECTIVES COVERED IN THIS CHAPTER MAY INCLUDE, BUT ARE NOT LIMITED TO, THE FOLLOWING:

✓ **Domain 1: Design Resilient Architectures**

- 1.1 Design a multi-tier architecture solution
- 1.2 Design highly available and/or fault-tolerant architectures
- 1.3 Design decoupling mechanisms using AWS services
- 1.4 Choose appropriate resilient storage

✓ **Domain 2: Design High Performing Architectures**

- 2.1 Identify elastic and scalable compute solutions for a workload
- 2.3 Select high-performing networking solutions for a workload
- 2.4 Choose high-performing database solutions for a workload

Introduction

Most applications and services depend on at least one database. A database allows an application to store, organize, and quickly retrieve data. Although you could use flat files to store data, they become increasingly slow to search as the amount of data grows. By relying on a database to perform these tasks, application developers can focus on the application without having to directly interact with a filesystem to store and retrieve data.

Consequently, the availability and performance of a database-backed application depend on the database you choose and how you configure it. Databases come in two flavors: relational and nonrelational. Each of these differs in the way it stores, organizes, and lets you retrieve data, so the type of database you choose depends on the needs of the application.

In this chapter, you'll learn the differences between these two database types and how to select the right one for your application. You'll also learn how to use the managed database services that AWS provides to get the level of performance and reliability your applications require, as well as how to protect your data and recover it in the event of a database failure. This chapter will introduce three different managed database services provided by AWS:

- Amazon Relational Database Service (RDS)
- Amazon Redshift
- DynamoDB

Relational Databases

A *relational database* contains at least one *table*, which you can visualize as a spreadsheet with columns and rows. In a relational database table, columns may also be called *attributes*, and rows may also be called *records* or *tuples*.

Columns and Attributes

Before you can add data to a relational database table, you must predefine each column's name and what data types it can accept. Columns are ordered, and you can't change the order after you create the table. The ordering creates a relationship between attributes in the table, which is where the term *relational database* comes from. Refer to Table 5.1 for an example of a relational database table containing employee records.

TABLE 5.1 The Employees table

Employee ID (Number)	Department (String)	Last name (String)	First name (String)	Birthdate (Date)
101	Information technology	Smith	Charlotte	07-16-1987
102	Marketing	Colson	Thomas	07-04-2000

Data must match the type defined under each column. For example, the Employee ID can't be a letter, because it's defined as a numeric data type. One advantage of a relational database is that it allows for flexible queries—you don't have to understand up front how you're going to query the data. As long as the data is there in a consistent format, you can craft queries to get the data you want, the way you want. This makes relational databases good for applications that need to query data in arbitrary columns and customize how that data is presented. For example, you could query the database to return the birthdate of every employee whose first name is Charlotte. Or you could query the database for every employee with a birthday in July. Relational databases make getting the data you want fast and easy.

Another advantage is that you can add more columns to a table after creating it. You can also delete columns, but deleting a column entails deleting all the data stored under the column. If you delete the First Name column, it will remove first name data for all employees in the table.

Using Multiple Tables

Storing all data in a single table can lead to unnecessary duplication, needlessly increasing the size of the database and making queries slower. Hence, it's common for applications to use multiple related tables. Using the preceding example, if 50 employees work in the information technology department, the string "Information technology" appears in the table 50 times—once for each record. To avoid this wasted space, you may create a separate table to hold department names, as shown in Table 5.2.

TABLE 5.2 The Departments table

Department ID (Number)	Department name (String)
10	Information technology
20	Marketing

Instead of putting the department names in every employee record, you would create a single record for each department in the Departments table. The Employees table would then refer to each department using the Department ID, as shown in Table 5.3.

TABLE 5.3 The Employees table

Employee ID (Number)	Department (String)	Last name (String)	First name (String)	Birthdate (Date)
101	10	Smith	Charlotte	07-16-1987
102	20	Colson	Thomas	07-04-2000

In this relationship, the Departments table is the *parent table*, and the Employees table is the *child table*. Each value in the Department column in the Employees table refers to the Department ID in the Departments table. Notice that the data type for the Department column is still a string. Although you could change the data type to a number, it's not necessary. The Department ID in the Departments table is called the *primary key*, and it must be unique in the table so that it can uniquely identify a row. The Employees table refers to the Department ID as a *foreign key*.

You must define primary and foreign keys so that the database knows how the columns in different tables are related. The database will enforce *foreign key constraints* to ensure that when a child table references a foreign key, that key also exists in the parent.

> **NOTE** There's a common misconception that the "relational" in relational database refers to the relationship between tables. But the relational aspect refers to the relationship of attributes *within* a row. In other words, all the attributes in a row are related to each other in some way. To better understand this, imagine a relational database with only one table. Even though there are no relationships between tables, it's still a relational database.

Structured Query Language

You use the *Structured Query Language* (SQL) with relational databases to store and query data and perform database maintenance tasks. For this reason, relational databases are often called SQL databases.

SQL commands or statements differ slightly depending on the specific relational database management system (RDBMS) you're using. All major programming languages have libraries that construct SQL statements and interact with the database, so you don't need to master SQL. But you do need to understand the concepts behind a few common SQL terms.

Querying Data

The SELECT statement is used to query data from a SQL database. It allows you to query based on the value in any column, as well as specify the specific columns you want the database to return. Thanks to the predictable structure of tables and the enforcement of foreign key constraints, you can use a JOIN clause with a SELECT statement to join together data from different tables.

Storing Data

The INSERT statement allows you to insert data directly into a table. For example, when adding new customer information to a customer relationship management (CRM) application, the application is probably using a SQL INSERT on the backend to add the customer data to a table. If you need to load a large number of records into a database, you can use the COPY command to copy data from a properly formatted file into the table you specify.

Online Transaction Processing vs. Online Analytic Processing

A relational database can be optimized for fast, frequent transactions, or it can be optimized for large, complex queries. Depending on its configuration, a relational database can fall into one of two categories: *online transaction processing* (OLTP) or *online analytic processing* (OLAP).

OLTP

OLTP databases are suited to applications that read and write data frequently, on the order of multiple times per second. They are optimized for fast queries, and those queries tend to be regular and predictable. Depending on the size of the database and its performance requirements, an OLTP database may have intense memory requirements so that it can store frequently accessed portions of tables in memory for quick access. Generally, a single server with ample memory and compute power handles all writes to an OLTP database. An OLTP database would be a good candidate for backing an online ordering system that processes hundreds of orders a minute.

OLAP

OLAP databases are optimized for complex queries against large data sets. As a result, OLAP databases tend to have heavy compute and storage requirements. In *data warehousing* applications, it's common to aggregate multiple OLTP databases into a single OLAP database. For example, in an OLTP database for an employee management system, employee data may be spread out across multiple tables. At regular but infrequent intervals, a data warehouse would aggregate these tables into a single table in an OLAP database. This makes it easier to structure SQL queries against the data. It also reduces the amount of time it takes to process such a computationally intensive query by offloading it to the high-powered OLAP database. With a large OLAP database, it's common for multiple database servers to share the computational load of complex queries. In a process called *partitioning* or *sharding*, each server gets a portion of the database for which it's responsible.

Amazon Relational Database Service

Amazon Relational Database Service (RDS) is a managed database service that lets you run relational database systems in the cloud. RDS takes care of setting up the database system, performing backups, ensuring high availability, and patching the database software and the underlying operating system. RDS also makes it easy to recover from database failures, restore data, and scale your databases to achieve the level of performance and availability that your application requires.

To deploy a database using RDS, you start by configuring a *database instance*, which is an isolated database environment. A database instance exists in a virtual private cloud (VPC) that you specify, but unlike an EC2 instance, AWS fully manages database instances. You can't establish an SSH session to them, and they don't show up under your EC2 instances.

Database Engines

A database engine is simply the software that stores, organizes, and retrieves data in a database. Each database instance runs only one database engine. RDS offers the following six database engines to choose from:

MySQL   MySQL is designed for OLTP applications such as blogs and e-commerce. RDS offers the latest MySQL Community Edition versions. MySQL offers two storage engines—MyISAM and InnoDB—but you should use the latter as it's the only one compatible with RDS-managed automatic backups.

MariaDB   MariaDB is a drop-in binary replacement for MySQL. It was created over concerns about MySQL's future after Oracle acquired the company that developed it. MariaDB supports the XtraDB and InnoDB storage engines, but AWS recommends using the latter for maximum compatibility with RDS.

Oracle   Oracle is one of the most widely deployed relational database management systems. Some applications expressly require an Oracle database. RDS provides the following Oracle Database editions:

- Standard Edition One (SE1)
- Standard Edition Two (SE2)
- Standard Edition (SE)
- Enterprise Edition (EE)

PostgreSQL   PostgreSQL advertises itself as the most Oracle-compatible open source database. This is a good choice when you have in-house applications that were developed for Oracle but want to keep costs down.

Amazon Aurora   Aurora is Amazon's drop-in binary replacement for MySQL and PostgreSQL. Aurora offers better write performance than both by using a virtualized storage layer that reduces the number of writes to the underlying storage. It provides two editions:

- MySQL compatible
- PostgreSQL compatible

Depending on the edition you choose, Aurora is compatible with PostgreSQL or MySQL import and export tools and snapshots. Aurora is designed to let you seamlessly migrate from an existing deployment that uses either of those two open source databases. For MySQL-compatible editions, Aurora supports only the InnoDB storage engine. Also, the Aurora Backtrack feature for MySQL lets you, within a matter of seconds, restore your database to any point in time within the last 72 hours.

Microsoft SQL Server   RDS offers multiple Microsoft SQL Server versions ranging from 2012 to the present. For the edition, you can choose Express, Web, Standard, or Enterprise. The variety of flavors makes it possible to migrate an existing SQL Server database from an on-premises deployment to RDS without having to perform any database upgrades.

Licensing Considerations

RDS provides two models for licensing the database engine software you run. The *license included* model covers the cost of the license in the pricing for an RDS instance. The *bring your own license* (BYOL) model requires you to obtain a license for the database engine you run.

License Included   MariaDB and MySQL use the GNU General Public License (GPL) v2.0, and PostgreSQL uses the PostgreSQL license, all of which allow for free use of the respective software.

All versions and editions of Microsoft SQL Server that you run on RDS include a license, as do Oracle Database Standard Edition One (SE1) and Standard Edition Two (SE2).

Bring Your Own License   Only the Oracle database engine supports this licensing model. The following Oracle Database editions allow you to bring your own license:

- Enterprise Edition (EE)
- Standard Edition (SE)
- Standard Edition One (SE1)
- Standard Edition Two (SE2)

Database Option Groups

Different database engines offer various features or *options* to help you manage your databases and improve security. *Option groups* let you specify these features and apply them to one or more instances. Options require more memory, so make sure your instances have ample memory and enable only the options you need.

The options available for a database option group depend on the engine. Oracle offers Amazon S3 integration. Both Microsoft SQL Server and Oracle offer *transparent data encryption* (TDE), which causes the engine to encrypt data before writing it to storage. MySQL and MariaDB offer an audit plug-in that lets you log user logons and queries run against your databases.

Database Instance Classes

When launching a database instance, you must decide how much processing power, memory, network bandwidth, and disk throughput it needs. RDS offers a variety of database instance classes to meet the diverse performance needs of different databases. If you get it wrong or if your needs change, you can switch your instance to a different class. RDS divides database instance classes into the following three types.

Standard

Standard instance classes meet the needs of most databases. The latest-generation instance class is db.m5, which provides up to:

- 384 GB of memory
- 96 vCPU
- 25 Gbps network bandwidth
- 19,000 Mbps (2,375 MBps) disk throughput

Memory Optimized

Memory-optimized instance classes are for databases that have hefty performance requirements. Providing more memory to a database allows it to store more data in memory, which can result in faster query times. There are three latest-generation instance classes: db.x1e, db.z1d, and db.r5. The most powerful of these is db.x1e, and it provides up to:

- 3,904 GB of memory
- 128 vCPU
- 25 Gbps network bandwidth
- 14,000 Mbps (1,750 MBps) disk throughput

Database instances use EBS storage. Both the standard and memory-optimized instance class types are EBS optimized, meaning they provide dedicated bandwidth for transfers to and from EBS storage.

Burstable Performance

Burstable performance instances are for development, test, and other nonproduction databases. The latest burstable performance instance class available is db.t3, and it gives you up to:

- 32 GB of memory
- 8 vCPU
- 5 Gbps network bandwidth
- 2,048 Mbps (256 MBps) disk throughput

The db.t3, db.m5, and db.r5 classes are based on the AWS Nitro System, accounting for significantly improved performance over older generation instance classes. Note that disk reads and writes count against the maximum disk throughput on these instance classes.

Storage

Selecting the right storage for your database instance is about more than just ensuring you have enough disk space. You also have to decide how fast the storage must be to meet the performance requirements of your database-backed application.

Understanding Input/Output Operations Per Second

AWS measures storage performance in *input/output operations per second* (IOPS). An input/output (I/O) operation is either a read from or write to storage. All things being equal, the more IOPS you can achieve, the faster your database can store and retrieve data.

RDS allocates you a number of IOPS depending on the type of storage you select, and you can't exceed this threshold. The speed of your database storage is limited by the number of IOPS allocated to it. The amount of data you can transfer in a single I/O operation depends on the page size that the database engine uses. To understand how many IOPS you need, you first need to understand how much disk throughput you need.

Before we look at a few examples, a little warning: the math can get tricky, but you don't need to memorize all the details. Just make sure you understand what IOPS actually measure. MySQL and MariaDB have a page size of 16 KB. Hence, writing 16 KB of data to disk would constitute one I/O operation. Oracle, PostgreSQL, and Microsoft SQL Server use a page size of 8 KB. Writing 16 KB of data using one of those database engines would consume two I/O operations. The larger the page size, the more data you can transfer in a single I/O operation.

Assuming a 16 KB page size, suppose your database needed to read 102,400 KB (100 MB) of data every second. To achieve this level of performance, your database would have to be able to read 6,400 16 KB pages every second. Because each page read counts as one I/O operation, your storage and instance class would need to be able to sustain 6,400 IOPS. Notice the inverse relationship between IOPS and page size: the larger your page size, the fewer IOPS you need to achieve the same level of throughput.

> Things get interesting when you move beyond a 32 KB page size. If your database engine writes more than 32 KB in a single I/O operation, AWS counts that as more than one I/O operation. For example, reading or writing a 64 KB page would count as two I/O operations. A 128 KB page would count as four I/O operations.

The number of IOPS you can achieve depends on the type of storage you select. RDS offers the following three different types of storage.

General-Purpose SSD

For most databases, *general-purpose SSD* (gp2) storage is sufficient. It's fast, giving you single-digit millisecond latency. You can allocate a volume of up to 64 TB. For each gigabyte of data that you allocate to a volume, RDS allocates that volume a baseline performance of three IOPS, up to a total of 16,000 IOPS per volume. A 20 GB volume would get 60 IOPS, whereas a 100 GB volume would get 300 IOPS. A 5,334 GB volume would get 16,000 IOPS. This means that the larger your volume, the better performance you'll get. Note that the minimum storage volume you can create is 20 GB.

The maximum throughput the gp2 storage type offers is 2,000 Mbps (250 MBps). To achieve this, two things have to fall into place. First, your instance must support a disk throughput of at least that much. For example, all `db.m5` instance classes support a throughput of 4,750 Mbps, so any one of those would suffice. Also, you must allocate a sufficient number of IOPS to sustain this throughput. Suppose you're running MariaDB with a page size of 16 KB (128 Kb or 0.128 Mb). To determine the number of IOPS you'll need to sustain 2,000 Mbps of disk throughput, you'd divide the bandwidth by the page size, as follows:

2000 Mbps / 0.128 Mb = 15,625 IOPS

To achieve 2,000 Mbps disk throughput to a volume, it would need 16,000 IOPS allocated. Note that you must specify provisioned IOPS in increments of 1,000. Also, as stated earlier, this is the maximum number of IOPS possible with gp2. To achieve this many IOPS, your volume would have to be 5,334 GB or about 5.33 TB.

If you think you might occasionally need up to 3000 IOPS but don't need a lot of storage, you don't have to overallocate storage just to get your desired number of IOPS. Volumes smaller than 1 TB can temporarily burst to 3,000 IOPS. The duration of the burst is determined by the following formula:

$$\text{Burst duration in seconds} = (\text{Credit balance}) / \left[3,000 - 3 * (\text{storage size in GB})\right]$$

When you initially boot a database instance, you get a credit balance of 5,400,000 IOPS. Whenever your instance uses IOPS above and beyond its baseline, it will dip into the credit balance. After the credit balance is depleted, you can no longer burst. For example, with a 200 GB volume, the burst duration would be 2,250 seconds, or 37.5 minutes.

The credit balance is replenished at a rate of one baseline IOPS every second. For example, if you have a 200 GB volume with a baseline IOPS of 600, your credit balance increases by 600 IOPS per second, up to the maximum of 5,400,000.

To get some practice with these concepts, complete Exercise 5.1 to create an instance using gp2 storage.

EXERCISE 5.1

Create an RDS Database Instance

In this exercise, you'll create an RDS database instance using MariaDB as the database engine. Free Tier accounts won't incur any charges.

1. Go to the RDS Dashboard.
2. Click Create Database.
3. Under Choose A Database Creation Method, select Standard Create.
4. Under Engine Options, select MariaDB. Keep the default version.
5. Under Templates, select Free Tier.
6. Enter a master username and master password of your choice.
7. Scroll down to the Additional Configuration section and expand it.
8. For Initial Database Name, enter a database name of your choice.
9. Click Create Database.

Provisioned IOPS SSD (io1)

If you're not excited about the rather confusing math involved with using gp2 storage, RDS provides a more straightforward option. *Provisioned IOPS SSD* lets you simply allocate the number of IOPS you need when you create your instance. There's no concept of bursting in io1 storage. The number of IOPS you provision is what you get and what you pay for, whether or not you use it. This makes it useful for databases that require consistent low-latency performance.

If you use a standard or memory-optimized instance class, RDS guarantees that you'll achieve within 10 percent of the provisioned IOPS for 99.9 percent of the year. That means you may get less than your specified number of IOPS for only about 2 hours and 45 minutes out of the year.

You can provision up to 64,000 IOPS per io1 volume. The maximum number of IOPS you can achieve and how much storage you can allocate depend on the database engine you select. Oracle, PostgreSQL, MariaDB, MySQL, Microsoft SQL Server, and Aurora let you choose 4 GB to 16 TB of storage and allocate up to 64,000 provisioned IOPS. The ratio of storage in gigabytes to IOPS must be at least 50:1. For example, if you want 32,000 IOPS, you must provision at least 640 GB of storage.

Throughput-Optimized HDD (st1)

Throughput-optimized HDD volumes use magnetic storage. Volume sizes range from 500 MB to 16 TB. The base throughput depends on the volume size, with 40 MBps of throughput per terabyte. The maximum throughput tops out at 500 MBps for a 16 TB volume.

Cold HDD (sc1)

Cold HDD volumes also use magnetic storage and come in sizes ranging from 500 MB to 16 TB. The difference is that sc1 volumes offer lower throughput. You get only 12 MBps of throughput per terabyte, with a maximum of 192 MBps for a 16 TB volume. As you might expect, sc1 is cheaper than st1 storage.

> **NOTE** The st1 and sc1 volume types are appropriate or frequent, sequential reads and writes, such as you might see with data warehousing, extract, transform, and load (ETL), and Elastic MapReduce (EMR) applications. These volume types are *not* appropriate for random I/O. For that, gp2 is a better choice.

Magnetic Storage (Standard)

RDS offers magnetic storage for backward compatibility with older instances. It's limited to a maximum size of 1 TB and approximately 100 IOPS. AWS is phasing out this storage type, so don't use it.

Read Replicas

If your database instance doesn't meet your performance requirements, you have the option to scale up (vertically) or scale out (horizontally), depending on where the bottleneck is.

Scaling Vertically

As mentioned previously, if your memory, compute, network speed, or disk throughput are the issue, you can simply throw more resources at your instance without having to make any changes to your application or databases. The way you do that is by upgrading to a larger instance class. This is called *scaling vertically* or *scaling up*.

Scaling Horizontally

Scaling horizontally, also known as *scaling out*, entails creating additional database instances called *read replicas*. All database engines except for Microsoft SQL Server support read replicas. Aurora exclusively supports a specific type of read replica called an *Aurora replica*.

A read replica is another database instance that services only queries against the database. A read replica takes some of the query load off the *master database instance*, which remains solely responsible for writing data to the database. Read replicas are useful for read-heavy applications.

You can have up to five read replicas for RDS and up to 15 Aurora replicas. Data from the master is asynchronously replicated to each read replica, meaning that there's a delay between when the data is written to the database by the master and when it shows up on the replica. Depending on how much data you can stand to lose, this can make read replicas unsuitable for disaster recovery. If the master database instance fails before the replication is complete, you'll lose the unsynchronized data.

When you create a read replica, RDS gives you a read-only endpoint, which is a domain name that resolves only to your read replica. If you have multiple replicas, RDS will load-balance the connection to each one of them. If you have reporting and analysis tools that only need to read data, you would point them to a read-only endpoint. Complete Exercise 5.2 to create a read replica for the database instance you created earlier.

EXERCISE 5.2

Create a Read Replica

1. In the RDS Dashboard, click Databases.
2. Select the instance you created earlier, click Actions, and then click Create Read Replica.
3. Under the Settings section, in the DB Instance Identifier field, enter a name for your read replica.
4. Click the Create Read Replicas button.

A read replica and the master may be in different availability zones, and even in different regions. In the event of the master instance failing, you can promote a read replica to the master. But keep in mind that because of the asynchronous nature of replication, you may lose data this way. Complete Exercise 5.3 to promote your read replica to a stand-alone master database instance.

EXERCISE 5.3

Promote the Read Replica to a Master

In this exercise, you'll promote the read replica you just created to master. Before beginning, wait for the replica to enter the Available state.

1. In the RDS console, click Databases.
2. Select the read replica you created.
3. Click Actions and then select Promote.
4. Click Continue.
5. Click Promote Read Replica.

High Availability (Multi-AZ)

To keep your database continuously available in the event of a database instance outage, you can deploy multiple database instances in different availability zones using what RDS calls a *multi-AZ deployment*. In a multi-AZ deployment, you have a *primary database instance* in one availability zone that handles reads and writes to the database, and you have a *standby database instance* in a different availability zone. If the primary instance experiences an outage, it will fail over to the standby instance, usually within two minutes.

Here are a few possible causes for a database instance outage:

- Availability zone outage
- Changing a database instance type
- Patching of the instance's operating system

You can configure multi-AZ when you create a database instance or later. All database engines support multi-AZ but implement it slightly differently.

If you enable multi-AZ after creating your instance, you'll experience a significant performance hit, so be sure to do it during a maintenance window.

Multi-AZ with Oracle, PostgreSQL, MariaDB, MySQL, and Microsoft SQL Server

In this multi-AZ deployment, all instances must reside in the same region. RDS synchronously replicates data from the primary to the standby instance. This replication can introduce some latency, so be sure to use EBS-optimized instances and provisioned IOPS SSD storage.

Your application connects to the endpoint domain name for the primary. Note that the standby instance is not a read replica and cannot serve read traffic. When a failover occurs, RDS changes the DNS record of the endpoint to point to the standby. The only thing your application has to do is reconnect to the endpoint.

For MySQL and MariaDB, you can create a multi-AZ read replica in a different region. This lets you fail over to a different region. Multi-region failover isn't supported for Microsoft SQL Server, PostgreSQL, or Oracle.

> **Note** that if you configure multi-AZ using the bring your own license model for Oracle, you must possess a license for both the primary and standby instances.

Multi-AZ with Amazon Aurora

Amazon Aurora gives you two options for multi-AZ: single-master and multi-master.

Single-Master

An Amazon Aurora single-master cluster consists of a primary instance. Aurora gives you a cluster endpoint that always points to the primary instance. An Aurora cluster also may include Aurora replicas.

The primary and all replicas share a single cluster volume, which is synchronously replicated across three availability zones. This cluster volume automatically expands as needed, up to 64 TB.

In the event the primary instance fails, one of two things will happen. If no Aurora replicas exist, Aurora will create a new primary instance to replace the failed one. If an Aurora replica does exist, Aurora will promote the replica to the primary. The entire process typically takes less than two minutes.

Multi-Master

In a multi-master cluster, all instances can write to the database. Thus, when one instance fails, no failover occurs because all instances can continue to write to the shared cluster volume that stores the database. Amazon refers to this as continuous availability, rather than high availability, because as long as at least one database instance is running, you can read from and write to the database.

Backup and Recovery

With RDS you can take EBS volume snapshots of your database instances. Snapshots include all databases on the instance and are stored in S3, just like regular EBS snapshots. Snapshots are kept in multiple zones in the same region for redundancy.

Taking a snapshot suspends all I/O operations for a few seconds, unless you're using multi-AZ with a database engine other than Microsoft SQL Server. Be sure to take your snapshots during off-peak times.

When considering your backup and recovery needs, you should understand two metrics. The *recovery time objective* (RTO) is the maximum acceptable time to recover data and resume processing after a failure. The *recovery point objective* (RPO) is the maximum period of acceptable data loss. Consider your own RTO and RPO requirements when choosing your RDS backup options.

When you restore from a snapshot, RDS restores it to a new instance. The time to restore a snapshot can take several minutes, depending on its size. The more provisioned IOPS you allocate to your new instance, the faster the recovery time.

Automated Snapshots

RDS can automatically create snapshots of your instances daily during a 30-minute backup window. You can customize this window or let RDS choose it for you. Because taking a snapshot impacts performance, choose a time when your database is least busy. If you let RDS select the backup window, it will randomly select a 30-minute window within an 8-hour block that varies by region.

Enabling automatic backups enables *point-in-time recovery*, which archives database change logs to S3 every 5 minutes. In the event of a failure, you'll lose only up to 5 minutes' worth of data. Restoring to a point-in-time can take hours, depending on how much data is in the transaction logs.

RDS keeps automated snapshots for a limited period of time and then deletes them. You can choose a *retention period* between one day and 35 days. The default is seven days. To disable automated snapshots, set the retention period to 0. Note that disabling automated snapshots immediately deletes all existing automated snapshots and disables point-in-time recovery. Also, if you change the retention period from 0 to any other value, it will trigger an immediate snapshot.

You can also manually take a snapshot of your database instance. Unlike automated snapshots, manual snapshots stick around until you delete them. If you delete an instance, RDS will prompt you to take a final snapshot. It will also prompt you to retain automated snapshots. RDS will keep the final snapshot and all manual snapshots. If you choose not to retain automated backups, it will immediately delete any automated snapshots.

Maintenance Items

Because RDS is a managed service, it's the responsibility of AWS to handle patching and upgrades. AWS routinely performs these maintenance items on your database instances.

Maintenance items include operating system security and reliability patches. These generally occur once every few months. Database engine upgrades also may occur during a maintenance window. When AWS begins to support a new version of a database engine, you can choose to upgrade to it. Major version upgrades may contain database changes that aren't backward compatible. As such, if you want a major version upgrade, you must apply it manually. AWS may automatically apply minor version changes that are nonbreaking.

You can determine when these maintenance tasks take place by specifying a 30-minute weekly maintenance window. The window cannot overlap with the backup window. Even though the maintenance window is 30 minutes, it's possible for tasks to run beyond this.

Amazon Redshift

Recall from earlier in the chapter that OLAP databases are optimized for data warehousing applications wherein multiple databases are fed into a single monster database optimized for complex, computationally intensive queries. *Redshift* is Amazon's managed data warehouse service. Although it's based on PostgreSQL, it's not part of RDS. Redshift uses columnar storage, meaning that it stores the values for a column close together. This improves storage speed and efficiency and makes it faster to query data from individual columns. Redshift supports Open Database Connectivity (ODBC) and Java Database Connectivity (JDBC) database connectors.

Redshift uses compression encodings to reduce the amount of size each column takes up in storage. You can apply compression manually on a column-by-column basis. Or if you use the COPY command to import data from a file into a Redshift database, Redshift will determine which columns to compress.

Compute Nodes

A Redshift cluster contains one or more *compute nodes* that are divided into two categories. *Dense compute* nodes can store up to 326 TB of data on magnetic storage. *Dense storage* nodes can store up to 8,192 TB of data on fast SSDs.

If your cluster contains more than one compute node, Redshift also includes a *leader node* to coordinate communication among the compute nodes, as well as to communicate with clients. A leader node doesn't incur any additional charges.

Data Distribution Styles

Rows in a Redshift database are distributed across compute nodes. How the data is distributed depends on the distribution style. In *EVEN* distribution, the leader node spreads the data out evenly across all compute nodes. This is the default style. *KEY* distribution spreads the data according to the value in a single column. Columns with the same value are stored on the same node. In the *ALL* distribution, every table is distributed to every compute node.

Redshift Spectrum

Redshift Spectrum is a service that allows you to query data from files stored in S3 without having to import the data into your cluster. You simply define the structure of the data and then query it as you wish. The bucket containing the data you want to query must be in the same region as your cluster.

AWS Database Migration Service

The AWS Database Migration Service (DMS) can automatically copy an existing database and its schema (if applicable) to another database. What makes DMS particularly powerful is its ability to migrate data between different database engines and between relational and nonrelational databases. DMS can migrate data between the following:

- Aurora
- DynamoDB
- IBM DB2
- MariaDB
- MongoDB
- MySQL
- Oracle
- PostgreSQL
- Redshift
- S3
- SAP

DMS provisions an EC2 instance called a DMS instance that's the brains behind DMS. It initiates connections to the source and target databases and performs the replication and any necessary schema conversions between them.

> If you need to migrate a database that's too large to transfer over the network, you can use Snowball Edge instead. For more information on Snowball, see Chapter 3, "AWS Storage."

Nonrelational (NoSQL) Databases

Nonrelational databases are designed to consistently handle tens of thousands of transactions per second. Although they can store the same data you'd find in a relational database, they're optimized for so-called unstructured data. Unstructured data is an unfortunate term, as all data you store in any database has some structure. A more accurate description would be multistructured data. The data you store in a nonrelational database can vary in structure, and that structure can change over time.

Nonrelational and relational databases have many elements in common. Nonrelational databases—also known as NoSQL databases—consist of collections that are confusingly also sometimes called *tables*. Within a table, you store items, which are similar to rows or tuples in a relational database. Each item consists of at least one attribute, which is analogous to a column in a SQL database. An attribute consists of a unique name called a *key*, a data type, and a value. Attributes are sometimes called *key/value pairs*.

Storing Data

One of the biggest differences between a relational and a nonrelational database is that nonrelational databases are *schemaless* and don't require all items in a table to have the same attributes. Each item requires a primary key attribute whose value must be unique within the table. The purpose of the primary key is to uniquely identify an item and provide a value by which to sort items. Nonrelational databases are flexible when it comes to the type of data you can store. With the exception of the primary key attribute, you don't have to define attributes when you create a table. You create attributes on the fly when you create or modify an item. These attributes are unordered and hence have no relation to each other, which is why they're called nonrelational.

Nonrelational databases do not give you a way to split data across tables and then merge it together at query time. Therefore, an application will generally keep all its data in one table. This can lead to the duplication of data, which in a large database can incur substantial storage costs.

Querying Data

The trade-off for having flexibility to store unstructured data comes in terms of being more limited in your queries. Nonrelational databases are optimized for queries based on the primary key. Queries against other attributes are slower, making nonrelational databases inappropriate for complex or arbitrary queries. Prior to creating a table, you need to understand the exact queries that you're going to need to perform against the data. Consider the item represented in Table 5.4.

TABLE 5.4 Item in an unstructured database

Key	Type	Value
Employee ID (primary key)	Number	101
Department	String	Information technology
Last Name	String	Smith
First Name	String	Charlotte

Imagine a database with millions of such items. If you wanted to list every department that has an employee named Charlotte, it would be difficult to obtain this information using a nonrelational database. Because items are sorted by Employee ID, the system would have to scan through every item to locate all items that have an attribute First Name with a value of Charlotte. And because the data in each item is unstructured, it may require searching through every attribute. It would then have to determine which of these items contain a Department attribute. Such a query would be slow and computationally expensive.

Types of Nonrelational Databases

You may hear nonrelational databases divided into categories such as key/value stores, document-oriented stores, and graph databases. But all nonrelational databases are key/value store databases.

A document-oriented store is a particular application of a nonrelational database that analyzes the contents of a document stored as a value and extracts metadata from it.

A graph database, such as Amazon Neptune, analyzes relationships between attributes in different items. This is different than a relational database that enforces relationships between records. A graph database discovers these relationships in unstructured data.

DynamoDB

DynamoDB is a managed nonrelational database service that can handle thousands of reads and writes per second. It achieves this level of performance by spreading your data across multiple *partitions*. A partition is an allocation of storage for a table, and it's backed by solid-state drives in multiple availability zones.

Partition and Hash Keys

When you create a table, you must specify a primary key and a data type. Because the primary key uniquely identifies an item in the table, its value must be unique within the table. There are two types of primary keys you can create.

A *partition key*, also known as a *hash key*, is a primary key that contains a single value. When you use only a partition key as a primary key, it's called a *simple primary key*. Good candidates for a partition key would be an email address, a unique username, or even a randomly generated identifier. A partition key can store no more than 2,048 bytes.

A primary key can also be a combination of two values: a partition key and a *sort* (or *range*) key. This is called a *composite primary key*. The partition key doesn't have to be unique, but the combination of the partition key and sort key must be unique. For example, a person's last name could be the partition key, whereas the first name could be the sort key. Using this approach, you could use the values in Table 5.5 for a composite primary key for a table.

TABLE 5.5 Composite primary keys

Last name (partition key)	First name (sort key)
Lewis	Clive
Lewis	Warren
Williams	Warren
Williams	Clive

Neither the last name Lewis nor the first name Warren is unique in the table. But combining the partition and sort keys together creates a unique primary key.

DynamoDB distributes your items across partitions based on the primary key. Using the preceding example, items with the last name Lewis would all be stored together on the same partition. DynamoDB would arrange the items in ascending order by the sort key. Note that a sort key can store no more than 1,024 bytes.

When a lot of read or write activity occurs against items stored in the same partition, the partition is said to be a *hot partition*. Hot partitions can negatively affect performance. To avoid hot partitions, try to make your partition keys as specific as possible. For example, if you're storing log entries, consider using a timestamp that changes frequently as the partition key.

Attributes and Items

Each key/value pair composes an attribute, and one or more attributes make up an item. DynamoDB can store an item size of up to 400 KB, which is roughly equivalent to 50,000 English words!

At a minimum, every item contains a primary key and corresponding value. When you create an attribute, you must define the data type. Data types fall into the following three categories:

Scalar A *scalar data type* can have only one value. You can have string, number, binary, Boolean, and null data types.

The string data type can store up to 400 KB of Unicode data with UTF-8 encoding. A string must always be greater than zero.

The number data type stores positive or negative numbers up to 38 significant digits. DynamoDB trims leading and trailing zeros.

The binary data type stores binary data in Base-64 encoded format. Like the string type, it's limited by the maximum item size to 400 KB.

The Boolean data type can store a value of either true or false.

The null data type is for representing an attribute with an undefined or unknown value. Oddly, it must contain a value of null.

Set A *set data type* holds an unordered list of scalar values. The values must be unique within a set, and a set must contain at least one value. You can create number sets, string sets, and binary sets.

Document *Document data types* are designed to hold different types of data that fall outside the constraints of scalar and set data types. You can nest document types together up to 32 levels deep.

A list document type can store an ordered collection of values of any type. For example, you could include the following in the value of a list document:

```
Chores:
["Make coffee",
Groceries: ["milk", "eggs", "cheese"],
"Pay bills",
Bills: [water: [60], electric: [100]]]
```

Notice that the Chores list contains string data, numeric data, and nested lists.

A map data type can store an unordered collection of key/value pairs in a format similar to JavaScript Object Notation (JSON). As with a list, there are no restrictions on the type of data you can include. The following is an example of a map that contains a nested list and a nested map:

```
{
  Day: "Friday",
  Chores: [
    "Make coffee",
    "Groceries", {
      milk: { Quantity: 1 },
      eggs: { Quantity: 12 }
    }
    "Mow the lawn"],
}
```

Throughput Capacity

When creating a table, you have the option of having DynamoDB operate in on-demand mode or provisioned mode. In on-demand mode, DynamoDB automatically scales to accommodate your workload. This is useful if you don't know what your workload demand is going to be or want to pay only for the capacity you use.

In provisioned mode, you specify the number of reads and writes per second your application will require. This is called *provisioned throughput.* DynamoDB reserves partitions based on the number of *read capacity units* (RCUs) and *write capacity units* (WCUs) you specify when creating a table.

When you read an item from a table, that read may be *strongly consistent* or *eventually consistent.* A strongly consistent read always gives you the most up-to-date data, whereas an eventually consistent read may produce stale data that does not reflect data from a recent write operation. Whether you use strongly or eventually consistent reads depends on whether your application can tolerate reading stale data. You need to understand whether you need strongly or eventually consistent reads when deciding how much throughput to provision.

> **NOTE** You can switch between provisioned and on-demand mode only once every 24 hours.

For an item up to 4 KB in size, one RCU buys you one strongly consistent read per second. To read an 8 KB item every second using a strongly consistent read, you'd need two RCUs.

If you use an eventually consistent read, one RCU buys you two eventually consistent reads per second. To read an 8 KB item every second using an eventually consistent read, you'd need only one RCU.

When it comes to writing data, one WCU gives you one write per second for an item up to 1 KB in size. This means if you need to write 100 items per second, each item being less than 1 KB, you'd have to provision 100 WCUs. If you have to write 10 items per second, each item being 2 KB, then you'd need 20 WCUs.

The throughput capacity you specify is an upper limit of what DynamoDB delivers. If you exceed your capacity, DynamoDB may throttle your request and yield an "HTTP 400 (Bad request)" error. AWS SDKs have built-in logic to retry throttled requests, so having a request throttled won't prevent your application from reading or writing data, but it will slow it down.

Complete Exercise 5.4 to create a table in DynamoDB.

EXERCISE 5.4

Create a Table in DynamoDB Using Provisioned Mode

1. Use the following command to create a table named Authors with a partition key named LastName and sort key named FirstName. Both keys should use the string data type. Provision the table with a WCU and an RCU of 1.

   ```
   aws dynamodb create-table --table-name Authors --attribute-definitions
   AttributeName=LastName,AttributeType=S
   AttributeName=FirstName,AttributeType=S
   --keyschema AttributeName=LastName,KeyType=HASH
   AttributeName=FirstName,KeyType=RANGE
   --provisioned-throughput ReadCapacityUnits=1,WriteCapacityUnits=1
   ```

2. Go to the DynamoDB service console to view the table you just created.

Auto Scaling

If you're unsure exactly how much throughput you have to provision for a table or if you anticipate your throughput needs will vary over time, you can configure Auto Scaling to automatically increase your provisioned throughput when it gets close to hitting a defined threshold. Note that this is different than on-demand mode where you don't have to specify an RCU and a WCU.

To configure Auto Scaling, you specify a minimum and maximum RCU and WCU. You also specify a desired utilization percentage. DynamoDB will automatically adjust your RCU and WCU to keep your utilization at this percentage. For example, suppose you set a utilization of 70 percent, a minimum RCU of 10, and a maximum RCU of 50. If you consume 21 RCU, Auto Scaling will adjust your provisioned capacity to around 30 RCU. If your consumption drops to 14, Auto Scaling will reduce your provisioned throughput to 20 RCU.

Setting the right utilization is a balancing act. The higher you set your utilization, the more likely you are to exceed your provisioned capacity. If that happens, your requests may get throttled. On the other hand, if you set your utilization too low, you will end up paying for capacity you don't need.

Reserved Capacity

If you need 100 or more WCU or RCU, you can purchase reserved throughput capacity to save money. You must reserve RCU and WCU separately, and you're limited to 100,000 units of each. You have to pay a one-time fee and commit to a period of one or three years.

Reading Data

DynamoDB provides two different operations to let you read data from a table. A *scan* lists all items in a table. It's a read-intensive operation and can potentially consume all of your provisioned capacity units. A *query* returns an item based on the value of the partition key. When performing a query, the value of the partition key you search for must exactly match that of an item. If your table contains a sort key, you may optionally query by the sort key as well. For the sort key, you have more flexibility. You can search by exact value, a value greater than or less than the key, a range of values, or the beginning of the value.

Secondary Indexes

Secondary indexes solve two issues with querying data from DynamoDB. When you query for a particular item, you must specify a partition key exactly. For example, the Author table you created earlier has LastName as the partition key and FirstName as the sort key. Secondary indexes let you look up data by an attribute other than the table's primary key. Think of a secondary index as a copy of some of the attributes in a table. The table that the index gets its data from is called the *base table*.

When you create a secondary index, you can choose which attributes get copied from the base table into the index. These are called *projected attributes*. A secondary index always includes the partition and sort key attributes from the base table. You can choose to copy just the partition and sort keys and their values, the keys plus other attributes, or everything. This lets you extract only the data you need. There are two types of secondary indexes.

Global Secondary Index

You can create a *global secondary index* (GSI) any time after creating a table. In a global secondary index, the partition and hash keys can be different than the base table. The same rules for choosing a primary key still apply. You want the primary key of your index to be as unique as possible. If you use a composite primary key, items having partition keys with the same value will be stored on the same partition.

When reading from a global secondary index, reads are always eventually consistent. If you add an item to a table, it may not immediately get copied to the secondary index.

Local Secondary Index

A local secondary index (LSI) must be created at the same time as the base table. You also cannot delete a local secondary index after you've created it. The partition key must always be the same as the base table, but the sort key can be different. For example, if the base table has LastName as the partition key and FirstName as the sort key, you can create a local secondary index with a partition key of LastName and a sort key of BirthYear. Reads from a local secondary index can be strongly or eventually consistent, depending on what you specify at read time.

Global Tables

To improve availability, you can use global tables to replicate a table across multiple regions. To use global tables, your table must be configured in on-demand mode or provisioned mode with Auto Scaling enabled. A global table is a collection of replica tables, and a global table can have only one replica table per region. Whenever you write an item to a replica table, it's replicated to replica tables in other regions. Global tables don't support strongly consistent read across regions.

Backups

You can back up a table at any time without consuming any RCUs or impacting the performance of DynamoDB. You can take an unlimited number of backups. You can restore a backup to the same region or a different region than the table the backup was taken from. When you restore a table, you can specify whether the destination table should operate in on-demand or provisioned mode, and you can specify options for indexes and encryption.

Summary

Whether you implement a relational or nonrelational database depends solely on the application that will use it. Relational databases have been around a long time, and many application developers default to modeling their data to fit into a relational database. Applications use database-specific SDKs to interact with the database, so often the needs of the application mandate the specific database engine required. This is why AWS RDS offers six of the most popular database engines and sports compatibility with a wide range of versions. The idea is to let you take an existing database and port it to RDS without having to make any changes to the application.

The nonrelational database is a more recent invention. DynamoDB is Amazon's proprietary nonrelational database service. Unlike applications designed for relational databases, an application designed for a nonrelational database generally cannot be ported from an on-premises deployment to DynamoDB without some code changes. You're therefore more likely to find cloud-native applications using DynamoDB. When developing or refactoring an application to use DynamoDB, there's a good chance that developers will need to consult with you regarding how to design the database. It's critical in this case that you understand how to choose partition and sort keys and data types and how to allocate throughput capacity to meet the performance needs of the application.

Regardless of which database you use, as with any AWS service, you as an AWS architect are responsible for determining your performance and availability requirements and implementing them properly.

Exam Essentials

Understand the differences between relational and nonrelational databases.   A relational database requires you to specify attributes up front when you create a table. All data you insert into a table must fit into the predefined attributes. It uses the Structured Query Language (SQL) to read and write data, and so it's also called a SQL database. A nonrelational database only requires you to specify a primary key attribute when creating a table. All items in a table must include a primary key but can otherwise have different attributes. Nonrelational—also called NoSQL databases—store unstructured data.

Know the different database engines RDS supports.   RDS supports all of the most popular database engines—MySQL, MariaDB, Oracle, PostgreSQL, Amazon Aurora, and Microsoft SQL Server. Understand the difference between the bring-your-own-license and license-included licensing models. Know which database engines support which licensing models.

Be able to select the right instance class and storage type given specific storage requirements.   Memory and storage tend to be the constraining factors for relational databases, so it's crucial that you know how to choose the right instance class and storage type based on the performance needs of a database. Know the three different instance classes: standard, memory optimized, and burstable. Also know how these relate to the different storage types: general-purpose SSD (gp2), provisioned IOPS SSD (io1), throughput optimized (st1), cold HDD (sc1), and magnetic.

Understand the differences between multi-AZ and read replicas.   Both multi-AZ and read replicas involve creating additional database instances, but there are some key differences. A read replica can service queries, whereas a standby instance in a multi-AZ deployment cannot. A master instance asynchronously replicates data to read replicas, whereas in a multi-AZ configuration, the primary instance synchronously replicates data to the standby. Understand how Aurora replicas work and how Aurora multi-AZ differs from multi-AZ with other database engines.

Be able to determine the appropriate primary key type for a DynamoDB table.
DynamoDB tables give you two options for a primary key. A simple primary key consists of just a partition key and contains a single value. DynamoDB distributes items across partitions based on the value in the partition key. When using a simple primary key, the partition key must be unique within a table. A composite primary key consists of a partition key and a sort key. The partition key does not have to be unique, but the combination of the partition key and the sort key must be.

Know how DynamoDB throughput capacity works. When you create a table, you must specify throughput capacity in write capacity units and read capacity units. How many read capacity units a read consumes depends on two things: whether the read is strongly or eventually consistent and how much data you read within one second. For an item of up to 4 KB in size, one strongly consistent read consumes one read capacity unit. Eventually consistent reads consume half of that. When it comes to writes, one write capacity unit lets you write up to one 1 KB item per second.

Review Questions

1. In a relational database, a row may also be called what? (Choose two.)
   A. Record
   B. Attribute
   C. Tuple
   D. Table

2. What must every relational database table contain?
   A. A foreign key
   B. A primary key
   C. An attribute
   D. A row

3. Which SQL statement would you use to retrieve data from a relational database table?
   A. QUERY
   B. SCAN
   C. INSERT
   D. SELECT

4. Which relational database type is optimized to handle multiple transactions per second?
   A. Offline transaction processing (OLTP)
   B. Online transaction processing (OLTP)
   C. Online analytic processing (OLAP)
   D. key/value store

5. How many database engines can an RDS database instance run?
   A. Six
   B. One
   C. Two
   D. Four

6. Which database engines are compatible with existing MySQL databases? (Choose all that apply.)
   A. Microsoft SQL Server
   B. MariaDB
   C. Aurora
   D. PostgreSQL

7. Which storage engine should you use with MySQL, Aurora, and MariaDB for maximum compatibility with RDS?
   A. MyISAM
   B. XtraDB
   C. InnoDB
   D. PostgreSQL

8. Which database engine supports the bring-your-own-license (BYOL) model? (Choose all that apply.)
   A. Oracle Standard Edition Two
   B. Microsoft SQL Server
   C. Oracle Standard Edition One
   D. PostgreSQL

9. Which database instance class provides dedicated bandwidth for storage volumes?
   A. Standard
   B. Memory optimized
   C. Storage optimized
   D. Burstable performance

10. If a MariaDB database running in RDS needs to write 200 MB of data every second, how many IOPS should you provision using io1 storage to sustain this performance?
    A. 12,800
    B. 25,600
    C. 200
    D. 16

11. Using general-purpose SSD storage, how much storage would you need to allocate to get 600 IOPS?
    A. 200 GB
    B. 100 GB
    C. 200 TB
    D. 200 MB

12. If you need to achieve 12,000 IOPS using provisioned IOPS SSD storage, how much storage should you allocate, assuming that you need only 100 GB of storage?
    A. There is no minimum storage requirement.
    B. 200 GB
    C. 240 GB
    D. 12 TB

13. What type of database instance only accepts queries?
    A. Read replica
    B. Standby database instance
    C. Primary database instance
    D. Master database instance

14. In a multi-AZ deployment using Oracle, how is data replicated?
    A. Synchronously from the primary instance to a read replica
    B. Synchronously using a cluster volume
    C. Asynchronously from the primary to a standby instance
    D. Synchronously from the primary to a standby instance

15. Which of the following occurs when you restore a failed database instance from a snapshot?
    A. RDS restores the snapshot to a new instance.
    B. RDS restores the snapshot to the failed instance.
    C. RDS restores only the individual databases to a new instance.
    D. RDS deletes the snapshot.

16. Which Redshift distribution style stores all tables on all compute nodes?
    A. EVEN
    B. ALL
    C. KEY
    D. ODD

17. Which Redshift node type can store up to 326 TB of data?
    A. Dense memory
    B. Leader
    C. Dense storage
    D. Dense compute

18. Which is true regarding a primary key in a nonrelational database? (Choose all that apply.)
    A. It's required to uniquely identify an item.
    B. It must be unique within the table.
    C. It's used to correlate data across different tables.
    D. Its data type can vary within a table.

19. In a DynamoDB table containing orders, which key would be most appropriate for storing an order date?
    A. Partition key
    B. Sort key

C. Hash key
D. Simple primary key

20. When creating a DynamoDB table, how many read capacity units should you provision to be able to sustain strongly consistent reads of 11 KB per second?
    A. 3
    B. 2
    C. 1
    D. 0

21. Which Redshift node type can provide the fastest read access?
    A. Dense compute
    B. Dense storage
    C. Leader
    D. KEY

22. Which DynamoDB index type allows the partition and hash key to differ from the base table?
    A. Eventually consistent index
    B. Local secondary index
    C. Global primary index
    D. Global secondary index

23. To ensure the best performance, in which of the following situations would you choose to store data in a NoSQL database instead of a relational database?
    A. You need to perform a variety of complex queries against the data.
    B. You need to query data based on only one attribute.
    C. You need to store JSON documents.
    D. The data will be used by different applications.

24. What type of database can discover how different items are related to each other?
    A. SQL
    B. Relational
    C. Document-oriented store
    D. Graph

Chapter 6

Authentication and Authorization—AWS Identity and Access Management

THE AWS CERTIFIED SOLUTIONS ARCHITECT ASSOCIATE EXAM OBJECTIVES COVERED IN THIS CHAPTER MAY INCLUDE, BUT ARE NOT LIMITED TO, THE FOLLOWING:

✓ **Domain 3: Design Secure Applications and Architectures**

- 3.1 Design secure access to AWS resources
- 3.2 Design secure application tiers
- 3.3 Select appropriate data security options

Introduction

Your AWS resources are probably your company's crown jewels, so you definitely don't want to leave them unprotected. But you also can't lock them down so tightly that even your admins and customers can't get in.

Finding the perfect balance is possible. Getting there will have a lot to do with the way you *authenticate* user requests to confirm they're legitimate and then *authorize* no more and no less than the exact access they'll need. On AWS, authentication and authorization are primarily handled by Identity and Access Management (IAM).

In this chapter, you're going to learn about IAM identities—which are sometimes described as *principals*. An identity represents an AWS user or a role. Roles are identities that can be temporarily assigned to an application, service, user, or group.

Identities can also be federated—that is, users or applications without AWS accounts can be authenticated and given temporary access to AWS resources using an external service such as Kerberos, Microsoft Active Directory, or the Lightweight Directory Access Protocol (LDAP).

Identities are controlled by attaching policies that precisely define the way they'll be able to interact with all the resources in your AWS account. You can attach policies to either principals (identity-based policies) or resources (resource-based policies).

This chapter will describe the following:

- Creating policies to closely control what your principals can do on your account
- Managing the various kinds of keys or tokens used by principals to prove their identities
- Providing single sign-on solutions for integrating IAM with external providers using identity federation
- Implementing best practices for configuring accounts and roles to properly secure your resources

IAM Identities

The one identity that comes with every new AWS account is the root user. By default, root has full rights over all the services and resources associated with your account. This makes sense, because otherwise there would be no way for you to get things done with the services that lie beyond root control. At the same time, having everything in the hands of this one user makes root an attractive target for hackers: they only need to get the root password or access keys to compromise the entire account.

To reduce your exposure to this vulnerability, AWS suggests that you heavily protect your root account and delegate specific powers for day-to-day operations to other users. Figure 6.1 shows the checklist of Amazon's recommended actions in the Security Status section of a typical account's IAM home page.

FIGURE 6.1 The Security Status checklist from the IAM page of an AWS account

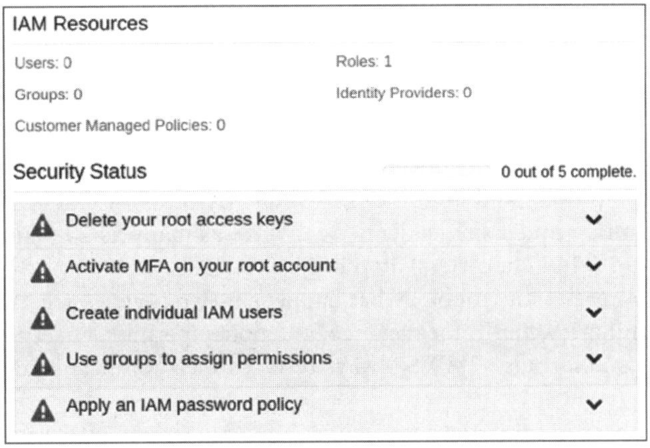

Understanding how—and why—you can satisfy those recommendations is an important first step for successfully managing your AWS account.

IAM Policies

Your first job should be to understand how policies are used to control the behavior of IAM identities. An IAM policy is a document that identifies one or more *actions* as they relate to one or more AWS *resources*. The policy document also determines the *effect* permitted by the action on the resource. The value of an effect will be either Allow or Deny.

A policy might, for instance, allow (the effect) the creation of buckets (the action) within S3 (the resource). Of course, the way resources and actions are defined will vary according to the particular service you're working with.

IAM provides hundreds of preset policies that you can view from the Policies page accessed from the IAM Dashboard. You can filter policies using key words to narrow down your search. You can also create your own policy using the tools available through the Create policy page in the Dashboard or by manually crafting your own using JSON-formatted text.

The following example is a JSON AdministratorAccess policy document from the IAM Dashboard interface. Note how it *allows* the identity holding the policy to perform any (*) action on any (*) resource in your account.

```
{
    "Version": "2012-10-17",
    "Statement": [
```

```
        {
            "Effect": "Allow",
            "Action": "*",
            "Resource": "*"
        }
    ]
}
```

Any action that's not explicitly allowed by a policy will be denied. But wait, if all actions are implicitly denied, why would you ever need to explicitly invoke "Deny" within a policy? Such policies could be useful in cases where a user requires access to most—but not all—resources within a domain. "Deny" can single out exactly those resources that should remain off-limits.

When you associate a policy document with an IAM identity, that identity will then be bound to the policy's powers and limitations. A single IAM policy can be associated with any number of identities, and a single identity can have as many as 10 managed policies (each no greater than 6,144 characters) attached to it.

This raises the potential for trouble: what happens if two policies are associated with a single identity conflict? What if, for instance, one policy permits a user to create new S3 buckets and the second forbids it? AWS always resolves such conflicts by denying the action in question.

You should also bear in mind that, in general, an explicit deny effect will always overrule an allow.

User and Root Accounts

The best way to protect your root account is to lock it down by doing the following:

- Delete any access keys associated with root.
- Assign a long and complex password and store it in a secure password vault.
- Enable multifactor authentication (MFA) for the root account.
- Wherever possible, don't use root to perform administration operations.

Before all that, however, you must create at least one new user and assign it enough authority to get the necessary work done. Typically, that means you give the main admin user the AdministratorAccess policy, and use that user to create other users, groups, and roles—each with just enough power to perform its specific task.

> **NOTE** You may be forgiven for wondering why giving a user the AdministratorAccess policy is any safer than leaving your root account in active service. After all, both seem to have complete control over all your resources, right? Wrong. There are some powers that even an AdministratorAccess holder doesn't have, including the ability to create or delete account-wide budgets and enable MFA Delete on an S3 bucket.

Once all that's done, you're ready to lock down root. You can do that right now using Exercise 6.1.

EXERCISE 6.1

Lock Down the Root User

1. If necessary, create a regular user and then assign it the AdministratorAccess policy.

2. Make sure there are no active access keys associated with your root account.

3. Enable MFA for the root account, where short-lived authentication codes are sent to software applications on preset mobile devices (including smartphones) to confirm a user's identity.

4. Update your root login to a password that's long and complex and that includes non-alphanumeric characters.

5. Confirm that you can still log in as root and then store the password safely.

Perhaps now is also a good time to work through Exercise 6.2 to create a new user for your own account, assign it a policy, and confirm that the policy has the intended effect.

EXERCISE 6.2

Assign and Implement an IAM Policy

1. Create a new user in the IAM Dashboard.

2. Attach the AmazonS3FullAccess policy that will permit your user to create, edit, and delete S3 buckets. (Hint: You can search IAM policies using s3 to display a much shorter list.)

3. Note the user login instructions that will be displayed.

4. Log in as your new user and try creating a new S3 bucket.

5. Just to prove everything is working, try launching an EC2 instance. Your request should be denied.

When you create a new IAM user, you'll have the option of applying a password policy. This policy can be set to enforce minimum lengths and complexity along with the maximum lapsed time between password resets. Getting your team members to use only higher-quality passwords is an important security precaution.

Users can open the My Security Credentials page (accessed via the pull-down menu beneath the user's name in the console) to manage their security settings. As you can see in Figure 6.2, that page includes sections for the following:

- Updating a password
- Activating or managing MFA

- Generating or deleting access keys for managing your AWS resources through the AWS CLI or programming SDKs
- Generating key pairs for authenticating signed URLs for your Amazon CloudFront distributions
- Generating X.509 certificates to encrypt Simple Object Access Protocol (SOAP) requests to those AWS services that allow it

> SOAP requests to S3 and Amazon Mechanical Turk are an exception to this rule, since they use regular access keys rather than X.509 certificates.

- Retrieving your 12-digit AWS Account ID and, for use with legacy S3 ACLs, your canonical user ID

FIGURE 6.2 The six action items displayed on the Your Security Credentials page

```
Your Security Credentials

Use this page to manage the credentials for your AWS account. To manage credentials for AWS Identity and Access
Management (IAM) users, use the IAM Console.
To learn more about the types of AWS credentials and how they're used, see AWS Security Credentials in AWS General
Reference.

  + Password
  + Multi-factor authentication (MFA)
  + Access keys (access key ID and secret access key)
  + CloudFront key pairs
  + X.509 certificate
  + Account identifiers
```

Obviously, not all of those items will be relevant to all your users.

Access Keys

Access keys provide authentication for programmatic or CLI-based access. Rather than having to find a way for your application to input a traditional username and password, you can make your local environment aware of the access key ID and secret access key. Using that information, your application or command can pass authentication data along with each request.

You learned about access keys in action in Chapter 2, "Amazon Elastic Compute Cloud and Amazon Elastic Block Store." You should also become familiar with the AWS access key lifecycle and how it should be managed.

Deactivating Unused Keys

Each active access key is a potential source of account vulnerability. Audit the keys associated with each of your users from time to time, and if you can confirm that any of these keys are not currently being used, deactivate them. If you have no plans to use them in the future, delete them altogether.

Key Rotation

Best practices require that you regularly retire older access keys because the longer a key has been in use, the greater the chance that it's been compromised. Therefore, it makes good sense to set a limit—perhaps 30 days—beyond which keys must be deleted and replaced by new ones.

Key rotation is automated for IAM roles used by EC2 resources to access other AWS services. But if your keys are designated for your own applications, things can be more complicated. It's a good idea to follow this protocol:

1. Generate a new access key for each of your users. Users can also be given the ability to manage their own keys.
2. Update your application settings to point to the new keys.
3. Deactivate (but don't delete) the old keys.
4. Monitor your applications for a few days to make sure all the updates were successful. You can use the CLI command
   `aws iam get-access-key-last-used --access-key-id ABCDEFGHIJKLMNOP`
   to determine whether any applications are still using an old key.
5. When you're sure you have everything, delete the old keys.

You can enforce key rotation by including rotation in the password policy associated with your IAM user accounts.

Exercise 6.3 guides you through a simple access key lifecycle.

EXERCISE 6.3

Create, Use, and Delete an AWS Access Key

1. Create a new AWS access key, and save both the access key ID and secret access key somewhere secure.
2. Enter **aws configure** at your local command line to add the key to your AWS CLI configuration.

   If you already have a different access key configured on your system, you can create and use multiple keys in parallel. By adding the --profile argument to the aws configure command, you can create separate profiles. You'll be prompted to enter configuration details for each new profile. Here's an example:

   ```
   $ aws configure --profile account2
   ```

EXERCISE 6.3 (continued)

You can then invoke a profile by adding the argument to a regular command:

```
$ aws s3 ls --profile account2
```

3. Try performing some operation—such as listing your S3 buckets—and then uploading a local file using the AWS CLI and your new key.

4. Disable (select Make Inactive) or delete the key you just created from the IAM Dashboard.

5. Confirm that you are now unable to administer your S3 buckets using the key.

Groups

Creating individual users with just the permissions they'll need to perform their duties can be a secure and efficient way to manage AWS operations. But as your account becomes busier and more complex, it can also become a nightmare to manage.

Say, for instance, you've given some level of access to a half a dozen admins and a similar number of developers. Manually assigning appropriate access policies to each user can be time-consuming. But just imagine how much fun you'll have if you need to *update* permissions across the entire team—perhaps applying one set of changes for just the developers and another for the admins.

The solution for such scenarios is to create a separate IAM group for each class of user and then associate each of your users with the group that fits their job description. You could create one group for developers, another for admins, and a third for your design team.

Now whenever you need to make changes to the access profile for a particular class, rather than having to edit each user account, you just update the appropriate group. If you're adding an Elastic Beanstalk workload, you can open it up for the entire developers' group with a single action. If you're no longer running your video content through Amazon Elastic Transcoder, you can remove access from the design group.

Exercise 6.4 is a good practical introduction to working with groups.

EXERCISE 6.4

Create and Configure an IAM Group

1. Make sure you have at least two IAM users in your account.

2. Create a new IAM group and attach at least one policy—perhaps IAMUserChangePassword.

3. Add your two users to the group.

4. Confirm that your users can now change their own passwords.
5. Delete the group or change its policies and then confirm that your users can no longer update their passwords.

Roles

An IAM *role* is a temporary identity that a user or service seeking access to your account resources can request. This kind of authorization can solve a lot of logistical problems.

You might have IAM users who will occasionally need to shut down and restart EC2 instances after an update. To prevent accidental shutdowns, however, you'd rather they didn't normally have that power. Much the way `sudo` works on a Linux or macOS machine—or the `runas` command syntax on Windows—you can allow your user to assume an authorizing role for only as long as it's needed.

Similarly, you might want to run an AWS service like Elastic Container Service (ECS) that will launch resources associated with other AWS services in your account. An ECS task, for instance, will probably require access to the Elastic Container Registry (ECR) in order to pull container images. You give ECS the authority to pull data from ECR through a role. In this case, a prebuilt managed role that'll do the job already exists: AmazonECSTaskExecutionRolePolicy.

You might also want to temporarily give IAM users from another AWS account—or users who sign in using a federated authentication service—access to resources on this account. An IAM role (which by default expires after 12 hours) is often the best mechanism for making this work.

You create a new role by defining the *trusted entity* you want given access. There are four categories of trusted entity: an AWS service; another AWS account (identified by its account ID); a web identity who authenticates using a login with Amazon, Amazon Cognito, Facebook, or Google; and Security Assertion Markup Language (SAML) 2.0 federation with a SAML provider you define separately.

Once your entity is defined, you give it permissions by creating and attaching your own policy document or assigning one or more preset IAM policies. When a trusted entity assumes its new role, AWS issues it a time-limited security token using the AWS Security Token Service (STS).

Authentication Tools

AWS provides a wide range of tools to meet as many user and resource management needs as possible. The IAM functionality that you've already seen is only part of the story. The Amazon Cognito, AWS Managed Microsoft AD, and AWS Single Sign-On services are for handling user authentication, whereas AWS Key Management Service (KMS), AWS Secrets Manager, and AWS CloudHSM simplify the administration of encryption keys and authentication secrets.

Amazon Cognito

Cognito provides mobile and web app developers with two important functions:

- Through Cognito's *user pools*, you can add user sign-up and sign-in to your applications.
- Through Cognito's *identity pools*, you can give your application users temporary, controlled access to other services in your AWS account.

Building a new user pool involves defining how you want your users to identify themselves when they sign up (attributes such as address or birth date) and sign in (username or email address). You can also set minimum requirements for password complexity, multi-factor authentication, and email verification.

When you set up an identity pool, you define the pool from which your users can come (using Cognito, AWS, federated, or even unauthenticated identities). You then create and assign an IAM role to the pool. Once your pool is live, any user whose identity matches your definition will have access to the resources specified in the role.

AWS Managed Microsoft AD

Managed Microsoft AD is actually accessed through the AWS Directory Service, as are a number of directory management tools like Amazon Cloud Directory and Cognito. (Cloud Directory is a way to store and leverage hierarchical data like lists of an organization's users or hardware assets.) What the Directory Service tools all share in common is the ability to handle large stores of data and integrate them into AWS operations.

Technically, Managed Microsoft AD is called *AWS Directory Service for Microsoft Active Directory*. But whichever way you refer to it, the goal is to have Active Directory control the way Microsoft SharePoint, .NET, and SQL Server–based workloads running in your VPC connect to your AWS resources. It's also possible to connect your AWS services to an on-premises Microsoft Active Directory using AD Connector.

Managed Microsoft AD domain controllers run in two VPC availability zones. As a managed service, AWS automatically takes care of all necessary infrastructure administration, including data replication and software updates.

AWS Single Sign-On

AWS Single Sign-On (SSO) allows you to provide users with streamlined authentication and authorization through an existing Microsoft Active Directory configured within AWS Directory Service. The service works across multiple AWS accounts within AWS Organizations. SSO also supports access to popular applications such as Salesforce, Box, and Office 365 in addition to custom apps that support SAML 2.0.

AWS Organizations, by the way, is a service that can manage policy-based controls across multiple AWS accounts. Companies with more than one AWS account can use AWS Organizations to unify and integrate the way their assets are exposed and consumed no matter how distributed they might be.

AWS Key Management Service

KMS deeply integrates with AWS services to create and manage your encryption keys, which you saw regarding encryption keys for EBS volumes in Chapter 2 and regarding both server-side and client-side encryption for S3 buckets in Chapter 3, "AWS Storage."

The value of KMS lies in how it provides fully managed and centralized control over your systemwide encryption. The service lets you create, track, rotate, and delete the keys that you'll use to protect your data. For regulatory compliance purposes, KMS is integrated with AWS CloudTrail, which records all key-related events.

Key creation and administration happens through the console, AWS CLI, or SDKs. Key administration powers can be assigned to individual IAM users, groups, or roles.

AWS Secrets Manager

You already know that you can manage identity authentication to AWS services using IAM roles. However, roles won't help you securely pass credentials—referred to as *secrets* in this context—to third-party services or databases.

Instead, the passwords and third-party API keys for many of the resources your applications might need can be handled by the AWS Secrets Manager. Rather than having to hardcode secrets into your code and then having to regularly update them when they change, with Secrets Manager you can deliver the most recent credentials to applications on request. The manager will even automatically take care of credential rotation.

AWS CloudHSM

CloudHSM (where the HSM stands for "hardware security module") launches virtual compute device clusters to perform cryptographic operations on behalf of your web server infrastructure. One typical goal is to off-load the burden of generating, storing, and managing cryptographic keys from your web servers so that their resources can be focused exclusively on serving your applications.

CloudHSM provides a service that's similar to AWS KMS, but according to AWS documentation (aws.amazon.com/cloudhsm/faqs), it is particularly useful for the following:

- Keys stored in dedicated, third-party validated HSMs under your exclusive control
- Federal Information Processing Standards (FIPS) 140-2 compliance
- Integration with applications using Public Key Cryptography Standards (PKCS)#11, Java JCE (Java Cryptography Extension), or Microsoft CNG (Cryptography API: Next Generation) interfaces
- High-performance in-VPC cryptographic acceleration (bulk crypto)

You activate an HSM cluster by running the CloudHSM client as a daemon on each of your application hosts. The client is configured to fully encrypt communication with the HSM.

AWS CLI Example

The following commands create a new user named steve and confirm that the user now exists:

```
$ aws iam create-user --user-name steve
$ aws iam get-user --user-name steve
```

The list-policies command will return a long list of the preset policies IAM provides. Among them is AmazonEC2ReadOnlyAccess, which permits its assignee only descriptions of running EC2 resources. You can attach a policy to a user by supplying the policy's Amazon Resource Name (ARN) as follows:

```
$ aws iam list-policies
$ aws iam attach-user-policy \
    --policy-arn arn:aws:iam::aws:policy/AmazonEC2ReadOnlyAccess \
    --user-name steve
```

Here's the AmazonEC2ReadOnlyAccess policy in JSON format:

```
{
    "Version": "2012-10-17",
    "Statement": [
        {
            "Effect": "Allow",
            "Action": "ec2:Describe*",
            "Resource": "*"
        },
        {
            "Effect": "Allow",
            "Action": "elasticloadbalancing:Describe*",
            "Resource": "*"
        },
        {
            "Effect": "Allow",
            "Action": [
                "cloudwatch:ListMetrics",
                "cloudwatch:GetMetricStatistics",
                "cloudwatch:Describe*"
            ],
            "Resource": "*"
        },
        {
```

```
            "Effect": "Allow",
            "Action": "autoscaling:Describe*",
            "Resource": "*"
        }
    ]
}
```

The `list-access-keys` command will return the names of any existing keys associated with the specified username. If you don't specify a name, keys belonging to root will be returned. The `create-access-key` command will create a new key (make sure you copy and save the key data that's returned), and `delete-access-key` will delete the specified key.

```
$ aws iam list-access-keys --user-name steve
$ aws iam create-access-key --user-name steve
$ aws iam delete-access-key --user-name steve --access-key-id AKIAJAP<...>
```

Summary

The IAM root user that's automatically enabled on a new AWS account should ideally be locked down and not used for day-to-day account operations. Instead, you should give individual users the precise permissions they'll need to perform their jobs.

All user accounts should be protected by strong passwords, multifactor authentication, and the use of encryption certificates and access keys for resource access.

Once authenticated, a user can be authorized to access a defined set of AWS resources using IAM policies. It's a good practice to associate users with overlapping access needs into IAM groups, where their permissions can be centrally and easily updated. Users can also be assigned temporary IAM roles to give them the access they need, when they need it.

Access keys should be regularly audited to ensure that unused keys are deleted and active keys are rotated at set intervals.

Identities (including users, groups, and roles) can be authenticated using a number of AWS services, including Cognito, Managed Microsoft AD, and Single Sign-On. Authentication secrets are managed by services such as AWS Key Management Service (KMS), AWS Secrets Manager, and AWS CloudHSM.

Exam Essentials

Understand how to work with IAM policies. You can build your own custom IAM policies or select preset policies and apply them to identities to control their access to AWS resources.

Understand how to protect your AWS account's root user. You should lock down your root user and instead delegate day-to-day tasks to specially defined users.

Understand how to effectively and securely manage user access. This includes the efficient use of IAM groups and roles and appropriately applying both preset and custom IAM policies.

Understand how to optimize account access security. You can use IAM administration tools to enforce the use of strong passwords and MFA, along with properly managed (rotated) access keys and tokens.

Be familiar with the various AWS authentication and integration tools. Cognito lets you manage your application's users, and Managed Microsoft AD applies Active Directory domains to compatible applications running in your VPC. Both permit federated identities (where users using external authentication services can be authenticated for AWS resources) through external providers.

Review Questions

1. Which of the following is the *greatest* risk posed by using your AWS account root user for day-to-day operations?
   A. There would be no easy way to control resource usage by project or class.
   B. There would be no effective limits on the effect of an action, making it more likely for unintended and unwanted consequences to result.
   C. Since root has full permissions over your account resources, an account compromise at the hands of hackers would be catastrophic.
   D. It would make it difficult to track which account user is responsible for specific actions.

2. You're trying to create a custom IAM policy to more closely manage access to components in your application stack. Which of the following syntax-related statements is a correct description of IAM policies?
   A. The Action element refers to the way IAM will react to a request.
   B. The * character applies an element globally—as broadly as possible.
   C. The Resource element refers to the third-party identities that will be allowed to access the account.
   D. The Effect element refers to the anticipated resource state after a request is granted.

3. Which of the following will—when executed on its own—prevent an IAM user with no existing policies from launching an EC2 instance? (Choose three.)
   A. Attach no policies to the user.
   B. Attach two policies to the user, with one policy permitting full EC2 access and the other permitting IAM password changes but denying EC2 access.
   C. Attach a single policy permitting the user to create S3 buckets.
   D. Attach the AdministratorAccess policy.
   E. Associate an IAM action statement blocking all EC2 access to the user's account.

4. Which of the following are important steps for securing IAM user accounts? (Choose two.)
   A. Never use the account to perform any administration operations.
   B. Enable multifactor authentication (MFA).
   C. Assign a long and complex password.
   D. Delete all access keys.
   E. Insist that your users access AWS resources exclusively through the AWS CLI.

5. To reduce your exposure to possible attacks, you're auditing the active access keys associated with your account. Which of the following AWS CLI commands can tell you whether a specified access key is still being used?
   A. `aws iam get-access-key-used -access-key-id <key_ID>`
   B. `aws iam --get-access-key-last-used access-key-id <key_ID>`

C. `aws iam get-access-key-last-used access-last-key-id <key_ID>`

D. `aws iam get-access-key-last-used --access-key-id <key_ID>`

6. You're looking to reduce the complexity and tedium of AWS account administration. Which of the following is the greatest benefit of organizing your users into groups?

    A. It enhances security by consolidating resources.

    B. It simplifies the management of user permissions.

    C. It allows for quicker response times to service interruptions.

    D. It simplifies locking down the root user.

7. During an audit of your authentication processes, you enumerate a number of identity types and want to know which of them might fit the category of "trusted identity" and require deeper investigation. Which of these is *not* considered a trusted entity in the context of IAM roles?

    A. A web identity authenticating with Google

    B. An identity coming through a SAML-based federated provider

    C. An identity using an X.509 certificate

    D. A web identity authenticating with Amazon Cognito

8. Your company is bidding for a contract with a U.S. government agency that demands any cryptography modules used on the project be compliant with government standards. Which of the following AWS services provides virtual hardware devices for managing encryption infrastructure that's FIPS 140-2 compliant?

    A. AWS CloudHSM

    B. AWS Key Management Service

    C. AWS Security Token Service

    D. AWS Secrets Manager

9. Which of the following is the best tool for authenticating access to a VPC-based Microsoft SharePoint farm?

    A. Amazon Cognito

    B. AWS Directory Service for Microsoft Active Directory

    C. AWS Secrets Manager

    D. AWS Key Management Service

10. What is the function of Amazon Cognito identity pools?

    A. Gives your application users temporary, controlled access to other services in your AWS account

    B. Adds user sign-up and sign-in to your applications

    C. Incorporates encryption infrastructure into your application lifecycle

    D. Delivers up-to-date credentials to authenticate RDS database requests

11. An employee with access to the root user on your AWS account has just left your company. Since you can't be 100 percent sure that the former employee won't try to harm your company, which of the following steps should you take? (Choose three.)
    A. Change the password and MFA settings for the root account.
    B. Delete and re-create all existing IAM policies.
    C. Change the passwords for all your IAM users.
    D. Delete the former employee's own IAM user (within the company account).
    E. Immediately rotate all account access keys.

12. You need to create a custom IAM policy to give one of your developers limited access to your DynamoDB resources. Which of the following elements will not play any role in crafting an IAM policy?
    A. Action
    B. Region
    C. Effect
    D. Resource

13. Which of the following are necessary steps for creating an IAM role? (Choose two.)
    A. Define the action.
    B. Select at least one policy.
    C. Define a trusted entity.
    D. Define the consumer application.

14. Which of the following uses authentication based on AWS Security Token Service (STS) tokens?
    A. Policies
    B. Users
    C. Groups
    D. Roles

15. What format must be used to write an IAM policy?
    A. HTML
    B. Key/value pairs
    C. JSON
    D. XML

16. If you need to allow a user full control over EC2 instance resources, which two of the following must be included in the policy you create?
    A. "Target": "ec2:*"
    B. "Action": "ec2:*"
    C. "Resource": "ec2:*"

D. `"Effect": "Allow"`

E. `"Effect": "Permit"`

17. What is the function of Amazon Cognito user pools?

    A. Gives your application users temporary, controlled access to other services in your AWS account

    B. Adds user sign-up and sign-in to your applications

    C. Incorporates encryption infrastructure into your application lifecycle

    D. Delivers up-to-date credentials to authenticate RDS database requests

18. Which of the following best describe the "managed" part of AWS Managed Microsoft AD? (Choose two.)

    A. Integration with on-premises AD domains is possible.

    B. AD domain controllers are launched in two availability zones.

    C. Data is automatically replicated.

    D. Underlying AD software is automatically updated.

19. Which of the following steps are part of the access key rotation process? (Choose three.)

    A. Monitor the use of your new keys.

    B. Monitor the use of old keys.

    C. Deactivate the old keys.

    D. Delete the old keys.

    E. Confirm the status of your X.509 certificate.

20. What tool will allow an Elastic Container Service task to access container images it might need that are being maintained in your account's Elastic Container Registry?

    A. An IAM role

    B. An IAM policy

    C. An IAM group

    D. An AIM access key

Chapter 7

CloudTrail, CloudWatch, and AWS Config

THE AWS CERTIFIED SOLUTIONS ARCHITECT ASSOCIATE EXAM OBJECTIVES COVERED IN THIS CHAPTER MAY INCLUDE, BUT ARE NOT LIMITED TO, THE FOLLOWING:

✓ **Domain 1: Design Resilient Architectures**

- 1.2 Design highly available and/or fault-tolerant architectures

✓ **Domain 2: Design High-Performing Architectures**

- 2.1 Identify elastic and scalable compute solutions for a workload

✓ **Domain 3: Design Secure Applications and Architectures**

- 3.1 Design secure access to AWS resources
- 3.3 Select appropriate data security options

Introduction

CloudTrail, CloudWatch, and AWS Config are three services that can help you ensure the health, performance, and security of your AWS resources and applications. These services collectively help you keep an eye on your AWS environment by performing the following operational tasks:

Tracking Performance   Understanding how your AWS resources are performing can tell you if they're powerful enough to handle the load you're throwing at them or if you need to scale up or out. By tracking performance over time, you can identify spikes in usage and determine whether those spikes are temporarily exhausting your resources. For example, if an EC2 instance maxes out its CPU utilization during times of high usage, you may need to upgrade to a larger instance type or add more instances.

Detecting Application Problems   Some application problems may be latent or hidden and go unreported by users. Checking application logs for warnings or errors can alert you to latent problems early on. For example, the words *exception* or *warning* in a log may indicate a data corruption, a timeout, or other issue that needs to be investigated before it results in a catastrophic failure.

Detecting Security Problems   Users having excessive permissions—or worse, accessing resources they're not supposed to—pose a security risk. Tracking user activities and auditing user permissions can help you reduce your risk and improve your security posture.

Logging Events   Maintaining a log of every single action that occurs against your AWS resources can be invaluable in troubleshooting and security investigations. Problems and breaches are inevitable and sometimes go undiscovered until months later. Having a detailed record of who did what and when can help you understand the cause, impact, and scope of an issue. Just as importantly, you can use this information to prevent the same problem in the future.

Maintaining an Inventory of AWS Resources   Understanding your existing resources, how they're configured, and their relationships and dependencies can help you understand how proposed changes will impact your current environment. Maintaining an up-to-date inventory can also help you ensure compliance with baselines your organization has adopted. Tracking your resource configurations over time makes it easy to satisfy auditing requirements that demand documentation on how a resource was configured at a point in time in the past.

CloudTrail, CloudWatch, and AWS Config are separate services that can be configured independently. They can also work together to provide a comprehensive monitoring solution for your AWS resources, applications, and even on-premises servers:

- CloudTrail keeps detailed logs of every read or write action that occurs against your AWS resources, giving you a trail that includes what happened, who did it, when, and even their IP address.

- CloudWatch collects numeric performance metrics from AWS and non-AWS resources such as on-premises servers. It collects and stores log files from these resources and lets you search them easily, and it provides alarms that can send you a notification or take an action when a metric crosses a threshold. CloudWatch can also take automatic action in response to events or on a schedule.

- AWS Config tracks how your AWS resources are configured and how they change over time. You can view how your resources are related to one another and how they were configured at any time in the past. You can also compare your resource configurations against a baseline that you define and have AWS Config alert you when a resource falls out of compliance.

CloudTrail

An *event* is a record of an action that a principal (a user or role; see Chapter 12, "The Security Pillar") performs against an AWS resource. CloudTrail logs read and write actions against AWS services in your account, giving you a detailed record including the action, the resource affected and its region, who performed the action, and when.

CloudTrail logs both API and non-API actions. API actions include launching an instance, creating a bucket in S3, and creating a virtual private cloud (VPC). Non-API actions include logging into the management console. Despite the name, API actions are API actions regardless of whether they're performed in the AWS management console, with the AWS command-line interface (CLI), with an AWS SDK, or by another AWS service. CloudTrail classifies events into *management events* and *data events*.

Management Events

Management events include operations that a principal executes (or attempts to execute) against an AWS resource. AWS also calls management events *control plane operations*. Management events are further grouped into write-only and read-only events.

Write-Only Events   Write-only events include API operations that modify or might modify resources. For example, the `RunInstances` API operation may create a new EC2 instance and would be logged, regardless of whether the call was successful. Write-only events also include logging into the management console as the root or an IAM user. CloudTrail does not log unsuccessful root logins.

Read-Only Events   Read-only events include API operations that read resources but can't make changes, such as the `DescribeInstances` API operation that returns a list of EC2 instances.

Data Events

Data events track two types of data plane operations that tend to be high volume: S3 object-level activity and Lambda function executions. For S3 object-level operations, CloudTrail distinguishes read-only and write-only events. `GetObject`—the action that occurs when you download an object from an S3 bucket—is a read-only event, whereas `DeleteObject` and `PutObject` are write-only events.

Event History

By default, CloudTrail logs 90 days of management events and stores them in a viewable, searchable, and downloadable database called the *event history*. The event history does not include data events.

CloudTrail creates a separate event history for each region containing only the activities that occurred in that region. So, if you're missing events, you're probably looking in the wrong region! But events for global services such as IAM, CloudFront, and Route 53 are included in the event history of every region.

Trails

If you want to store more than 90 days of event history or if you want to customize the types of events CloudTrail logs—for example, by excluding specific services or actions, or including data events such as S3 downloads and uploads—you can create a *trail*.

A trail is a configuration that records specified events and delivers them as CloudTrail log files to an S3 bucket of your choice. A log file contains one or more log entries in JavaScript Object Notation (JSON) format. A log entry represents a single action against a resource and includes detailed information about the action, including, but not limited to, the following:

eventTime   The date and time of the action, given in universal coordinated time (UTC). Log entries in a log file are sorted by timestamp, but events with the same time-stamp are not necessarily in the order in which the events occurred.

userIdentity Detailed information about the principal that initiated the request. This may include the type of principal (e.g., IAM role or user), its Amazon resource name (ARN), and IAM username.

eventSource The global endpoint of the service against which the action was taken (e.g., ec2.amazonaws.com).

eventName The name of the API operation (e.g., RunInstances).

awsRegion The region the resource is located in. For global services, such as Route 53 and IAM, this is always us-east-1.

sourceIPAddress The IP address of the requester.

Creating a Trail

You can choose to log events from a single region or all regions. If you apply the trail to all regions, then whenever AWS launches a new region, it automatically adds it to your trail.

You can create up to five trails for a single region. A trail that applies to all regions will count against this per-region limit. For example, if you create a trail in us-east-1 and then create another trail that applies to all regions, CloudTrail will count this as two trails in the us-east-1 region.

After you create a trail, it can take up to 15 minutes between the time CloudTrail logs an event and the time it writes a log file to the S3 bucket. This also applies to events written to CloudTrail event history. If you don't see a recent event, just be patient!

Logging Management and Data Events

When you create a trail, you can choose whether to log management events, data events, or both. If you log management events, you must choose whether to log read-only events, write-only events, or both. This allows you to log read-only and write-only events to separate trails. Complete Exercise 7.1 to create a trail that logs write-only events in all regions.

EXERCISE 7.1

Create a Trail

In this exercise, you'll configure CloudTrail to log write-only management events in all regions.

1. Browse to the CloudTrail service console and click the Create Trail button.

2. Under Trail Name, enter a trail name of your choice. Names must be at least three characters and can't contain spaces.

3. Under the Storage Location heading, select Create New S3 Bucket. Enter the name of the S3 bucket you want to use. Remember that bucket names must be globally unique.

EXERCISE 7.1 (continued)

4. Under Log File SSE-KMS Encryption, clear the box next to Enabled.
5. Click the Next button.
6. Under Event Types, select the box next to Management Events. Don't select any other boxes.
7. Under Management Events, make sure only Write is selected.
8. Click Next.
9. Click the Create Trail button.

The following list explains the difference between management events and data events.

Logging Management Events   If you create a trail using the web console and log management events, the trail will automatically log global service events also. These events are logged as occurring in the us-east-1 region. This means that if you create multiple trails using the web console, global events will be logged to each trail. To avoid these duplicate events, you can disable logging global service events on an existing trail using the AWS CLI command `aws cloudtrail update-trail --name mytrail --no-include-global-service-events`.

Alternately, if a trail is configured to log for all regions and you reconfigure it to log only for a single region, CloudTrail will disable global event logging for that trail. Another option is to forego the web console and create your trails using the AWS CLI, including the `--no-include-global-service-events` flag.

Logging Data Events   You're limited to selecting a total of 250 individual objects per trail, including Lambda functions and S3 buckets and prefixes. If you have more than 250 objects to log, you can instruct CloudTrail to log all Lambda functions or all S3 buckets. If you create a single-region trail to log Lambda events, you're limited to logging functions that exist in that region. If you create a trail that applies to all regions, you can log Lambda functions in any region.

> **NOTE**   Don't log data events on the bucket that's storing your CloudTrail logs. Doing so would create an infinite loop!

Log File Integrity Validation

CloudTrail provides a means to ensure that no log files were modified or deleted after creation. When an attacker hacks into a system, it's common for them to delete or modify log files to cover their tracks. During quiet periods of no activity, CloudTrail also gives you assurance that no log files were delivered, as opposed to being delivered and then maliciously deleted. This is useful in forensic investigations where someone with access to the S3 bucket may have tampered with the log file.

With log file integrity validation enabled, every time CloudTrail delivers a log file to the S3 bucket, it calculates a cryptographic hash of the file. This hash is a unique value derived from the contents of the log file itself. If even one byte of the log file changes, the entire hash changes. Hashes make it easy to detect when a file has been modified.

Every hour, CloudTrail creates a separate file called a *digest file* that contains the cryptographic hashes of all log files delivered within the last hour. CloudTrail places this file in the same bucket as the log files but in a separate folder. This allows you to set different permissions on the folder containing the digest file to protect it from deletion. CloudTrail also cryptographically signs the digest file using a private key that varies by region and places the signature in the file's S3 object metadata.

Each digest file also contains a hash of the previous digest file, if it exists. If there are no events to log during an hourlong period, CloudTrail still creates a digest file. This lets you know that no log files were delivered during the quiet period.

You can validate the integrity of CloudTrail log and digest files by using the AWS CLI. You must specify the ARN of the trail and a start time. The AWS CLI will validate all log files from the starting time to the present. For example, to validate all log files written from January 1, 2021, to the present, you'd issue the following command:

```
aws cloudtrail validate-logs --trail-arn arn:aws:cloudtrail:
us-east-1:account-id:trail/benpiper-trail --start-time
2021-01-01T00:00:00Z.
```

> **NOTE** By default, log and digest files are encrypted using Amazon server-side encryption with Amazon S3-managed encryption keys (SSE-S3). You can choose instead to use server-side encryption with AWS KMS-managed keys (SSE-KMS) for log files. The customer master key (CMK) you use must be in the same region as the bucket. Digest files are always encrypted with SSE-S3.

CloudWatch

CloudWatch lets you collect, retrieve, and graph numeric performance metrics from AWS and non-AWS resources. All AWS resources automatically send their metrics to CloudWatch. These metrics include EC2 instance CPU utilization, EBS volume read and write

IOPS, S3 bucket sizes, and DynamoDB consumed read and write capacity units. Optionally, you can send custom metrics to CloudWatch from your applications and on-premises servers. CloudWatch Alarms can send you a notification or take an action based on the value of those metrics. CloudWatch Logs lets you collect, store, view, and search logs from AWS and non-AWS sources. You can also extract custom metrics from logs, such as the number of errors logged by an application or the number of bytes served by a web server.

CloudWatch Metrics

CloudWatch organizes metrics into *namespaces*. Metrics from AWS services are stored in AWS namespaces and use the format AWS/service to allow for easy classification of metrics. For example, AWS/EC2 is the namespace for metrics from EC2, and AWS/S3 is the namespace for metrics from S3.

You can think of a namespace as a container for metrics. Namespaces help prevent metrics from being confused with similar names. For example, CloudWatch stores the WriteOps metric from the Relational Database Service (RDS) in the AWS/RDS namespace, whereas the EBS metric VolumeWriteOps goes in the AWS/EBS namespace. You can create custom namespaces for custom metrics. For example, you can store metrics from an Apache web server under the custom namespace Apache. Metrics exist only in the region in which they were created.

A metric functions as a variable and contains a time-ordered set of *data points*. Each data point contains a timestamp, a value, and optionally a unit of measure. Each metric is uniquely defined by a namespace, a name, and optionally a *dimension*. A dimension is a name/value pair that distinguishes metrics with the same name and namespace from one another. For example, if you have multiple EC2 instances, CloudWatch creates a CPUUtilization metric in the AWS/EC2 namespace for each instance. To uniquely identify each metric, AWS assigns it a dimension named InstanceId with the value of the instance's resource identifier.

Basic and Detailed Monitoring

How frequently an AWS service sends metrics to CloudWatch depends on the monitoring type the service uses. Most services support *basic monitoring*, and some support basic monitoring and *detailed monitoring*.

Basic monitoring sends metrics to CloudWatch every five minutes. EC2 provides basic monitoring by default. EBS uses basic monitoring for gp2 volumes.

EC2 collects metrics every minute but sends only the five-minute average to CloudWatch. How EC2 sends data points to CloudWatch depends on the hypervisor. For instances using the Xen hypervisor, EC2 publishes metrics at the end of the five-minute interval. For example, between 13:00 and 13:05, an EC2 instance has the following CPUUtilization metric values measured in percent: 25, 50, 75, 80, and 10. The average CPUUtilization over the five-minute interval is 48. Therefore, EC2 sends the CPUUtilization metric to CloudWatch with a timestamp of 13:00 and a value of 48.

For instances using the Nitro hypervisor, EC2 sends a data point every minute during a five-minute period, but a data point is a rolling average. For example, at 13:00, EC2 records a data point for the `CPUUtilization` metric with a value of 25. EC2 sends this data point to CloudWatch with a timestamp of 13:00. At 13:01, EC2 records another data point with a value of 50. It averages this new data point with the previous one to get a value of 37.5. It then sends this new data point to CloudWatch, but with a timestamp of 13:00. This process continues for the rest of the five-minute interval.

Services that use detailed monitoring publish metrics to CloudWatch every minute. More than 70 services support detailed monitoring, including EC2, EBS, RDS, DynamoDB, ECS, and Lambda. EBS defaults to detailed monitoring for io1 volumes.

Regular and High-Resolution Metrics

The metrics generated by AWS services have a timestamp resolution of no less than one minute. For example, a measurement of `CPUUtilization` taken at 14:00:28 would have a timestamp of 14:00. These are called *regular-resolution* metrics. For some AWS services, such as EBS, CloudWatch stores metrics at a five-minute resolution. For example, if EBS delivers a `VolumeWriteBytes` metric at 21:34, CloudWatch would record that metric with a timestamp of 21:30.

CloudWatch can store custom metrics with up to one-second resolution. Metrics with a resolution of less than one minute are *high-resolution metrics*. You can create your own custom metrics using the `PutMetricData` API operation. When publishing a custom metric, you can specify the timestamp to be up to two weeks in the past or up to two hours into the future. If you don't specify a timestamp, CloudWatch creates one based on the time it received the metric in coordinated universal time (UTC).

Expiration

You can't delete metrics in CloudWatch. Metrics expire automatically, and when a metric expires depends on its resolution. Over time, CloudWatch aggregates higher-resolution metrics into lower-resolution metrics.

A high-resolution metric is stored for three hours. After this, all the data points from each minute-long period are aggregated into a single data point at one-minute resolution. The high-resolution data points simultaneously expire and are deleted. After 15 days, five data points stored at one-minute resolution are aggregated into a single data point stored at five-minute resolution. These metrics are retained for 63 days. At the end of this retention period, 12 data points from each metric are aggregated into a single 1-hour resolution metric and retained for 15 months. After this, the metrics are deleted.

To understand how this works, consider a `VolumeWriteBytes` metric stored at five-minute resolution. CloudWatch will store the metric at this resolution for 63 days, after which it will convert the data points to one-hour resolution. After 15 months, CloudWatch will delete those data points permanently.

Graphing Metrics

CloudWatch can perform statistical analysis on data points over a period of time and graph the results as a time series. This is useful for showing trends and changes over time, such as spikes in usage. You can choose from the following *statistics*:

Sum   The total of all data points in a period

Minimum   The lowest data point in a period

Maximum   The highest data point in a period

Average   The average of all data points in a period

Sample count   The number of data points in a period

Percentile   The data point of the specified percentile. You can specify a percentile of up to two decimal places. For example, 50 would yield the median of the data points in the period. You must specify a percentile statistic in the format p50.

To graph a metric, you must specify the metric, the statistic, and the period. The period can be from one second to 30 days, and the default is 60 seconds. If you want CloudWatch to graph each data point as is, use the Sum statistic and set the period equal to the metric's resolution. For example, if you're using detailed monitoring to record the `CPUUtilization` metric from an EC2 instance, that metric will be stored at one-minute resolution. Therefore, you would graph the `CPUUtilization` metric over a period of one minute using the Sum statistic. To get an idea of how this would look, refer to Figure 7.1.

FIGURE 7.1   CPU utilization

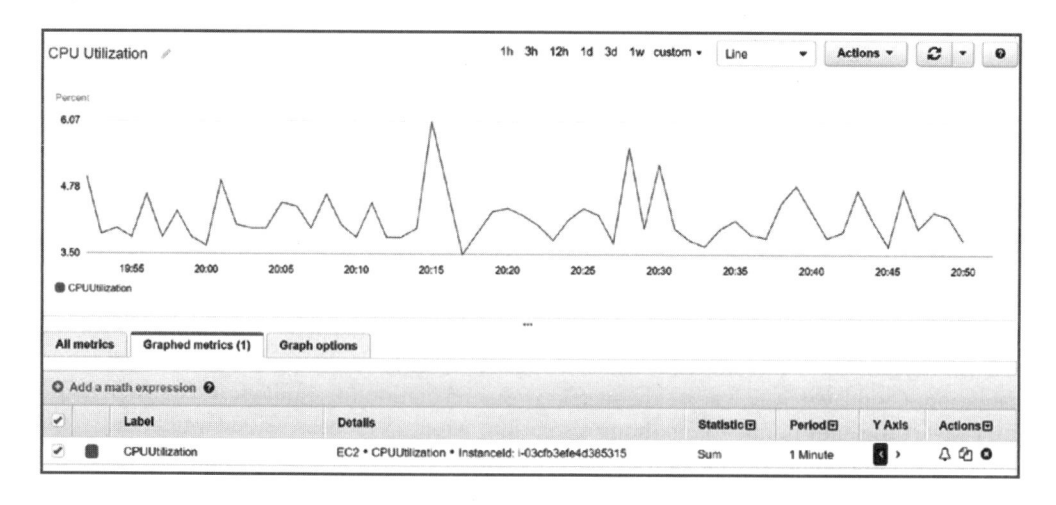

Note that the statistic applies only to data points within a period. In the preceding example, the Sum statistic adds all data points within a one-minute period. Because CloudWatch stores the metric at one-minute resolution, there is only one data point per minute. Hence, the resulting graph shows each individual metric data point.

The *time range* in the preceding graph is set to one hour, but you can choose a range of time between 1 minute and 15 months. Choosing a different range doesn't change the time series, but only how it's displayed.

Which statistic you should choose depends on the metric and what you're trying to understand about the data. CPU utilization fluctuates and is measured as a percentage, so it doesn't make much sense to graph the sum of CPU utilization over, say, a 15-minute period, because doing so would yield a result over 100 percent. It does, however, make sense to take the average CPU utilization over the same timeframe. Hence, if you were trying to understand long-term patterns of CPU utilization, you can use the Average statistic over a 15-minute period.

The `NetworkOut` metric in the `AWS/EC2` namespace measures the number of bytes sent by an instance during the collection interval. To understand peak hours for network utilization, you can graph the Sum statistic over a one-hour period and set the time range to one day, as shown in Figure 7.2.

FIGURE 7.2 The sum of network bytes sent out over a one-hour period

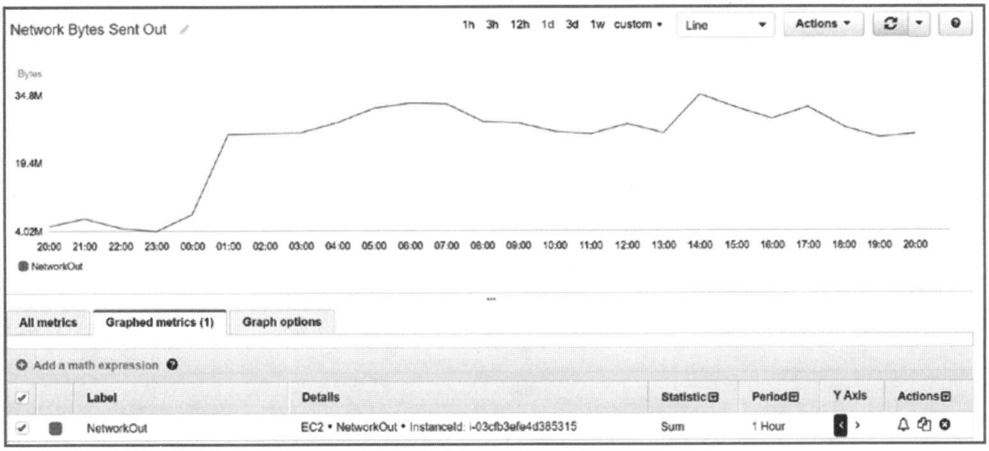

The Details column shows information that uniquely identifies the metric: the namespace, metric name, and the metric dimension, which in this case is `InstanceId`.

Metric Math

CloudWatch lets you perform various mathematical functions against metrics and graph them as a new time series. This capability is useful for when you need to combine multiple metrics into a single time series by using arithmetic functions, which include addition, subtraction, multiplication, division, and exponentiation. For example, you might divide the AWS/Lambda `Invocations` metrics by the `Errors` metric to get an error rate. Complete Exercise 7.2 to create a CloudWatch graph using *metric math*.

EXERCISE 7.2

Create a Graph Using Metric Math

In this exercise, you'll create a graph that plots the `NetworkIn` and `NetworkOut` metrics for an EC2 instance. You'll then use metric math to graph a new time series combining both metrics.

1. Browse to the CloudWatch service console and click Metrics on the navigation menu.

2. On the All Metrics tab, descend into the `EC2` namespace. Select Per-Instance Metrics; then locate and select the `NetworkIn` and `NetworkOut` metrics.

3. Click the Graphed Metrics tab.

4. For each metric, select Sum for Statistic and 5 Minutes for Period. Refer to Figure 7.2 as needed.

5. Click Math Expression and select Start With Empty Expression. CloudWatch will add an ID column next to each metric and assign the values m1 and m2. It will also add another row with an ID of e1.

6. In the Details column for the e1 row, enter the expression **m1+m2**. CloudWatch will add another time series to the graph representing the sum of the `NetworkIn` and `NetworkOut` metrics. Your graph should look similar to this one:

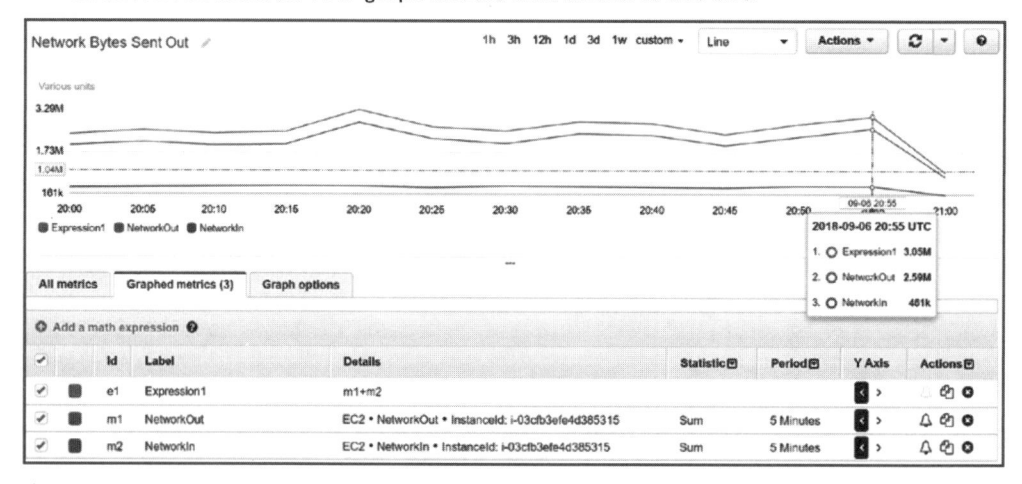

In addition to arithmetic functions, CloudWatch provides the following statistical functions that you can use in metric math expressions:

- `AVG`—Average
- `MAX`—Maximum
- `MIN`—Minimum
- `STDDEV`—Standard deviation
- `SUM`—Sum

Statistical functions return a scalar value, not a time series, so they can't be graphed. You must combine them with the `METRICS` function, which returns an array of time series of all selected metrics. For instance, referring to step 6 in Exercise 7.2, you could replace the expression m1+m2 with the expression `SUM(METRICS())` to achieve the same result. This would simply add up all the graphed metrics.

As another example, suppose you want to compare the CPU utilization of an instance with the standard deviation. You would first graph the `AWS/EC2` metric `CPUUtilization` for the instance in question. You'd then add the metric math expression `METRICS()/STDDEV(m1)`, where m1 is the time series for the `CPUUtilization` metric. See Figure 7.3 for an example of what such as graph might look like.

FIGURE 7.3 Combining metric math functions

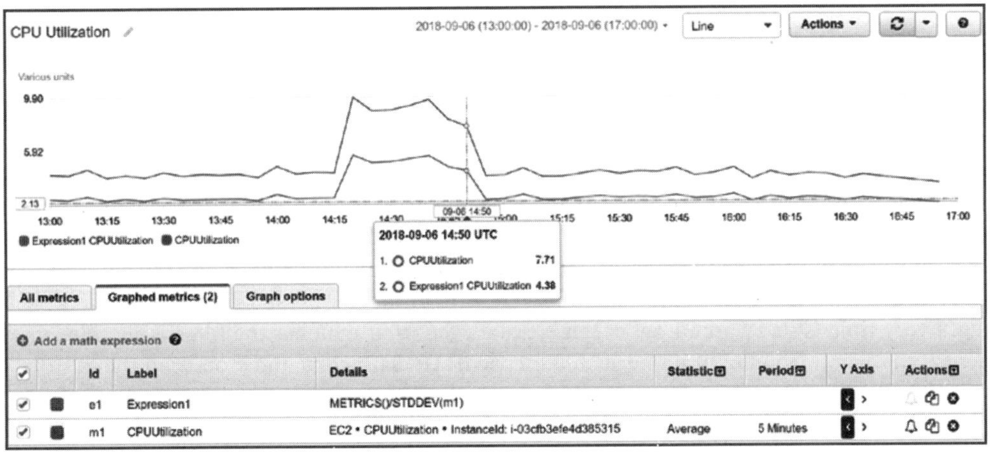

Keep in mind that the function `STDDEV(m1)` returns a scalar value—the standard deviation of all data points for the `CPUUtilization` metric. You must therefore use the `METRICS` function in the numerator to yield a time series that CloudWatch can graph.

CloudWatch Logs

CloudWatch Logs is a feature of CloudWatch that collects logs from AWS and non-AWS sources, stores them, and lets you search and even extract custom metrics from them. Some common uses for CloudWatch Logs include receiving CloudTrail logs, collecting application logs from an instance, and logging Route 53 DNS queries.

Log Streams and Log Groups

CloudWatch Logs stores *log events* that are records of activity recorded by an application or AWS resource. For CloudWatch Logs to understand a log event, the event must contain a timestamp and a UTF-8 encoded event message. In other words, you can't store binary data in CloudWatch Logs.

CloudWatch Logs stores log events from the same source in a *log stream*. The source may be an application or AWS resource. For example, if you have multiple instances running a web server that generates access logs, the logs from each instance would go into a separate log stream. You can manually delete log streams, but not individual log events.

CloudWatch organizes log streams into *log groups*. A stream can exist in only one log group. To organize related log streams, you can place them into the same log group. For example, you might stream logs from all instances in the same Auto Scaling group to the same log group. There's no limit to the number of streams in a group.

You may define the retention settings for a log group, choosing to keep log events from between one day to 10 years or indefinitely, which is the default setting. The retention settings apply to all log streams in a log group. You can manually export a log group to an S3 bucket for archiving.

Metric Filters

You can use *metric filters* to extract data from log streams to create CloudWatch metrics. Metric values must be numeric. You can use a metric filter to extract a numeric value, such as the number of bytes transferred in a request, and store that in a metric. But you can't use a metric filter to extract a non-numeric string, such as an IP address, from a log and store it as a metric. You can, however, increment a metric when the metric filter matches a specific string. You can create a metric filter to track the occurrences of a string at a particular position in a log file. For example, you may want to track the number of 404 Not Found errors that appear in an Apache web server application log. You would create a metric filter to track the number of times the string "404" appears in the HTTP status code section of the log. Every time CloudWatch Logs receives a log event that matches the filter, it increments a custom metric. You might name such a metric **HTTP404Errors** and store it in the custom Apache namespace. Of course, you can then graph this metric in CloudWatch.

Metric filters apply to entire log groups, and you can create a metric filter only after creating the group. Metric filters are not retroactive and will not generate metrics based on log events that CloudWatch recorded before the filter's creation.

CloudWatch Agent

The CloudWatch Agent is a command line–based program that collects logs from EC2 instances and on-premises servers running Linux or Windows operating systems. The agent can also collect performance metrics, including metrics that EC2 doesn't natively produce, such as memory utilization. Metrics generated by the agent are custom metrics and are stored in a custom namespace that you specify.

> **NOTE:** The CloudWatch Agent is different than the legacy CloudWatch Logs agent, which sends only logs and not metrics. AWS recommends using the CloudWatch agent.

Sending CloudTrail Logs to CloudWatch Logs

You can configure CloudTrail to send a trail log to a CloudWatch Logs log stream. Doing so lets you search and extract metrics from your trail logs. Remember that CloudTrail generates trail logs in JSON format and stores them in an S3 bucket of your choice, but it doesn't provide a way to search those logs. CloudWatch understands JSON format and makes it easy to search for specific events. For example, to search for failed console logins, you would filter the log stream in CloudWatch using the following syntax:

{$.eventSource = "signin.amazonaws.com" &&
$.responseElements.ConsoleLogin = "Failure" }

CloudTrail does not send log events larger than 256 KB to CloudWatch Logs. Hence, a single RunInstances call to launch 500 instances would exceed this limit. Therefore, make sure you break up large requests if you want them to be available in CloudWatch Logs. Complete Exercise 7.3 to configure your existing CloudTrail to deliver trail logs to CloudWatch Logs.

EXERCISE 7.3

Deliver CloudTrail Logs to CloudWatch Logs

In this exercise, you'll reconfigure the trail you created in Exercise 7.1 to stream events captured by CloudTrail to CloudWatch Logs.

1. Browse to the CloudTrail service console and click Trails.
2. Click the name of the trail you created in Exercise 7.1.
3. Under the heading CloudWatch Logs, click the Edit button.
4. Under CloudWatch Logs, select the Enabled check box.
5. CloudTrail prompts you to use a New or Existing Log Group. Select New and enter a log group name of your choice.
6. CloudTrail must assume an IAM role that will give it permissions to stream logs to CloudWatch Logs. CloudTrail can create the role for you. Just click the New radio button under IAM Role. Enter a custom role name of your choice.
7. Click the Save Changes button.

Delivery isn't instant, and it can take a few minutes before trail logs show up in CloudWatch Logs.

CloudWatch Alarms

A *CloudWatch alarm* watches over a single metric and performs an action based on a change in its value. The action CloudWatch takes can include tasks such as sending an email notification, rebooting an instance, or executing an Auto Scaling action.

To create an alarm, you first define the metric you want CloudWatch to monitor. In much the same way CloudWatch doesn't graph metrics directly but graphs metric *statistics* over a period, a CloudWatch alarm does not directly monitor metrics. Instead, it performs statistical analysis of a metric over time and monitors the result.

Data Point to Monitor

Suppose you want to monitor the average of the AWS/EBS VolumeReadOps metric over a 15-minute period. The metric has a resolution of 5 minutes. You would choose Average for the statistic and 15 minutes for the period. Every 15 minutes CloudWatch would take three metric data points—one every 5 minutes—and would average them together to generate a single *data point to monitor*.

You should set the period equal to or greater than the resolution of the metric. If you set a lower period, such as one minute, CloudWatch will look for a data point every minute, but because the metric updates only once every five minutes, it will count four of the five metrics as missing. This will result in the alarm not working properly.

If you use a percentile for the statistic, you must also select whether to ignore data points until the alarm collects a statistically significant number of data points. What constitutes statistically significant depends on the percentile. If you set the percentile to .5 (p50) or greater, you must have 10/1(1-percentile) data points to have a statistically significant sample. For instance, if you use the p80 statistic, a statistically significant number of data points would be 10/(1-.8) or 50. If the percentile is less than .5, you need 10/percentile data points to have a statistically significant sample. Supposing you were using the p25 statistic, you'd need 10/(.25), or 40 data points. If you choose to ignore data points before you have a statistically significant sampling, CloudWatch will not evaluate any of them. In other words, your alarm will be effectively disabled.

Threshold

The threshold is the value the data point to monitor must meet or cross to indicate something is wrong. There are two types of thresholds:

Static Threshold   You define a static threshold by specifying a value and a condition. If you want to trigger an alarm when CPUUtilization meets or exceeds 50 percent, you would set the threshold for that alarm to >= 50. Or if you want to know when CPUCreditBalance falls below 800, you would set the threshold to < 800.

Anomaly Detection   Anomaly detection is based on whether a metric falls outside of a range of values called a band. You define the size of the band based on the number of standard deviations. For example, if you set an anomaly detection threshold of 2, the alarm would trigger when a value is outside of two standard deviations from the average of the values.

Alarm States

The period of time a data point to monitor must remain crossing the threshold to trigger an alarm state change depends on the *data points to alarm*. An alarm can be in one of three states at any given time:

ALARM   The data points to alarm have crossed and remained past a defined threshold for a period of time.

OK   The data points to alarm have not crossed and remained past a defined threshold for a period of time.

INSUFFICIENT_DATA   The alarm hasn't collected enough data to determine whether the data points to alarm have crossed a defined threshold.

New alarms always start out in an INSUFFICIENT_DATA state. It's important to remember that an ALARM state doesn't necessarily indicate a problem, and an OK state doesn't necessarily indicate the absence of a problem. Alarm states track only whether the data points to alarm have crossed and remained past a threshold for a period of time. As an example, if the period is five minutes and the data points to alarm is three, then the data points to monitor must cross and remain crossing the threshold for 15 minutes before the alarm goes into an ALARM state.

Data Points to Alarm and Evaluation Period

There are cases where you may want to trigger the alarm if a data point to monitor crosses the threshold periodically but doesn't remain past it. For this, you can set an evaluation period that's equal to or greater than the data points to alarm. Suppose you want to trigger an alarm if the data points to monitor cross the threshold for three out of five data points. You would set the evaluation period to 5. The alarm would trigger if *any* three of the latest five data points exceed the threshold. The exceeding values don't have to be consecutive. This is called an *m out of n alarm*, where *m* is the data point to alarm and *n* is the evaluation period. The evaluation period can't exceed 24 hours.

To give an illustration, let's say you create an alarm with a threshold of >= 40. The data points to alarm is 2, and the evaluation period is 3, so this is a 2 out of 3 alarm. Now suppose CloudWatch evaluates the following three consecutive data points: 46, 39, and 41. Two of the three data points exceed the threshold, so the alarm will transition to the ALARM state.

Following that, CloudWatch evaluates the consecutive data points 45, 30, and 25. Two of the three data points fall below the threshold, so the alarm transitions to an OK state. Notice that CloudWatch must evaluate three data points (the evaluation period) before it changes the alarm state.

> When monitoring metrics for EC2 instances using the Nitro hypervisor, make sure the evaluation period is at least 2. Triggering an alarm based on a single data point may cause false positives because of the way EC2 delivers Nitro metrics to CloudWatch.

Missing Data

Missing data can occur during an evaluation period. This may happen if you detach an EBS volume from or stop an instance. CloudWatch offers the following four options for how it evaluates periods with missing data:

As Missing This treats the period missing data as if it never happened. For instance, if over four periods the data points are 41, no data, 50, and 25, CloudWatch will remove the period with no data from consideration. If the evaluation period is 3, it will evaluate only three periods with the data points 41, 50, and 25. It will not consider the missing data as occurring in the evaluation period. This is the default setting.

Not Breaching Missing data points are treated as not breaching the threshold. Consider a three out of four alarm with a threshold of <40, given the same data points: 41, no data, 50, and 25. Even though two values are breaching, the alarm would not trigger because the missing data is assumed to be not breaching.

Breaching CloudWatch treats missing data as breaching the threshold. Using the preceding illustration, the alarm would trigger, as three of the four values would exceed the threshold.

Ignore The alarm doesn't change state until it receives the number of consecutive data points specified in the data points to alarm setting.

Actions

You can configure an alarm to take an action when it transitions to a given state. You're not limited to just when an alarm goes into an ALARM state. You can also have CloudWatch take an action when the alarm transitions to an OK state. This is useful if you want to receive a notification when CPU utilization is abnormally high and another when it returns to normal. You can also trigger an action when an alarm monitoring an instance goes into an INSUFFICIENT_DATA state, which may occur when the instance shuts down. You can choose from the following actions:

Notification Using Simple Notification Service The Simple Notification Service (SNS) uses communication channels called *topics*. A topic allows a sender or *publisher* to send a notification to one or more recipients called *subscribers*.

A subscriber consists of a protocol and an endpoint. The protocol can be HTTP, HTTPS, Simple Queue Service (SQS), Lambda, a mobile push notification, email, email-JSON, or short message service (SMS, also known as text messages). The endpoint depends on the protocol. In the case of email or email-JSON, the endpoint would be an email address. In the case of SQS, it would be a queue. The endpoint for HTTP or HTTPS would be a URL.

When creating an alarm action, you specify an SNS topic. When the alarm triggers, it sends a notification to the topic. SNS takes care of relaying the notification to the subscribers of the topic.

Auto Scaling Action If you're using Auto Scaling, you can create a simple Auto Scaling policy to add or remove instances. The policy must exist before you can select it as an alarm action.

EC2 Action You can stop, terminate, reboot, or recover an instance in response to an alarm state change. You might choose to monitor the `AWS/EC2 StatusCheckFailed_Instance` metric, which returns 1 if there's a problem with the instance, such as memory exhaustion, filesystem corruption, or incorrect network or startup configuration. Such issues could be corrected by rebooting the instance.

You might also monitor the `StatusCheckedFailed_System` metric, which returns 1 when there's a problem that requires AWS involvement to repair, such as a loss of network connectivity or power, hypervisor problems, or hardware failure. In response to this, the recover action migrates the instance to a new host and restarts it with the same settings. In-memory data is lost during this process.

EC2 actions are available only if the metric you're monitoring includes the `InstanceId` as a dimension, as the action you specify will take place against that instance. Using an EC2 action requires a service-linked role named `AWSServiceRoleForCloudWatchEvents`, which CloudWatch can create for you when you create the alarm.

Although a single alarm can have multiple actions, it will take those actions only when it transitions to a single state that you specify when configuring the alarm. You cannot, for instance, set one action when an alarm transitions to an `ALARM` state and another action when it transitions to an `OK` state. Instead, you'd have to create two separate alarms.

Amazon EventBridge

EventBridge (formerly known as CloudWatch Events) monitors for and takes an action either based on specific events or on a schedule. For example, a running EC2 instance entering the stopped state would be an event. An IAM user logging into the AWS Management Console would be another event. EventBridge can automatically take immediate action in response to such events.

EventBridge differs from CloudWatch Alarms in that EventBridge takes some action based on specific events, not metric values. Hence, EventBridge can send an SNS notification as soon as an EC2 instance stops. Or it can execute a Lambda function to process an image file as soon as a user uploads it to an S3 bucket.

Event Buses

EventBridge monitors event buses. By default, every AWS account has one event bus that receives events for all AWS services. You can create a custom event bus to receive events from other sources, such as your applications or third-party services.

Rules and Targets

A rule defines the action to take in response to an event. When an event matches a rule, you can route the event to a target that takes action in response to the event. For example, if you want to receive an email whenever an EC2 Auto Scaling event occurs, you could create a rule to watch for instances in an Auto Scaling group being launched or terminated. For the target, you could select an SNS topic that's configured to send you an email notification.

A rule can also invoke a target on a schedule. This is useful if you want to take hourly EBS snapshots of an EC2 instance. Or to save money, you might create a schedule to shut down test instances every day at 7 p.m.

AWS Config

AWS Config tracks the configuration state of your AWS resources at a point in time. Think of AWS Config as a time machine. You can use it to see what a resource configuration looked like at some point in the past versus what it looks like now.

It can also show you how your resources are related to one another so that you can see how a change in one resource might impact another. For instance, suppose you create an EBS volume, attach it to an instance, and then later detach it. You can use AWS Config to see not only exactly when you created the volume but when it was attached to and detached from the instance.

Note that this is different from CloudTrail, which logs events, and from EventBridge, which can alert on events. Only AWS Config gives you a holistic view of your resources and how they were configured at any point in time. In other words, EventBridge deals with events or actions that occur against a resource, whereas AWS Config deals with the *state* of a resource. AWS Config can help you with the following objectives:

Security   AWS Config can notify you whenever a resource configuration changes, alerting you to potential breaches. You can also see what users had which permissions at a given time.

Easy Audit Reports   You can provide a configuration snapshot report showing you how your resources were configured at any point in time.

Troubleshooting   You can analyze how a resource was configured around the time a problem started. AWS Config makes it easy to spot misconfigurations and how a problem in one resource might impact another.

Change Management   AWS Config lets you see how a potential change to one resource could impact another. For example, if you plan to change a security group, you can use AWS Config to quickly see all instances that use that security group.

The Configuration Recorder

The *configuration recorder* is the workhorse of AWS Config. It discovers your existing resources, records how they're configured, monitors for changes, and tracks those changes over time. By default, it monitors all items in the region in which you configure it. It can also monitor the resources of global services such as IAM. If you don't want to monitor all resources, you can select specific resource types to monitor, such as EC2 instances, IAM users, S3 buckets, or DynamoDB tables. You can have only one configuration recorder per region.

Configuration Items

The configuration recorder generates a *configuration item* for each resource it monitors. The configuration item contains the specific settings for that resource at a point in time, as well as the resource type, its ARN, and when it was created. The configuration item also includes the resource's relationships to other resources. For example, a configuration item for an EBS volume would include the instance ID of the instance it was attached to at the time the item was recorded. AWS Config maintains configuration items for every resource it tracks, even after the resource is deleted. Configuration items are stored internally in AWS Config, and you can't delete them manually.

Configuration History

AWS Config uses configuration items to build a configuration history for each resource. A configuration history is a collection of configuration items for a given resource over time. A configuration history includes details about the resources, such as when it was created, how it was configured at different points in time, and when it was deleted, if applicable. It also includes any related API logged by CloudTrail.

Every six hours in which a change occurs to a resource, AWS Config delivers a configuration history file to an S3 bucket that you specify. The S3 bucket is part of what AWS calls the *delivery channel*. Configuration history files are grouped by resource type. The configuration history for all EC2 instances goes into one file, and the history for EBS volumes goes into another. The files are timestamped and kept in separate folders by date. You can also view configuration history directly from the AWS Config console. You can optionally add an SNS topic to the delivery channel to have AWS Config notify you immediately whenever there's a change to a resource.

Configuration Snapshots

A configuration snapshot is a collection of all configuration items from a given point in time. Think of a configuration snapshot as a configuration backup for all monitored resources in your account.

Using the AWS CLI, you can have AWS Config deliver a configuration snapshot to the bucket defined in your delivery channel. By default, the delivery channel is named default, so to deliver a configuration snapshot, you would manually issue the following command:
`aws configservice deliver-config-snapshot --delivery-channel-name default`

AWS Config can automatically deliver a configuration snapshot to the delivery channel at regular intervals. You can't set automatic delivery of this snapshot in the console but must configure the delivery using the CLI. To do this, you must also specify a JSON file containing at a minimum the following items:

- Delivery channel name (default)
- S3 bucket name
- Delivery frequency of the configuration snapshot

The file may also optionally contain an SNS ARN. Let's refer to the following file as deliveryChannel.json:

```
{
  "name": "default",
  "s3BucketName": "my-config-bucket-us-east-1",
   "snsTopicARN": "arn:aws:sns:us-east-1:account-id:config-topic",
  "configSnapshotDeliveryProperties": {
    "deliveryFrequency": "TwentyFour_Hours"
  }
}
```

The delivery frequency can be every hour or every 24, 12, 6, or 3 hours. To reconfigure the delivery channel according to the settings in the deliveryChannel.json file, you'd issue the following command:

`aws configservice put-delivery-channel --delivery-channel file://deliveryChannel.json`

To verify that the configuration change succeeded, issue the following command:
`aws configservice describe-delivery-channels`

If the output matches the configuration settings in the file, then the configuration change was successful.

Monitoring Changes

The configuration recorder generates at least one new configuration item every time a resource is created, changed, or deleted. Each new item is added to the configuration history for the resource as well as the configuration history for the account.

A change to one resource will trigger a new configuration item for the changed resource and related resources. For instance, removing a rule in a security group causes the configuration recorder to create a new item for the security group and every instance that uses that security group.

Although you can't delete configuration items manually, you can configure AWS Config to keep configuration items between 30 days and 7 years. Seven years is the default. Note that the retention period does not apply to the configuration history and configuration snapshot files AWS Config delivers to S3.

Starting and Stopping the Configuration Recorder

You can start and stop the configuration recorder at any time using the web console or the CLI. During the time the configuration recorder is stopped, it doesn't monitor or record changes. But it does retain existing configuration items. To stop it using the following CLI command, you must specify the configuration recorder's name, which is default.

```
aws configservice stop-configuration-recorder --configuration-
recorder-name default
```

To start the recorder, issue the following command:

```
aws configservice start-configuration-recorder --configuration-
recorder-name default
```

Recording Software Inventory

AWS Config can record software inventory changes on EC2 instances and on-premises servers. This includes the following:

- Applications
- AWS components such as the CLI and SDKs
- The name and version of the operating system
- IP address, gateway, and subnet mask
- Firewall configuration
- Windows updates

To have AWS Config track these changes, you must enable inventory collection for the server using the AWS Systems Manager. You also must ensure AWS Config monitors the SSM:ManagedInstanceInventory resource type.

Managed and Custom Rules

In addition to monitoring resources changes, AWS Config lets you specify rules to define the optimal baseline configuration for your resources. AWS Config also provides customizable, predefined rules that cover a variety of common scenarios. For example, you may want to verify that CloudTrail is enabled, that every EC2 instance has an alarm tracking the CPUUtilization metric, that all EBS volumes are encrypted, or that multifactor authentication (MFA) is enabled for the root account. If any resources are noncompliant, AWS Config flags them and generates an SNS notification.

When you activate a rule, AWS Config immediately checks monitored resources against the rules to determine whether they're compliant. After that, how often it reevaluates resources is based on how the rule is configured. A reevaluation can be triggered by configuration changes or it can be set to run periodically. Periodic checks can occur every hour or every 3, 6, 12, or 24 hours. Note that even if you turn off the configuration recorder, periodic rules will continue to run.

Summary

You must configure CloudWatch and AWS Config before they can begin monitoring your resources. CloudTrail automatically logs only the last 90 days of management events even if you don't configure it. It's therefore a good idea to configure these services early on in your AWS deployment.

CloudWatch, CloudTrail, and AWS Config serve different purposes, and it's important to know the differences among them and when each is appropriate for a given use case.

CloudWatch tracks performance metrics and can take some action in response to those metrics. It can also collect and consolidate logs from multiple sources for storage and searching, as well as extract metrics from them.

CloudTrail keeps a detailed record of activities performed on your AWS account for security or auditing purposes. You can choose to log read-only or write-only management or data events.

AWS Config records resource configurations and relationships past, present, and future. You can look back in time to see how a resource was configured at any point. AWS Config can also compare current resource configurations against rules to ensure that you're in compliance with whatever baseline you define.

Exam Essentials

Know how to configure the different features of CloudWatch. CloudWatch receives and stores performance metrics from various AWS services. You can also send custom metrics to CloudWatch. You can configure alarms to take one or more actions based on a metric. CloudWatch Logs receives and stores logs from various resources and makes them searchable.

Know the differences between CloudTrail and AWS Config. CloudTrail tracks events, while AWS Config tracks how those events ultimately affect the configuration of a resource. AWS Config organizes configuration states and changes by resource, rather than by event.

Understand how CloudWatch Logs integrates with and complements CloudTrail. CloudTrail can send trail logs to CloudWatch Logs for storage, searching, and metric extraction.

Understand how SNS works. SNS uses a push paradigm. CloudWatch and AWS Config send notifications to an Amazon SNS topic. The SNS topic passes these notifications on to a subscriber, which consists of a protocol and endpoint. Know the various protocols that SNS supports.

Know the differences between CloudWatch Alarms and EventBridge. CloudWatch Alarms monitors and alerts on metrics, whereas EventBridge monitors and takes action on events.

Review Questions

1. You've configured CloudTrail to log all management events in all regions. Which of the following API events will CloudTrail log? (Choose all that apply.)
   A. Logging into the AWS console
   B. Creating an S3 bucket from the web console
   C. Uploading an object to an S3 bucket
   D. Creating a subnet using the AWS CLI

2. You've configured CloudTrail to log all read-only data events. Which of the following events will CloudTrail log?
   A. Viewing all S3 buckets
   B. Uploading a file to an S3 bucket
   C. Downloading a file from an S3 bucket
   D. Creating a Lambda function

3. Sixty days ago, you created a trail in CloudTrail to log read-only management events. Subsequently someone deleted the trail. Where can you look to find out who deleted it? No other trails are configured.
   A. The IAM user log
   B. The trail logs stored in S3
   C. The CloudTrail event history in the region where the trail was configured
   D. The CloudTrail event history in any region

4. What uniquely distinguishes two CloudWatch metrics that have the same name and are in the same namespace?
   A. The region
   B. The dimension
   C. The timestamp
   D. The data point

5. Which type of monitoring sends metrics to CloudWatch every five minutes?
   A. Regular
   B. Detailed
   C. Basic
   D. High resolution

6. You update a custom CloudWatch metric with the timestamp of 15:57:08 and a value of 3. You then update the same metric with the timestamp of 15:57:37 and a value of 6. Assuming the metric is a high-resolution metric, which of the following will CloudWatch do?

   A. Record both values with the given timestamp.
   B. Record the second value with the timestamp 15:57:37, overwriting the first value.
   C. Record only the first value with the timestamp 15:57:08, ignoring the second value.
   D. Record only the second value with the timestamp 15:57:00, overwriting the first value.

7. How long does CloudWatch retain metrics stored at one-hour resolution?

   A. 15 days
   B. 3 hours
   C. 63 days
   D. 15 months

8. You want to use CloudWatch to graph the exact data points of a metric for the last hour. The metric is stored at five-minute resolution. Which statistic and period should you use?

   A. The Sum statistic with a five-minute period
   B. The Average statistic with a one-hour period
   C. The Sum statistic with a one-hour period
   D. The Sample count statistic with a five-minute period

9. Which CloudWatch resource type stores log events?

   A. Log group
   B. Log stream
   C. Metric filter
   D. CloudWatch Agent

10. The CloudWatch Agent on an instance has been sending application logs to a CloudWatch log stream for several months. How can you remove old log events without disrupting delivery of new log events? (Choose all that apply.)

    A. Delete the log stream.
    B. Manually delete old log events.
    C. Set the retention of the log stream to 30 days.
    D. Set the retention of the log group to 30 days.

11. You created a trail to log all management events in all regions and send the trail logs to CloudWatch logs. You notice that some recent management events are missing from the log stream, but others are there. What are some possible reasons for this? (Choose all that apply.)

    A. The missing events are greater than 256 KB in size.
    B. The metric filter is misconfigured.

C. There's a delay between the time the event occurs and the time CloudTrail streams the event to CloudWatch.

D. The IAM role that CloudTrail assumes is misconfigured.

12. Two days ago, you created a CloudWatch alarm to monitor the `VolumeReadOps` on an EBS volume. Since then, the alarm has remained in an `INSUFFICIENT_DATA` state. What are some possible reasons for this? (Choose all that apply.)

    A. The data points to monitor haven't crossed the specified threshold.

    B. The EBS volume isn't attached to a running instance.

    C. The evaluation period hasn't elapsed.

    D. The alarm hasn't collected enough data points to alarm.

13. You want a CloudWatch alarm to change state when four consecutive evaluation periods elapse with no data. How should you configure the alarm to treat missing data?

    A. As Missing

    B. Breaching

    C. Not Breaching

    D. Ignore

    E. As Not Missing

14. You've configured an alarm to monitor a metric in the AWS/EC2 namespace. You want CloudWatch to send you a text message and reboot an instance when an alarm is breaching. Which two actions should you configure in the alarm? (Choose two.)

    A. SMS action

    B. Auto Scaling action

    C. Notification action

    D. EC2 action

15. In a CloudWatch alarm, what does the EC2 recover action do to the monitored instance?

    A. Migrates the instance to a different host

    B. Reboots the instance

    C. Deletes the instance and creates a new one

    D. Restores the instance from a snapshot

16. You learn that an instance in the `us-west-1` region was deleted at some point in the past. To find out who deleted the instance and when, which of the following must be true?

    A. The AWS Config configuration recorder must have been turned on in the region at the time the instance was deleted.

    B. CloudTrail must have been logging write-only management events for all regions.

    C. CloudTrail must have been logging IAM events.

    D. The CloudWatch log stream containing the deletion event must not have been deleted.

17. Which of the following may be included in an AWS Config delivery channel? (Choose all that apply.)
    A. A CloudWatch log stream
    B. The delivery frequency of the configuration snapshot
    C. An S3 bucket name
    D. An SNS topic ARN

18. You configured AWS Config to monitor all your resources in the us-east-1 region. After making several changes to the AWS resources in this region, you decided you want to delete the old configuration items. How can you accomplish this?
    A. Pause the configuration recorder.
    B. Delete the configuration recorder.
    C. Delete the configuration snapshots.
    D. Set the retention period to 30 days and wait for the configuration items to age out.

19. Which of the following metric math expressions can CloudWatch graph? (Choose all that apply.)
    A. AVG(m1)-m1
    B. AVG(m1)
    C. METRICS()/AVG(m1)
    D. m1/m2

20. You've configured an AWS Config rule to check whether CloudTrail is enabled. What could prevent AWS Config from evaluating this rule?
    A. Turning off the configuration recorder
    B. Deleting the rule
    C. Deleting the configuration history for CloudTrail
    D. Failing to specify a frequency for periodic checks

21. Which of the following would you use to execute a Lambda function whenever an EC2 instance is launched?
    A. CloudWatch Alarms
    B. EventBridge
    C. CloudTrail
    D. CloudWatch Metrics

Chapter 8

The Domain Name System and Network Routing: Amazon Route 53 and Amazon CloudFront

THE AWS CERTIFIED SOLUTIONS ARCHITECT ASSOCIATE EXAM OBJECTIVES COVERED IN THIS CHAPTER MAY INCLUDE, BUT ARE NOT LIMITED TO, THE FOLLOWING:

✓ **Domain 1: Design Resilient Architectures**

- 1.1 Design a multi-tier architecture solution
- 1.2 Design highly available and/or fault-tolerant architectures

✓ **Domain 2: Design High-Performing Architectures**

- 2.1 Identify elastic and scalable compute solutions for a workload
- 2.3 Select high-performing networking solutions for a workload

✓ **Domain 4: Design Cost-Optimized Architectures**

- 4.3 Design cost-optimized network architectures

Introduction

Making your cloud resources accessible to the people and machines that will consume them is just as important as creating them in the first place. Facilitating network access to your content generally has three goals: availability, speed, and accurately getting the right content to the right users.

The Domain Name System (DNS) is primarily concerned with making your resources addressable using a human-readable name. The Internet's DNS infrastructure ensures that content living behind IP addresses can be reliably associated with human-readable domain names—like amazon.com.

Amazon's Route 53 is built to manage domain services, but it can also have a significant impact on the precision, reliability, and speed with which resources are delivered.

CloudFront, Amazon's global content delivery network (CDN), can take both speed and accuracy a few steps further.

You're going to learn about how AWS addresses all three of those functions in this chapter.

The Domain Name System

DNS is responsible for mapping human-readable domain names (like example.com) to one or more machine-readable IP addresses (like 93.184.216.34).

Whenever you launch a new network-facing service on AWS—or anywhere else—and want it to make it accessible through a readable domain name, you need to satisfy some configuration requirements. But before you can learn to do that, it's important to be familiar with the basic concepts on which domain name services are built.

In this section, we'll define the key elements of DNS infrastructure, particularly the way they're used within Amazon Route 53.

Namespaces

The addressing structure organizing the billions of objects making up the Internet is managed through naming conventions. If, for instance, there was more than one website called Amazon.com or more than one resource identified by the IP address

205.251.242.103, then things would quickly get chaotic. So, there's got to be a reliable, top-down administration authority.

The Internet naming system is maintained within the *domain name hierarchy* namespace, which controls the use of human-readable names. It's a *hierarchy* in the sense that the Internet can be segmented into multiple smaller namespaces through the assignment of blocks of public or private IP addresses or through the use of top-level domains (TLDs).

Both the Internet Protocol (IP) and the domain name hierarchy are administered through the Internet Corporation for Assigned Names and Numbers (ICANN).

Name Servers

Associating a domain name like amazon.com with its actual IP address is the job of a *name server*. All computers will have a simple name server database available locally. That database might contain entries associating *hostnames* (like localhost) with an appropriate IP address. The following code is an example of the /etc/hosts file from a typical Linux machine. It includes a line that allows you to enter **fileserver.com** in your browser to open the home page of a web server running within the local network with the IP 192.168.1.5.

```
127.0.0.1     localhost
127.0.1.1     MyMachine
192.168.1.5 fileserver.com

# The following lines are desirable for IPv6 capable hosts
::1     ip6-localhost ip6-loopback
fe00::0 ip6-localnet
ff00::0 ip6-mcastprefix
ff02::1 ip6-allnodes
ff02::2 ip6-allrouters
```

If a query isn't satisfied by a local name server, it will be forwarded to one of the external DNS name servers specified in your computer's network interface configuration. Such a configuration might point to a public DNS server like Google's at 8.8.8.8 or OpenDNS at 208.67.222.222. Their job is to provide an IP address matching the domain name you entered so that your application (a web browser, for instance) can complete your request.

Domains and Domain Names

In terms of Internet addressing, a *domain* is one or more servers, data repositories, or other digital resources identified by a single domain name. A *domain name* is a name that has been registered for the domain; that registered name is used to direct network requests to the domain's resources.

Domain Registration

Top-level name servers must be made aware of a new domain name before they can respond to related queries. Propagating domain name data among name servers is the job of a *domain name registrar*. Registrars are businesses that manage reservations of domain names. Registrars work with registry operators like VeriSign (that are responsible for all top-level domain activity) so that domain registrations are globally authoritative. Among its other roles, Amazon Route 53 acts as a domain name registrar.

Domain Layers

A domain name is made up of multiple parts. The rightmost text of every domain address (like .com or .org) indicates the *top-level domain* (TLD). The name to the left of the TLD (the *amazon* part of amazon.com) is called the *second-level domain* (SLD).

> **NOTE** This SLD designation would also refer to the unique second-level domains used by some countries, like the .co of .co.uk used for UK-based businesses.

A subdomain identifies a subset of a domain's resources. Web and email servers from the administration department of a college, for instance, might all use the administration.school.edu name. Thus, dean@administration.school.edu might be a valid email address, whereas administration.school.edu/apply.pdf could point to a file kept in the web root directory of the administration.school.edu server.

www.school.edu, api.school.edu, and ftp.school.edu are all common examples of subdomains (sometimes referred to as *hosts*). Figure 8.1 illustrates the parts of a simple subdomain.

FIGURE 8.1 A simple DNS domain broken down to its parts

aws.amazon.com
— SLD — TLD —
— Subdomain —

Fully Qualified Domain Names

Based on the default DNS settings on many systems, the system's default domain name will be automatically appended when resolving requests for partial domain names. As an example, a request for workstation might be resolved as workstation.localhost. If, however, you want to request a domain name as is, without anything being appended to it, you'll need to use a *fully qualified domain name* (FQDN).

An FQDN contains the absolute location of the domain including, at the least, a subdomain and the TLD. In addition, convention will often require a trailing dot after the TLD—which represents the domain root—to confirm that this is, indeed, an FQDN. Addresses in DNS zone files, for instance, will fail without a trailing dot. Here's how that might look:

administration.school.edu.

Zones and Zone Files

A *zone* (or *hosted zone* as Route 53 calls it) defines a DNS domain. A *zone file* is a text file that describes the way resources within the zone should be mapped to DNS addresses within the domain. The file consists of *resource records* containing the data fields listed in Table 8.1.

TABLE 8.1 The data categories contained in a resource record from a zone file

Directive	Purpose
Name	The domain or subdomain name being defined
TTL	The time to live before the record expires
Record Class	The namespace for this record—usually IN (Internet)
Record Type	The record type defined by this record (A, CNAME, etc.)

This following example is a typical resource record defining the name server (NS) of the example.com domain as an Internet class record (IN) with a one-hour time to live (TTL) (1h) whose value is ns-750.awsdns-30.net.:

example.com. 1h IN NS ns-750.awsdns-30.net.

Record Types

The *record type* you enter in a zone file's resource record will determine how the record's data is formatted and how it should be used. There are currently around 40 types in active use, but Table 8.2 explains only the more common ones that also happen to be offered by Route 53.

TABLE 8.2 Some common DNS record types

Type	Function
A	Maps a hostname to an IPv4 IP address
CNAME	*Canonical name*—allows you to define one hostname as an alias for another
MX	*Mail exchange*—maps a domain to specified message transfer agents

TABLE 8.2 Some common DNS record types *(continued)*

Type	Function
AAAA	Maps a hostname to an IPv6 IP address
TXT	*Text*—contains human or machine-readable text
PTR	*Pointer*—points to another location within the domain space (this type will not be processed)
SRV	A customizable record for service location
SPF	*Sender Policy Framework*—an email validation protocol (no longer widely supported as of RFC 7208)
NAPTR	*Name Authority Pointer*—allows regex-based domain name rewriting
CAA	*Certification Authority Authorization*—establishes authority to issue security certificates for a domain
NS	Identifies a name server to be used by a zone
SOA	*Start of authority*—Defines a zone's authoritative meta information

Alias Records

You can use alias records to route traffic to a resource—such as an elastic load balancer—without specifying its IP address. Although the use of alias records has not yet been standardized across providers, Route 53 makes them available within record sets, allowing you to connect directly with network-facing resources running on AWS.

Amazon Route 53

With those DNS basics out of the way, it's time to turn our attention back to AWS. Route 53 provides more than just basic DNS services. In fact, it focuses on four distinct areas:

- Domain registration
- DNS management
- Availability monitoring (*health checks*)
- Traffic management (*routing policies*)

In case you're curious, the "53" in Route 53 reflects the fact that DNS traffic uses TCP or UDP port 53.

Domain Registration

While there's nothing stopping you from registering your domains through any ICANN-accredited registrar—like GoDaddy—you can just as easily use Route 53. For domains that will be associated with AWS infrastructure, using Route 53 for registration can in fact help simplify your operations.

You can transfer registration of an existing domain from your current registrar by unlocking the domain transfer setting in the registrar's admin interface and then requesting an authorization code. You'll supply that code to Route 53 when you're ready to do the transfer.

> **NOTE** If you'd prefer to leave your domain with its current registrar, you can still use Route 53 to manage your DNS configuration. Simply copy the name server addresses included in your Route 53 record set and paste them as the new name server values in your registrar's admin interface.

DNS Management

However you registered your domain name, until you find some way to connect your domain with the resources it's supposed to represent, it won't do you a lot of good. On Route 53, the way you create and configure a hosted zone determines what your users will be shown when they invoke your domain name in a browser, email client, or programmatically.

Within Route 53, rather than importing a preconfigured zone file (which is certainly an option), you'll usually configure your hosted zones directly via the console or AWS CLI.

When you create a new hosted zone and enter your domain name, you'll need to tell Route 53 whether you'd like this zone to be a public hosted zone or a private hosted zone. Access for resources in a privately hosted zone will be available only from within the AWS VPCs that you specify. If you want your resources to be accessible to external users (which is true of most domains), then you'll create a publicly hosted zone.

Route 53 will automatically create a start of authority (SOA) record and provide four name server addresses. From there, it'll be your job to create new record sets defining the relationships between your domain and any subdomains you choose to create and the resources you want made available. Exercise 8.1 guides you through the process of setting up a functioning hosted zone.

EXERCISE 8.1

Create a Hosted Zone on Route 53 for an EC2 Web Server

1. Click the Create Hosted Zone button in the GUI.

2. Enter a valid domain name that's currently not managed and select the Public Hosted Zone type. If you don't already have an existing domain on Route 53 and don't want to create a new one just for this exercise, use a fake domain—even something like example.com. Of course, you won't be able to fully test the zone if you're not using a real domain name. Either way, remember that leaving the hosted zone up for more than 12 hours will incur a monthly hosting charge—that's currently only $0.50.

3. You'll be given name server and SOA record sets. Select those records one at a time and spend a few minutes exploring.

4. From the Amazon EC2 Dashboard, create an AWS Linux AMI-based instance the way you did in Exercise 2.1 in Chapter 2, "Amazon Elastic Compute Cloud and Amazon Elastic Block Store." You'll configure this instance as a simple web server so that you can test the domain. When configuring your security group, open HTTP port 80 so that web traffic can get in.

5. Log in to your instance (as you did in Exercise 2.1) and run the following commands to update the software repository, install the Apache web server, create a default index.html web page (add some text to the page when the text editor opens, and, when you're done, save and exit using Ctrl+X and then Y; this will later help you know that everything is working), and start up the Apache process.

   ```
   sudo yum update -y
   sudo yum install -y httpd
   sudo nano /var/www/html/index.html
   sudo systemctl start httpd
   ```

6. You can confirm your web server is working by pointing your browser to the instance's public IP address. You may want to keep this instance running, as it will be also helpful for the next exercise.

7. Back in the Route 53 Dashboard, click the Create Record Set button to create an A record to map your domain name to the IP address of a web server instance running on EC2. Leave the Name field to the left of the example.com value blank and enter your server's IP address (or any realistic fake address if your domain isn't real) in the Value field. Click Create.

8. Click Create Record Set once again to create a second A record. This time, you'll enter **www** in the Name field.

9. Click the Yes radio button next to Alias, click once inside the Alias Target field, and select the example.com value that should be among those that appear.

10. Domain propagation might take a few hours, but assuming you're working with an actual server, you should eventually be able to point your browser to example.com and your server's root web page should load.

Availability Monitoring

You should always be aware of the status of the resources you have running: closing your eyes and just hoping everything is fine has never been a winning strategy. Route 53 offers tools to monitor the health of the resources it's managing and ways to work around problems.

When you create a new record set, you're given the option of choosing a routing policy (which we'll discuss at length a bit later in this chapter). Selecting any policy besides Simple will make it possible to associate your policy with a *health check*. A health check, which you create, configure, and name separately, will regularly test the resource that's represented by your record set to confirm it's healthy.

If everything's okay, Route 53 will continue routing traffic to that resource. But if a preset number of tests are run without an acceptable response, Route 53 will assume the resource is offline and can be set to redirect traffic to a backup resource. Exercise 8.2 illustrates how you can configure a health check that you can use with record sets.

EXERCISE 8.2

Set Up a Health Check

1. In the Route 53 GUI, click Create Health Check.

2. Give your check a name and select the Endpoint radio button for What To Monitor. You can use the web server you created in Exercise 8.1 as the health check endpoint.

3. With the IP address radio button selected for Specify Endpoint By, choose either HTTP or HTTPS for the protocol (this will depend on how you configured Transport Layer Security [TLS] encryption for your website—HTTP is a good bet if you're not sure).

4. Enter your server's IP address, and enter either **80** or **443** for the Port value (again, this will depend on your site's TLS settings—80 is a good bet if you're not sure).

EXERCISE 8.2 (continued)

5. Enter the name of a file in the web root directory of your website as the Path value. This is the resource the health check will try to load to test your site's health. You could use `index.html` (or `index.php`, if you happen to have a PHP application running), but, over the long term, that will probably cause a lot of unnecessary overhead. Instead, consider creating a small file called something like `test.html` and entering that for Path.

6. You can create an alarm to alert you to failed tests or simply click Create Health Check.

Routing Policies

In a world where your users all happily access a single, reliable server instance hosting your application, complicated routing policies make little sense. But because that's not the way things usually work in the real world, you'll generally want to apply flexible protocols to ensure that you're always providing the most reliable and low-latency service possible.

Route 53 routing policies provide this kind of functionality at the domain level that can be applied globally across all AWS regions. To make the most of their power, you'll need to understand what choices are available and how they work.

The simplest of the policies is, predictably, called simple. It routes all requests to the IP address or domain name you assign it. Simple is the default routing policy for new record sets.

Weighted Routing

A weighted policy will route traffic among multiple resources according to the ratio you set. To explain that better, imagine you have three servers (or load balancers representing three groups of servers), all hosting instances of the same web application. One of the servers (or server groups) has greater unused compute and memory capacity and can therefore handle far more traffic. It would be inefficient to send equal numbers of users to each of the servers. Instead, you can assign the larger server a numeric weight of, say, 50 and then 25 to the other two. That would result in half of all requests being sent to the larger server and 25 percent to each of the others.

To configure a weighted policy in Route 53, you create a separate record set for each of your servers, give the same value for the set ID of each of the record sets, and then enter an instance-appropriate numeric value for the weight. The matching set IDs tell Route 53 that those record sets are meant to work together.

Latency Routing

Latency-based routing lets you leverage resources running in multiple AWS regions to provide service to clients from the instances that will deliver the best experience. Practically this means that, for an application used by clients in both Asia and Europe, you could place parallel resources in, say, the `ap-southeast-1` and `eu-west-1` regions. You then create a record set for each resource using latency-based policies in Route 53, with one pointing to your `ap-southeast-1` resource and the other to `eu-west-1`.

Assuming you gave both record sets the same value for Set ID, Route 53 will know to direct requests for those resources to the instance that will provide the lowest latency for each client.

Failover Routing

A failover routing policy will direct traffic to the resource you identify as *primary* as long as health checks confirm that the resource is running properly. Should the primary resource go offline, subsequent traffic will be sent to the resource defined within a second record set and designated as *secondary*. As with other policies, the desired relationship between record sets is established by using matching set ID values for each set.

Geolocation Routing

Unlike latency policies that route traffic to answer data requests as *quickly* as possible, geolocation uses the continent, country, or U.S. state where the request originated to decide *what resource to send*. This can help you focus your content delivery, allowing you to deliver web pages in customer-appropriate languages, restrict content to regions where it's legally permitted, or generate parallel sales campaigns.

You should be aware that Route 53 will sometimes fail to identify the origin of a requesting IP address. You might want to configure a default record to cover those cases.

Multivalue Answer Routing

It's possible to combine a health check configuration with multivalue routing to make a deployment more highly available. Each multivalue-based record set points to a single resource and can be associated with a health check.

As many as eight records can be pointed to parallel resources and connected to one another through matching set ID values. Route 53 will use the health checks to monitor resource status and randomly route traffic among the healthy resources.

Working through Exercise 8.3 will show you the way the failover routing policy process works.

EXERCISE 8.3

Configure a Route 53 Routing Policy

1. Make sure you have two separate web-facing resources running. You could create a second Apache web server instance in addition to the one you launched in Exercise 8.1, or perhaps create a simple static website in S3 the way you did in Exercise 3.4 in Chapter 3, "AWS Storage." In this case, though, your bucket name will have to exactly match the domain name. That might mean something like `secondary.example.com`.

EXERCISE 8.3 *(continued)*

2. Configure a health check for each of your resources. You create a health check for an S3 static website by selecting Domain Name for Specify Endpoint By. Note that you'll need to strip off the `http://` and trailing / characters from the endpoint before the health check configuration will accept it.

3. Assuming you have an active hosting zone (perhaps the one from Exercise 8.1), use the Route 53 GUI to create a record set for each of your instances. You might enter domain names like **server1.example.com** and **server2.example.com**—it's up to you.

4. Create a regular A record type if you're pointing to an EC2 instance, and enter the instance's IP address as the record's Value. If you're using an S3 static website for one or both of your resources, click Yes for Alias in the record set, click once within the Alias Target box that will appear, and select the appropriate S3 resource from among those that appear. Your static website must live in the same AWS account.

5. For Routing Policy, select Failover and choose either Primary or Secondary for the Failover Record Type radio buttons. The Set ID value will automatically change according to your choice. Make sure the record set for one of your resources is designated as Primary.

6. Select Yes for Evaluate Target Health, and select Yes for Associate With Health Check. Select the health check that matches the resource you're working with.

7. Test your configuration by first loading the primary website by its normal URL. Now, disable your primary website. You could do that by deleting or renaming its `index.html` file, by blocking HTTP access in the instance security group, or in the case of an S3, by disabling the static website setting. Point your browser to the same primary URL. If failover is working, you should see the `index.html` web page from your secondary resource.

Traffic Flow

Route 53 Traffic Flow is a console-based graphical interface that allows you to visualize complex combinations of routing policies as you build them. Figure 8.2 shows what a Traffic Flow configuration can look like.

Because it integrates routing policies with all the possible resource endpoints associated with your AWS account, Traffic Flow can make it simple to quickly build a sophisticated routing structure. You can also use and customize routing templates.

Traffic Flow offers *geoproximity routing*, which gives you the precision of geolocation routing but at a far finer level. Geoproximity routing rules can specify geographic areas by their relationship either to a particular longitude and latitude or to an AWS region. In both cases, setting a bias score will define how widely beyond your endpoint you want to apply your rule.

FIGURE 8.2 A sample Traffic Flow policy

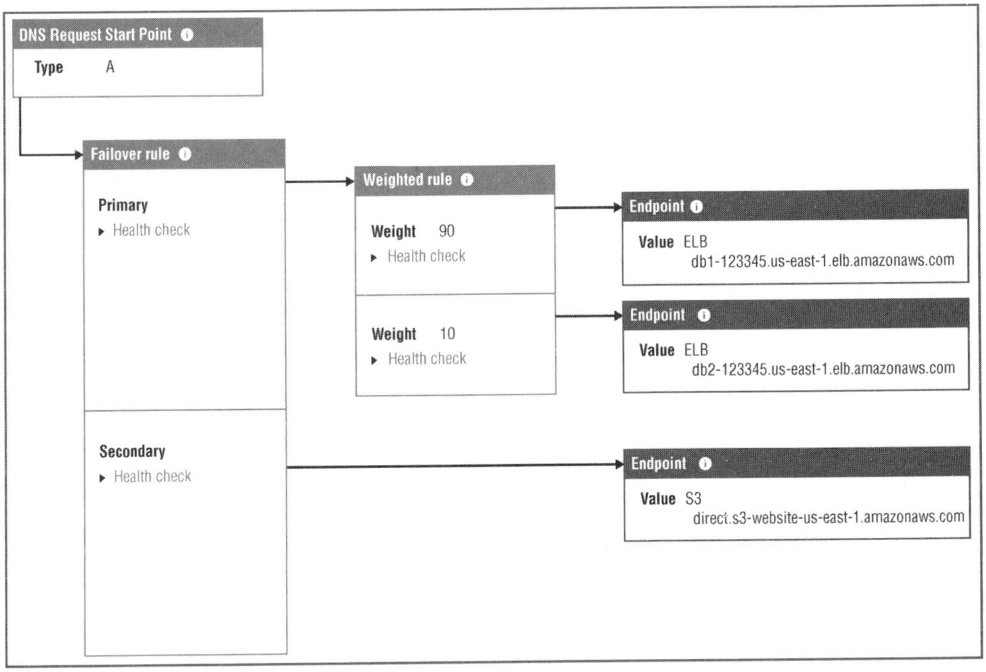

```
DNS Request Start Point
  Type    A

    ├─► Failover rule
    │      Primary
    │      ► Health check          ──► Weighted rule
    │                                    Weight  90           ──► Endpoint
    │                                    ► Health check            Value  ELB
    │                                                                     db1-123345.us-east-1.elb.amazonaws.com
    │
    │                                    Weight  10           ──► Endpoint
    │                                    ► Health check            Value  ELB
    │                                                                     db2-123345.us-east-1.elb.amazonaws.com
    │
    │      Secondary
    │      ► Health check                                     ──► Endpoint
                                                                   Value  S3
                                                                          direct.s3-website-us-east-1.amazonaws.com
```

Source: console.aws.amazon.com/route53/trafficflow/home#

Route 53 Resolver

You can now extend Route 53's powerful routing tools across your hybrid infrastructure using Route 53 Resolver. Resolver can manage bidirectional address queries between servers running in your AWS account and on-premises resources. This can greatly simplify workloads that are meant to seamlessly span private and public platforms.

Amazon CloudFront

Amazon's global *content delivery network* (CDN) CloudFront can help solve one of the primary problems addressed by Route 53: getting your content into the hands of your users as quickly as possible.

CloudFront maintains a network of physical edge locations placed geographically close to the end users who are likely to request content. When you configure how you want your content delivered—as part of what AWS calls a CloudFront distribution—you define how you want your content distributed through that network and how it should then be delivered to your users.

How do your users "know" to address their requests to your CloudFront endpoint address rather than getting it directly from your resources? Normally, they don't. But when you use Route 53 to configure your domain to direct incoming DNS requests to the CloudFront distribution, users will be automatically routed appropriately.

When a request is made, CloudFront assesses the user's location and calculates which endpoint will be available to deliver the content with the lowest latency. If this is the first time this content has been requested through the endpoint, the content will be copied from the origin server (an EC2 web server instance, perhaps, or the contents of an S3 bucket). Delivery of subsequent requests will—because the cached copy is still stored at the endpoint—be much faster.

The kind of CloudFront distribution you create will depend on the kind of media you're providing. For web pages and graphic content, you'll select a web distribution. For video content stored in S3 buckets that can use Adobe's Real-Time Messaging Protocol (RTMP), CloudFront's RTMP distribution makes the most sense.

As you briefly saw in Chapter 3, when you configure your distribution, you'll have the option of adding a free *AWS Certificate Manager* (ACM) SSL/TLS encryption certificate to your distribution. This will protect your content as it moves between CloudFront and your users' devices from network sniffers and man-in-the-middle attacks.

CloudFront currently supports three categories of content origin, described in Table 8.3.

TABLE 8.3 Permitted CloudFront origins

Category	Description
Amazon S3 bucket	Any accessible S3 bucket
AWS MediaPackage channel endpoint	Video packaging and origination
AWS MediaStore container endpoint	Media-optimized storage service

Exercise 8.4 will take you through the steps for creating a CloudFront distribution complete with an encryption certificate.

EXERCISE 8.4

Create a CloudFront Distribution for Your S3-Based Static Website

1. Create a static website in S3 the way you did in Exercise 3.4 in Chapter 3.
2. From the CloudFront dashboard, start a new distribution using the web delivery method.

3. Click once inside the Origin Domain Name box, and a list of all the available resources within your account will be displayed. The S3 bucket containing your website files should be included. Select it.

4. Select a Price Class in the Distribution Settings section. Using the All Edge Locations setting will provide the best performance for the greatest number of customers, but it will also cost more. By the way, if you're only creating this as an experiment and you're unlikely to get more than a few hundred requests, then don't worry about the cost; it'll be in the low pennies.

   The next three steps will apply only if you have a valid DNS domain to work with.

5. Add the DNS domain names you want to use for this distribution to the Alternate Domain Names (CNAMES) box. Those might include **server1.example.com** and **www.example.com**. If you don't do this, your encryption certificate might not work for all content requests.

6. Click the Request Or Import A Certificate With ACM button and follow the simple instructions to request a certificate for your registered domain name. Once your request has been granted, your certificate will appear in the Custom SSL Certificate box. You can select it.

7. When you're satisfied with your settings, click Create Distribution.

8. In the Route 53 console, create (or edit) a record set that points your domain name to the CloudFront distribution you just created. Wait for the new settings to propagate and confirm that you can access the S3 content through a domain using HTTPS (encryption).

AWS CLI Example

The following command will list all the hosted zones in Route 53 within your AWS account. Among the data that will be returned is an Id for each zone. You can take the value of Id and pass it to the get-hosted-zone command to return record set details for that zone.

```
$ aws route53 list-hosted-zones
HOSTEDZONES   C81362-0F3F-8F0-8F3-1432D81AB  /hostedzone/Z3GZZDB5SZ3
fakedomain.org.  5
CONFIG   False
$ aws route53 get-hosted-zone --id /hostedzone/Z38LGIZCB3CSZ3
NAMESERVERS  ns-2026.awsdns-63.co.uk
```

```
NAMESERVERS  ns-771.awsdns-32.net
NAMESERVERS  ns-1050.awsdns-03.org
NAMESERVERS  ns-220.awsdns-30.com
HOSTEDZONE   C81362-0F3F-8F0-8F3-1432D81AB     /hostedzone/Z3GZZDB5SZ3
fakedomain.org.       6
CONFIG       False
```

Summary

The DNS system manages Internet resource addressing by mapping IP addresses to human-readable domain names. ICANN administers name servers and domain name registrars (like Route 53) through registry operators (like VeriSign).

A fully qualified domain name consists of a TLD (like org or com) and an SLD (like amazon). FQDNs are also often given a trailing dot representing "root."

DNS configuration details are organized as hosted zones, where resource record sets are created to control the way you want inbound domain traffic to be directed. That behavior can be defined through any one of a number of record types, like A, CNAME, and MX.

A newly created Route 53 hosted zone will contain a start of authority (SOA) record set and a list of name servers to which requests can be directed. You'll need to create at least one record set so that Route 53 will know from which domain name to expect requests.

Route 53 can be configured to regularly monitor the run status of your resources using health checks. Besides alerting you to problems, health checks can also be integrated with Route 53 routing policies to improve application availability.

Weighted routing policies let you direct traffic among multiple parallel resources proportionally according to their ability to handle it. Latency routing policies send traffic to multiple resources to provide the lowest-latency service possible. Failover routing monitors a resource and, on failure, reroutes subsequent traffic to a backup resource. Geolocation routing assesses the location of a request source and directs responses to appropriate resources.

Amazon CloudFront is a CDN that caches content at edge locations to provide low-latency delivery of websites and digital media.

Exam Essentials

Understand how DNS services enable predictable and reliable network communication.   DNS registration ensures that domain names are globally unique and accessible. Domain requests are resolved using name servers—both local and remote.

Understand DNS naming conventions.   You should be familiar with "parsing" domain names. Each of the TLD, SLD, and subdomain/host sections will be read by DNS clients in a predictable way.

Be familiar with the key DNS record types. Recognize the function of key record types, including A and AAAAA (address records for IPv4 and IPv6), CNAME (canonical name record or alias), MX (mail exchange record), NS (name server record), and SOA (start of authority record).

Be familiar with Route 53 routing policies. Recognize the function of key routing policies, including simple (for single resources), weighted (routing among multiple resources by percentage), latency (low-latency content delivery), failover (incorporate backup resources for higher availability), and geolocation (respond to end user's location).

Understand how to create a CloudFront distribution. CloudFront distributions can be configured to deliver content through low-latency, geographically based content to end users.

Review Questions

1. Which of the following describes the function of a name server?
   A. Translating human-readable domain names into IP addresses
   B. Registering domain names with ICANN
   C. Registering domain names with VeriSign
   D. Applying routing policies to network packets

2. Your organization is planning a new website and you're putting together all the pieces of information you'll need to complete the project. Which of the following describes a domain?
   A. An object's FQDN
   B. Policies controlling the way remote requests are resolved
   C. One or more servers, data repositories, or other digital resources identified by a single domain name
   D. A label used to direct network requests to a domain's resources

3. You need to decide which kind of website name will best represent its purpose. Part of that task will involve choosing a top-level domain (TLD). Which of the following is an example of a TLD?
   A. amazon.com/documentation/
   B. aws.
   C. amazon.
   D. .com

4. Which of the following is the name of a record type— as used—in a zone file?
   A. CNAME (canonical name)
   B. TTL (time to live)
   C. Record type
   D. Record data

5. Which of the following DNS record types should you use to associate a domain name with an IP address?
   A. NS
   B. SOA
   C. A
   D. CNAME

6. Which of the following are services provided by Amazon Route 53? (Choose three.)
   A. Domain registration
   B. Content delivery network

C. Health checks
   D. DNS management
   E. Secure and fast direct network connections to an AWS VPC

7. For regulatory compliance, your application may only provide data to requests coming from the United States. Which of the following routing policies can be configured to do this?
   A. Simple
   B. Latency
   C. Geolocation
   D. Multivalue

8. Your web application is hosted within multiple AWS regions. Which of the following routing policies will ensure the fastest possible access for your users?
   A. Latency
   B. Weighted
   C. Geolocation
   D. Failover

9. You're testing three versions of a new application, with each version running on its own server and the current production version on a fourth server. You want to route 5 percent of your total traffic to each of the test servers and route the remaining 85 percent of traffic to the production server. Which routing policy will you use?
   A. Failover
   B. Weighted
   C. Latency
   D. Geolocation

10. You have production infrastructure in one region sitting behind one DNS domain, and for disaster recovery purposes, you have parallel infrastructure on standby in a second AWS region behind a second domain. Which routing policy will automate the switchover in the event of a failure in the production system?
    A. Latency
    B. Weighted
    C. Geolocation
    D. Failover

11. Which of the following kinds of hosted zones are real options within Route 53? (Choose two.)
    A. Public
    B. Regional
    C. VPC
    D. Private
    E. Hybrid

12. Which of the following actions will you need to perform to transfer a domain from an external registrar to Route 53? (Choose two.)
    A. Unlock the domain transfer setting on the external registrar admin page.
    B. Request an authorization code from the external registrar.
    C. Copy the name server addresses from Route 53 to the external registrar admin page.
    D. Create a hosted zone CNAME record set.

13. Which of the following actions will you need to perform to use Route 53 to manage a domain that's being hosted on an external registrar?
    A. Request an authorization code from the external registrar.
    B. Copy the name server addresses from Route 53 to the external registrar admin page.
    C. Create a hosted zone CNAME record set.
    D. Unlock the domain transfer setting on the external registrar admin page.

14. Your multiserver application has been generating quality-related complaints from users and your logs show some servers are underused and others have been experiencing intermittent failures. How do Route 53 health checks test for the health of a resource so that a failover policy can direct your users appropriately?
    A. It periodically tries to load the `index.php` page.
    B. It periodically tries to load the `index.html` page.
    C. It periodically tries to load a specified web page.
    D. It periodically tries to log into the resource using SSH.

15. Which of the following most accurately describes the difference between geolocation and geoproximity routing policies?
    A. Geoproximity policies specify geographic areas by their relationship either to a particular longitude and latitude or to an AWS region, whereas geolocation policies use the continent, country, or U.S. state where the request originated to decide what resource to send.
    B. Geolocation policies specify geographic areas by their relationship either to a particular longitude and latitude or to an AWS region, whereas geoproximity policies use the continent, country, or U.S. state where the request originated to decide what resource to send.
    C. Geolocation policies will direct traffic to the resource you identify as primary as long as health checks confirm that that resource is running properly, whereas geoproximity policies allow you to deliver web pages in customer-appropriate languages.
    D. Geolocation policies use a health check configuration routing to make a deployment more highly available, whereas geoproximity policies leverage resources running in multiple AWS regions to provide service to clients from the instances that will deliver the best experience.

16. Which of the following are challenges that CloudFront is well positioned to address? (Choose two.)
    A. A heavily used website providing media downloads for a global audience
    B. An S3 bucket with large media files used by workers on your corporate campus
    C. A file server accessed through a corporate VPN
    D. A popular website with periodically changing content

17. Which of the following is not a permitted origin for a CloudFront distribution?
    A. Amazon S3 bucket
    B. AWS MediaPackage channel endpoint
    C. API Gateway endpoint
    D. Web server

18. What's the best way to control the costs your CloudFront distribution incurs?
    A. Select a price class that maintains copies in only a limited subset of CloudFront's edge locations.
    B. Configure a custom SSL certificate to restrict access to HTTPS requests only.
    C. Disable the use of Alternate Domain Names (CNAMES) for your distribution.
    D. Enable Compress Objects Automatically for your distribution.

19. Which of the following is not a direct benefit of using a CloudFront distribution?
    A. User requests from an edge location that's recently received the same request will be delivered with lower latency.
    B. CloudFront distributions can be directly mapped to Route 53 hosted zones.
    C. All user requests will be delivered with lower latency.
    D. You can incorporate free encryption certificates into your infrastructure.

20. Which of the following content types is the best fit for a Real-Time Messaging Protocol (RTMP) distribution?
    A. Amazon Elastic Transcoder–based videos
    B. S3-based videos
    C. Streaming videos
    D. A mix of text and media-rich digital content

Chapter 9

Simple Queue Service and Kinesis

THE AWS CERTIFIED SOLUTIONS ARCHITECT ASSOCIATE EXAM OBJECTIVES COVERED IN THIS CHAPTER MAY INCLUDE, BUT ARE NOT LIMITED TO, THE FOLLOWING:

✓ **Domain 1: Design Resilient Architectures**

- 1.1 Design a multi-tier architecture solution
- 1.2 Design highly available and/or fault-tolerant architectures
- 1.3 Design decoupling mechanisms using AWS services
- 1.4 Choose appropriate resilient storage

✓ **Domain 2: Design High-Performing Architectures**

- 2.1 Identify elastic and scalable compute solutions for a workload
- 2.2 Select high-performing and scalable storage solutions for a workload
- 2.3 Select high-performing networking solutions for a workload

Introduction

In this chapter, you'll learn about the following services:
- Simple Queue Service
- Kinesis Video Streams
- Kinesis Data Streams
- Kinesis Data Firehose

Simple Queue Service

Application developers often use a design principle called *loose coupling* to make their applications more scalable and reliable. Without getting into unnecessary detail, the way they do this is by taking a single monolithic application that normally runs on one server and breaking it up into separate components, sometimes called microservices. These microservices run on different servers and must communicate with each other over the network by passing messages to each other. The Simple Queue Service (SQS) is a managed service that allows these different application components to pass messages back and forth. SQS is highly available and elastic, capable of handling hundreds of thousands of messages per second.

Queues

SQS lets you create queues to hold messages for processing. SQS uses the term *producer* to describe the component that places messages into the queue and *consumer* to describe the component that reads messages from the queue. A message can be up to 256 KB in size. Here's how the basic process works:

1. A producer places one or more messages into a queue using the `SendMessage` action. When a message is in the queue, it's said to be in-flight.
2. A consumer checks or polls the queue for new messages to process. It reads or consumes one or more messages from the queue using the `ReceiveMessage` action.

3. After the consumer processes a message, it deletes it from the queue using the `DeleteMessage` action.

Refer to Figure 9.1 for an illustration of the process.

FIGURE 9.1 SQS workflow

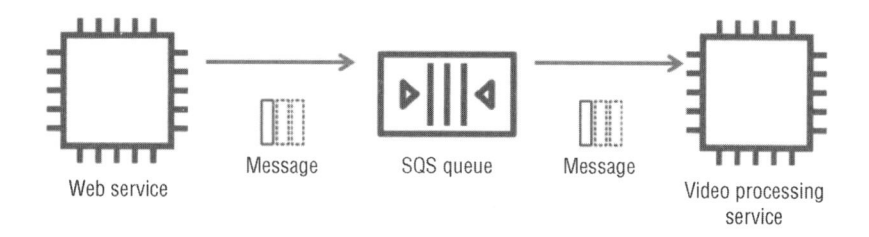

To reduce the number of API calls to SQS, you can use batching to perform actions on a batch of up to 10 messages at a time. If that were all there is to SQS, this would be a short chapter. But there's a lot of nuance in this seemingly simple process.

Visibility Timeout

When a consumer grabs a message from the queue, the message actually remains in the queue. It's up to the consumer to delete the message after processing it. Once a consumer receives a message, SQS doesn't allow any other consumers to read the same message for a period of time, called the *visibility timeout*. This is meant to prevent duplication of effort caused by other consumers processing the same message. However, as you'll see shortly, the visibility timeout doesn't always achieve this desired result. The default visibility timeout is 30 seconds but can range from 0 seconds to 12 hours.

Retention Period

Messages don't sit in a queue forever. The default retention period is 4 days, but you can configure it to be anywhere from 1 minute to 14 days.

Delay Queues and Message Timers

You can use the per-queue delay setting to messages that are added to the queue. The default per-queue delay is 0 seconds, but you can set it as long as 15 minutes.

You can also set a delay for individual messages by setting a message timer. The default message timer is 0 seconds, and you can set it as high as 15 minutes. If you set a message timer, it will override the delay queue settings.

Queue Types

When an application depends on a queue, the behavior and performance of the queue can make or break the application. Some applications may need to process thousands of messages a second. Others may require that messages always be delivered in the order in which they were received. To address these different needs, SQS offers two different types of queues: standard and first-in, first-out (FIFO).

Standard Queues

Standard queues offer almost unlimited throughput, allowing you to dump as many messages into the queue as fast as you need to. Messages may be delivered out-of-order, and occasionally may be delivered more than once. Any application that uses a standard queue should be written to handle duplicate messages. Standard is the default queue type. Standard queues can have up to about 120,000 in-flight messages.

First-In, First-Out (FIFO) Queues

First-in, first-out (FIFO) queues support sending about 3,000 messages per second into a queue. Messages are delivered in the order they were received, and each message is delivered exactly once. This eliminates the possibility of duplicate messages. FIFO queues can handle about 20,000 in-flight messages.

FIFO queues also let you divide a queue up into message groups to maintain message ordering within a subset of messages in the queue. When you have a lot of producers sending messages, using message groups can preserve the order of messages from each producer. For example, suppose you have two producers each sending temperature data from a different sensor every five seconds. To avoid mixing up the temperatures from the two sensors, you'd have each producer send its messages using a different message group ID. Then, when the consumer receives messages from the queue, it receives the temperature readings of each sensor in the correct order. In practice, using message groups is akin to having two separate queues.

Polling

When receiving messages from a queue, you can use short polling (which is the default) or long polling. The difference comes down to which you want more: getting an immediate response and potentially missing some messages or getting a delayed response with the assurance that you've received all the messages in the queue.

In short polling, SQS checks only a subset of its servers for waiting messages. Short polling gives you an immediate response by either returning a message or a response indicating that there are no messages. Even if there are messages in the queue, it's possible to get a false empty response. In other words, SQS may *tell* you there are no messages in the queue, even if there really are. Thus, to be sure you get all your messages, you may have to poll the queue multiple times.

In long polling, SQS ensures that you receive any messages waiting in the queue. Because long polling checks all the servers servicing your queue, it can take up to 20 seconds to return a response. Long polling is the cheaper option because you don't have to poll the queue as frequently to get all your messages.

Dead-Letter Queues

In some cases, you may have a message that your consumers can't process. In this case, a consumer may receive the message, attempt to process it, but leave it in the queue for another consumer to then attempt to process. The end result is that the message never leaves the queue. It becomes a dead letter.

You can have SQS automatically pull such a message from your source queue and place it into a dead-letter queue after it's been received so many times. To configure a dead-letter queue, you first create a queue of the same type (FIFO or standard); then you set the queue's `maxReceiveCount` to the maximum number of times a message can be received before it's moved to the dead-letter queue. The dead-letter queue must always be in the same region as the source queue.

A dead-letter queue has a retention period just like any other queue. One thing to keep in mind is that when a message is moved to a dead-letter queue, the message will be deleted based on its original creation date. For example, if the retention period for the dead-letter queue is seven days, and a message is already six days old when it's moved to the dead-letter queue, it will spend at most one day in the dead-letter queue before being deleted.

Kinesis

Amazon Kinesis is a collection of services that let you collect, process, store, and deliver streaming data. Kinesis can perform real-time ingestion of gigabytes per second from thousands of sources, making it perfect for things like audio and video feeds, application logs, and telemetry data.

AWS offers the following Kinesis services for different types of streaming data:

- Kinesis Video Streams
- Kinesis Data Streams
- Kinesis Data Firehose

Kinesis Video Streams

Kinesis Video Streams is designed to ingest and index almost unlimited amounts of streaming video data, such as from webcams, security cameras, and smartphones. Three popular use cases for Kinesis Video Streams include but aren't limited to:

- Computer vision applications, such as image recognition
- Streaming video
- Two-way videoconferencing

Kinesis Video Streams uses a producer-consumer model. The data source that feeds data into a Kinesis stream is called the *producer*. In addition to sending a video stream, a producer may also send non-video data, such as audio, subtitles, or GPS coordinates as part of the Kinesis video stream. Kinesis automatically indexes the stream by using timestamps, allowing for easy retrieval later on. By default, Kinesis Video Streams will store a video stream for 24 hours, but you can extend this retention period up to 7 days.

The application that reads data from the stream is called a consumer. A consumer pulls data from a Kinesis video stream for playback or processing. A consumer may be a video player, a custom application running on an EC2 instance, or another AWS service such as SageMaker or Rekognition Video. A consumer may receive data from a stream in real time or after-the-fact.

For video playback, Kinesis Video Streams supports HTTPS Live Streams (HLS) and Dynamic Adaptive Streaming Over HTTPS (DASH). For peer-to-peer videoconferencing, it supports Web Real-Time Communication (WebRTC).

Kinesis Data Streams

Kinesis Data Streams is a data pipeline that can aggregate, buffer, and reliably store data from producers until a consumer is ready to process it. Consumers might include big data analytics applications such as MapReduce. Kinesis Data Streams can rapidly ingest and store different types of binary data, including:

- Application logs
- Stock trades
- Social media feeds
- Financial transactions
- Location tracking information

Kinesis Data Streams works in a similar way to Kinesis Video Streams. Producers feed data into a Kinesis Data Stream, and that data is stored in a data record. How you get data into a stream depends on the type of data and the source. For example, if you want to stream server or application logs, you can use the Amazon Kinesis Agent, which is a Java-based application you can run on Linux servers. You can also integrate Kinesis into your applications using the Kinesis Producer Library (KPL) to send data directly from your application into a Kinesis stream.

Kinesis Data Streams stores data in a data record, along with a partition key and sequence number to uniquely identify each record. Unlike Kinesis Video Streams, which are time-indexed, Kinesis Data Streams are indexed by the partition key and sequence number. Using sequence numbers instead of timestamps makes Kinesis Data Streams appropriate for ordered data that may not necessarily be time-dependent. For example, suppose you have a smartphone application that tracks distance traveled by recording and streaming GPS coordinates at regular intervals. The application streams a series of data records containing GPS coordinates, using the same partition key for each record but incrementing the sequence number each time. When a consumer retrieves or "plays back" the stream, it will receive the records in the same order.

Multiple consumers can read from a stream concurrently, a process called *fan-out*. For example, you can have one consumer analyze application logs for suspicious activity, while having another consumer compress and store the logs in S3 for archival. The delay between the time data is put into the stream and the time a consumer can read it (the put-to-get delay) is typically less than one second.

The maximum throughput of a stream depends on the number of shards you configure. Each shard uniquely identifies a sequence of data records and has a fixed capacity. Each shard supports up to five read transactions per second, with a maximum data rate of 2 MB per second. For writes, you can push up to 1,000 records per second, with a data rate of 1 MB per second. If you need more capacity, you can increase the number of shards.

> You might notice that Kinesis Data Streams seems similar to SQS. SQS is typically used to allow an application component to pass small, short-lived messages to other components. SQS temporarily holds a small message in queue until a single consumer processes and deletes it. Kinesis Data Streams, on the other hand, is designed to provide durable storage and playback of large data streams—such as log files—to multiple consumers.

Kinesis Data Firehose

Kinesis Data Firehose can ingest streaming data and transform it before sending it to a destination. Data transformation may include cleaning data or converting it to a different format. For example, you may need to change JSON-formatted data to Apache Parquet format before sending it to Hadoop. Kinesis Data Firehose uses Lambda functions to perform the data transformation, giving you the flexibility to create custom transformations. Kinesis Data Firehose can send a copy of your untransformed data to an S3 bucket for safekeeping. It can also optionally buffer data before delivering it to its destination.

You can configure Kinesis Data Firehose to receive data from a Kinesis Data Stream. Doing so is especially useful if you need to retain data beyond the seven-day maximum. You can simply configure a Kinesis Data Firehose delivery stream, specify the Kinesis Data Stream as the source, and specify an S3 bucket as the destination.

Kinesis Data Firehose vs. Kinesis Data Streams

Kinesis Data Firehose seems confusingly similar to Kinesis Data Streams. Although all Kinesis services can ingest massive amounts of streaming data, each is tailored to a slightly different need.

Kinesis Data Streams uses a producer-consumer model. The data source that feeds data into a Kinesis stream is called the *producer*, and the application that reads data from the stream is called a *consumer*. Kinesis Video Streams and Kinesis Data Streams use a one-to-many model, allowing multiple consumers to subscribe to a single data stream.

Kinesis Data Firehose, on the other hand, is not an open-ended producer-consumer model where a consumer can simply subscribe to a stream. Instead, you must specify one or more destinations for the data.

Kinesis Data Firehose is tightly integrated with managed AWS services and third-party applications, so it's generally more appropriate for streaming data to services such as Redshift, S3, or Splunk. Kinesis Data Streams, on the other hand, is usually the better choice for streaming data to a custom application. Refer to Table 9.1 for a comparison of SQS and Kinesis services.

TABLE 9.1 Comparison of SQS and Kinesis services

Service	Transforms data	Maximum retention	Model
Simple Queue Service	No	14 days	Producer-consumer
Kinesis Video Streams	No	7 days	Producer-consumer
Kinesis Data Streams	No	7 days	Producer-consumer
Kinesis Data Firehose	Yes	24 hours	Source-destination

Summary

SQS acts as a highly available buffer for data that must move between different application components. On the other hand, the Kinesis services collect, process, and store streaming, real-time binary data such as audio and video. Both SQS and the Kinesis services enable loose coupling.

Exam Essentials

Understand the difference between a standard and FIFO queue in SQS. A standard queue can hold more messages than a FIFO queue but may deliver messages out of order or more than once. A FIFO queue holds fewer messages but always delivers them in order and only once.

Know the different use cases for Kinesis Video Streams, Kinesis Data Streams, and Kinesis Data Firehose. Kinesis Video Streams is appropriate for any time-encoded data, including video, audio, and radar images. Kinesis Data Streams can ingest and store sequences of any binary data. Kinesis Data Firehose ingests, transforms, and delivers streaming data to a specific destination.

Be able to choose the right Kinesis service for a particular application. Which Kinesis service is right for you comes down to two things: the type of data and whether you need to manipulate it. If you have time-indexed data, such as video or radar images, Kinesis Video Streams is probably the best choice. For streaming other types of binary data to one or more consumers, Kinesis Data Streams is your best bet. If you need to stream data into an Amazon service such as Redshift or S3, or if you need to transform data, use Kinesis Data Firehose.

Review Questions

1. When a consumer grabs a message from an SQS queue, what happens to the message? (Select two.)
   A. It is immediately deleted from the queue.
   B. It remains in the queue for 30 seconds and is then deleted.
   C. It remains in the queue for the remaining duration of the retention period.
   D. It becomes invisible to other consumers for the duration of the visibility timeout.

2. What is the default visibility timeout for an SQS queue?
   A. 0 seconds
   B. 30 seconds
   C. 12 hours
   D. 7 days

3. What is the default retention period for an SQS queue?
   A. 30 minutes
   B. 1 hour
   C. 1 day
   D. 4 days
   E. 7 days
   F. 14 days

4. You want to make sure that only specific messages placed in an SQS queue are not available for consumption for 10 minutes. Which of the following settings can you use to achieve this?
   A. Delay queue
   B. Message timer
   C. Visibility timeout
   D. Long polling

5. Which of the following SQS queue types can handle over 50,000 in-flight messages?
   A. FIFO
   B. Standard
   C. Delay
   D. Short

6. What SQS queue type always delivers messages in the order they were received?
   A. FIFO
   B. Standard

C. LIFO
D. FILO
E. Basic

7. You have an application that uses long polling to retrieve messages from an SQS queue. Occasionally, the application crashes because of duplicate messages. Which of the following might resolve the issue?
   A. Configure a per-queue delay.
   B. Use a standard queue.
   C. Use a FIFO queue.
   D. Use short polling.

8. A producer application places messages in an SQS queue, and consumer applications poll the queue every 5 seconds using the default polling method. Occasionally, when a consumer polls the queue, SQS reports there are no messages in the queue, even though there are. When the consumer subsequently polls the queue, SQS delivers the messages. Which of the following may explain the missing messages?
   A. Using long polling
   B. Using short polling
   C. Using a FIFO queue
   D. Using a standard queue

9. Which of the following situations calls for a dead-letter queue?
   A. A message sits in the queue for too long and gets deleted.
   B. Different consumers receive and process the same message.
   C. Messages are mysteriously disappearing from the queue.
   D. A consumer repeatedly fails to process a particular message.

10. A message that's 6 days old is sent to a dead-letter queue. The retention period for the dead-letter queue and the source queue is 10 days. What will happen to the message?
    A. It will sit in the dead-letter queue for up to 10 days.
    B. It will be immediately deleted.
    C. It will be deleted after four days.
    D. It will sit in the dead-letter queue for up to 20 days.

11. You're developing an application to predict future weather patterns based on RADAR images. Which of the following Kinesis services is the best choice to support this application?
    A. Kinesis Data Streams
    B. Kinesis Video Streams
    C. Kinesis Data Firehose
    D. Kinesis ML

12. You're streaming image data to Kinesis Data Streams and need to retain the data for 30 days. How can you do this? (Choose two.)
   A. Create a Kinesis Data Firehose delivery stream.
   B. Increase the stream retention period to 14 days.
   C. Specify an S3 bucket as the destination.
   D. Specify CloudWatch Logs as the destination.

13. Which of the following Kinesis services requires you to specify a destination for the stream?
   A. Kinesis Video Streams
   B. Kinesis Data Streams
   C. Kinesis Data Firehose
   D. Kinesis Data Warehouse

14. You're running an on-premises application that frequently writes to a log file. You want to stream this log file to a Kinesis Data Stream. How can you accomplish this with the least effort?
   A. Use the CloudWatch Logs Agent.
   B. Use the Amazon Kinesis Agent.
   C. Write a script that uses the Kinesis Producer Library.
   D. Move the application to an EC2 instance.

15. When deciding whether to use SQS or Kinesis Data Streams to ingest data, which of the following should you take into account?
   A. The frequency of data
   B. The total amount of data
   C. The number of consumers that need to receive the data
   D. The order of data

16. You want to send streaming log data into Amazon Redshift. Which of the following services should you use? (Choose two.)
   A. SQS with a standard queue
   B. Kinesis Data Streams
   C. Kinesis Data Firehose
   D. SQS with a FIFO queue

17. Which of the following is not an appropriate use case for Kinesis?
   A. Stock feeds
   B. Facial recognition
   C. Static website hosting
   D. Videoconferencing

18. You need to push 2 MB per second through a Kinesis Data Stream. How many shards do you need to configure?
    A. 1
    B. 2
    C. 4
    D. 8

19. Multiple consumers are receiving a Kinesis Data Stream at a total rate of 3 MB per second. You plan to add more consumers and need the stream to support reads of at least 5 MB per second. How many shards do you need to add?
    A. 1
    B. 2
    C. 3
    D. 4

20. Which of the following does Kinesis Data Firehose *not* support?
    A. Videoconferencing
    B. Transforming video metadata
    C. Converting CSV to JSON
    D. Redshift

PART II

The Well-Architected Framework

Chapter 10

The Reliability Pillar

THE AWS CERTIFIED SOLUTIONS ARCHITECT ASSOCIATE EXAM OBJECTIVES COVERED IN THIS CHAPTER MAY INCLUDE, BUT ARE NOT LIMITED TO, THE FOLLOWING:

✓ **Domain 1: Design Resilient Architectures**

- 1.1 Design a multi-tier architecture solution
- 1.2 Design highly available and/or fault-tolerant architectures
- 1.3 Design decoupling mechanisms using AWS service
- 1.4 Choose appropriate resilient storage

Introduction

Reliability—sometimes also called *resiliency*—is the ability of an application to avoid failure and recover from it quickly when it occurs. The degree of reliability your application will achieve is proportional to the amount of effort you put into the design up front, as well as the money you spend to operate it. Achieving a high degree of reliability for your application is not a trivial matter and entails significant cost and complexity. Therefore, before you begin architecting your AWS deployment, you have to decide how much reliability you actually need. The reliability of an application is quantified in terms of *availability*—the percentage of time the application is performing as expected.

After you decide how much availability you need, you can begin designing your AWS environment. The level of availability your application can achieve depends in part on the availability of the AWS resources it uses, including networking, compute, and storage. As you might expect, even AWS resources aren't going to be available 100 percent of the time. They'll fail, so how you plan for and handle these inevitable failures when they happen will have a significant impact on your application's availability.

In this chapter, you'll learn how to design your AWS environment to tolerate resource failures so that the failure of an instance or even an entire availability zone doesn't cause your entire application to become unavailable. You'll also learn how to recover from unavoidable failures. If all of your application's instances go down or if your application crashes because of a bug, security breach, database corruption, or just human error, you'll know how to recover quickly.

Calculating Availability

A common way of quantifying and expressing reliability is in terms of availability. Availability is the percentage of time an application is working as expected. Note that this measurement is subjective, and "working as expected" implies that you have a certain expectation of how the application should work.

Generally, you want application availability to be very high, 99 percent or greater. Table 10.1 lists the relationship between percentage of availability and the amount of time an application is unavailable during the year.

TABLE 10.1 The relationship between annual availability percentage and time unavailable

Availability percentage	Time unavailable
99%	3 days, 15 hours, 39 minutes
99.9%	8 hours, 45 minutes
99.95%	4 hours, 22 minutes
99.99%	52 minutes
99.999%	5 minutes

You may hear availability measured in nines. For example, "2 nines" is 99 percent, "3 nines" is 99.9 percent, and so on. Some call 99.95 percent "3-1/2 nines," which isn't quite correct, but it's easier and less confusing than saying "3 and 5/9 nines!"

Availability Differences in Traditional vs. Cloud-Native Applications

As an AWS architect, you must understand how application design decisions affect reliability. The applications you run in the cloud will fall into one of two broad categories: traditional or cloud-native.

Traditional Applications

Traditional applications are those written to run on and use the capabilities of traditional Linux or Windows servers. To deploy such an application on AWS, you'll need to run it on one or more EC2 instances. If the application uses a database, you'll either run and manage your own database software on an EC2 instance or use the AWS-managed equivalent. If your application requires a relational database, you can use Amazon Relational Database Service (RDS). If it uses a nonrelational database such as Redis, you can use ElastiCache for Redis. Traditional applications operate the same whether they're running on the cloud or in a data center. If you can "lift and shift" an application from the data center to the cloud without making any changes to the application's code, then it's a traditional application.

Suppose you have a traditional application running on a single EC2 instance backed by a multi-availability zone (AZ) RDS deployment. The application's availability depends on the availability of both the EC2 instance and the RDS instance. Both of these are called *hard dependencies*. To calculate the total availability of the application, you simply multiply the availability of those hard dependencies. AWS advertises the availability of a single EC2 instance as 90 percent and multi-AZ RDS as 99.95 percent. If you multiply these availabilities together (.9 × .9995), you get 89.955 percent, which is about 36 days of downtime per year. Not very good!

To increase availability, you can use redundant components. Let's start with EC2. Suppose that instead of one EC2 instance, you use three EC2 instances each in a different availability zone. Each EC2 instance runs the application, and an application load balancer (ALB) distributes connections to the instances in a *target group*. If the application on one instance fails or if the EC2 instance itself fails, the ALB will route connections to the remaining healthy instance. In this sense, the EC2 instances are redundant. To calculate the availability of these redundant components, you take 100 percent minus the product of the instance failure rate. If the availability of an EC2 instance is 90 percent, then the failure rate of that instance is 10 percent. Calculating the availability for three EC2 instances would look like this:

$$100\% - (10\% \times 10\% \times 10\%) = 99.9\%$$

This is about 9 hours of downtime per year. Quite an improvement over 36 days! But that's just for EC2. You also have to consider the database and the ALB. These represent hard dependencies, so you must multiply the database's availability—99.95 percent—and the ALB's availability—99.99 percent—by the availability of the EC2 instances—99.9 percent. The product would be about 99.84 percent, which is about 14 hours of downtime a year. To improve this number, you could add more EC2 instances or a redundant database. Later in this chapter, we'll show you how to do that.

Cloud-Native Applications

Cloud-native applications are written to use the resources of a specific cloud platform like AWS. These may be serverless applications written to use Lambda functions. Or they may run on EC2 instances and store objects in S3 or use DynamoDB instead of a relational database for storing data. This is especially likely to be true if the application requires low-latency access to data stored in a nonrelational data format such as JavaScript Object Notation (JSON). DynamoDB is also a popular choice for storing session state data.

Suppose you have a Linux application that runs on a single EC2 instance—just as a traditional app would—but is designed to use DynamoDB for database storage. In a single region, DynamoDB has an availability of 99.99 percent. Again, three EC2 instances have an availability of 99.9 percent, and the ALB has an availability of 99.99 percent—a combined availability of 99.89 percent. So considering just these factors, the total application availability is as follows:

$$99.89\% \times 99.9\% = 99.79\%$$

This is about 18 hours of downtime per year. But suppose you want even greater availability. Instead of running three EC2 instances, you run six, each in a different AZ. Again, you'll use ALB to distribute traffic to these instances. To calculate the availability of these instances, you'd do this:

$$100\% - (10\% \times 10\% \times 10\% \times 10\% \times 10\% \times 10\% = 99.9999\%$$

To get the total application availability, multiply 99.9999 percent by the availability of DynamoDB—99.99 percent—and the availability of the ALB—99.99 percent—which would give you 99.979 percent, or 2 hours of downtime per year.

Believe it or not, you can make this even better by using two regions. In one region, you would still have six instances, each in a different AZ, and an ALB to distribute traffic to the application instances. But now you replicate this setup in a different region and then use a Route 53 weighed routing policy to distribute traffic between the two regions. To calculate the total availability, start with the failure rate of the application in each region:

$$100\% - 99.979\% = .021\%$$

Now multiply the failure rate by itself and subtract from 100%:

$$100\% - (.021\% \times .021\%)$$

This is about 99.999 percent, approximately 5 minutes of downtime per year.

> **Note:** Route 53 advertises an availability of 100 percent, so its failure rate is zero.

Since you'd be running the application across regions, you could take advantage of DynamoDB global tables, which replicate your DynamoDB tables across regions. DynamoDB global tables have an availability of 99.999 percent. So, to calculate the total availability for the application, you'd multiply 99.999 percent by 99.999 percent to get about 99.998 percent, or 10 minutes of downtime per year.

Building Serverless Applications with Lambda

Although as an AWS architect you don't need to know how to program, it's important to understand how serverless applications that run on Lambda differ from executables that run on EC2 instances. Lambda allows you to create a function written in one of a variety of languages, including the following:

- C#
- Go
- Java
- JavaScript
- Node.js

- PowerShell
- Python
- Ruby

Lambda functions are useful for performing intermittent tasks that don't warrant keeping a running EC2 instance around. For example, you can create a Lambda function that retrieves an image from an S3 source bucket, resizes it, and then uploads the processed image to a destination bucket. When a new image is uploaded to the source bucket, S3 triggers the Lambda function. You're billed only for the time the function runs.

Lambda executes functions using a highly available distributed computing platform. Unlike an EC2 instance, which can fail or be stopped or terminated, Lambda is always available.

Know Your Limits

Although one of the big selling points of the cloud is the ability to grow with you, cloud capacity is not unlimited. AWS imposes limits to prevent anyone from accidentally or intentionally consuming all resources, effectively resulting in a denial of service for other customers. The limits depend on the service and include things such as network throughput, S3 PUT requests per second, the number of instances per region, the number of elastic IP addresses per region, and so on. Many of these limits can be increased upon request.

Use the AWS Trusted Advisor to see what service limits apply to your account. You may also consider setting CloudWatch Alarms to let you know when you're getting close to hitting a limit so that you can react to avoid hitting it. Avoiding hitting a limit may involve requesting a limit increase by contacting AWS support, adding another availability zone, or even shifting some of the load to another region.

Increasing Availability

The availability numbers you've looked at thus far are based only on the availability of AWS services. But the actual availability of your application could be worse. Many factors can impact availability. For example, your app can crash, perhaps because of a bug, memory leak, or data corruption.

The best way to maximize availability is to avoid failure in the first place. Instead of having a single monster instance hosting a web application, have multiple smaller instances and spread them out across different availability zones. This way, a failure of a single instance or even an entire availability zone won't render your application unavailable. Rather than depending on one instance to be highly available, distributed application design distributes the work across multiple smaller resources that are not dependent on one another.

But that's not the whole story. If an instance fails, it means the other instances have to pick up the slack. If enough instances fail, only a few instances will be servicing a lot of traffic, resulting in poor performance and likely in those remaining instances also crashing. Therefore, you need a way to re-create failed instances when such a crash occurs.

Another issue that could arise is that increased demand could place such a load on your instances that they become unusably slow, or even crash altogether. An advantage of using a distributed application design is that it makes it easier to add capacity with disruption. Instead of having to upgrade to a more powerful instance class, which requires downtime, you simply add more instances. When you have a distributed system, getting more resources just involves scaling out- –adding more of the same.

EC2 Auto Scaling

The *EC2 Auto Scaling* service offers a way to both avoid application failure and recover from it when it happens. Auto Scaling works by provisioning and starting on your behalf a specified number of EC2 instances. It can dynamically add more instances to keep up with increased demand. And when an instance fails or gets terminated, Auto Scaling will automatically replace it.

EC2 Auto Scaling uses either a *launch configuration* or a *launch template* to automatically configure the instances that it launches. Both perform the same function of defining the basic configuration parameters of the instance as well as what scripts (if any) run on it at launch time. Launch configurations have been around longer and are more familiar to you if you've been using AWS for a while. You're also more likely to encounter them if you're going into an existing AWS environment. Launch templates are newer, and they are what AWS now recommends. You'll learn about both, but the decision about which to use is up to you.

Launch Configurations

When you create an instance manually, you have to specify many configuration parameters, including an Amazon Machine Image (AMI), instance type, SSH key pair, security group, instance profile, block device mapping, whether it's EBS optimized, placement tenancy, and user data, such as custom scripts to install and configure your application. A launch configuration is essentially a named document that contains the same information you'd provide when manually provisioning an instance.

You can create a launch configuration from an existing EC2 instance. Auto Scaling will copy the settings from the instance for you, but you can customize them as needed. You can also create a launch configuration from scratch.

Launch configurations are for use only with EC2 Auto Scaling, meaning you can't manually launch an instance using a launch configuration. Also, after you create a launch configuration, you can't modify it. If you want to change any of the settings, you have to create an entirely new launch configuration.

Launch Templates

Launch templates are similar to launch configurations in that you can specify the same settings. But the uses for launch templates are more versatile. You can use a launch template with Auto Scaling, of course, but you can also use it for spinning up one-off EC2 instances or even creating a spot fleet.

Launch templates are also versioned, allowing you to change them after creation. Any time you need to make changes to a launch template, you create a new version of it. AWS keeps all versions, and you can then flip back and forth between versions as needed. This approach makes it easier to track your launch template changes over time. Complete Exercise 10.1 to create your own launch template.

> If you have an existing launch configuration, you can copy it to a launch template using the AWS web console. There's no need to create launch templates from scratch!

EXERCISE 10.1

Create a Launch Template

In this exercise, you'll create a launch template that installs and configures a simple web server. You'll then use the launch template to manually create an instance.

1. In the EC2 service console, click Launch Templates.
2. Click the Create Launch Template button.
3. Give the launch template a name such as **MyTemplate**.
4. Click the Search For AMI link to locate one of the Ubuntu Server LTS AMIs. If you're in the us-east-1 region, you can use **ami-0ac019f4fcb7cb7e6**.
5. For Instance Type, select t2.micro.
6. Under Security Groups, select a security group that allows inbound HTTP access. Create a new security group if necessary.
7. Expand the Advanced Details section and enter the following in the User Data text field:

    ```
    #!/bin/bash
    apt-get update
    apt-get install -y apache2
    echo "Welcome to my website" > index.html
    cp index.html /var/www/html
    ```

8. Click the Create Launch Template button.

9. Click the Launch Instance From This Template link.

10. Under Source Template Version, select 1 (Default).

11. Click the Launch Instance From Template button.

12. After the instance boots, browse to its public IP address. You should see a web page that says "Welcome to my website."

13. Terminate the instance when you're done with it.

Auto Scaling Groups

An *Auto Scaling group* is a group of EC2 instances that Auto Scaling manages. When creating an Auto Scaling group, you must first specify either the launch configuration or launch template you created. You must also specify how many running instances you want Auto Scaling to provision and maintain using the launch configuration or template you created. Specifically, you have to specify the minimum and maximum size of the Auto Scaling group. You also have the option to set the desired number of instances you want Auto Scaling to provision and maintain.

Minimum   Auto Scaling will ensure the number of healthy instances never goes below the minimum. If you set this to 0, Auto Scaling will not spawn any instances and will terminate any running instances in the group.

Maximum   Auto Scaling will make sure the number of healthy instances never exceeds this number. This might seem strange, but remember that AWS imposes service limits on how many instances you can run simultaneously. Setting your maximum to less than or equal to your limit ensures you never exceed it.

Desired Capacity   Desired Capacity is an optional setting that must lie within the minimum and maximum values. If you don't specify a desired capacity, Auto Scaling will launch the number of instances as the minimum value. If you specify a desired capacity, Auto Scaling will add or terminate instances to stay at the desired capacity. For example, if you set the minimum to 1, maximum to 10, and desired capacity to 4, then Auto Scaling will create four instances. If one of those instance gets terminated—for example, because of human action or a host crash—Auto Scaling will replace it to maintain the desired capacity of 4. In the web console, desired capacity is also called the *group size*.

Specifying an Application Load Balancer Target Group

If you want to use an application load balancer to distribute traffic to instances in your Auto Scaling group, just plug in the name of the ALB target group when creating the Auto

Scaling group. Whenever Auto Scaling creates a new instance, it will automatically add it to the ALB target group.

Health Checks Against Application Instances

When you create an Auto Scaling group, Auto Scaling will strive to maintain the minimum number of instances, or the desired number if you've specified it. If an instance becomes unhealthy, Auto Scaling will terminate and replace it.

By default, Auto Scaling determines an instance's health based on EC2 health checks. Recall from Chapter 7, "CloudTrail, CloudWatch, and AWS Config," that EC2 automatically performs system and instance status checks. These checks monitor for instance problems such as memory exhaustion, filesystem corruption, or an incorrect network or startup configuration, as well as system problems that require AWS involvement to repair. Although these checks can catch a variety of instance and host-related problems, they won't necessarily catch application-specific problems.

If you're using an application load balancer to route traffic to your instances, you can configure health checks for the load balancer's target group. Target group health checks can check for HTTP response codes from 200 to 499. You can then configure your Auto Scaling group to use the results of these health checks to determine whether an instance is healthy.

If an instance fails the ALB health check, it will route traffic away from the failed instance, ensuring users don't reach it. At the same time, Auto Scaling will remove the instance, create a replacement, and add the new instance to the load balancer's target group. The load balancer will then route traffic to the new instance.

> A good design practice is to have a few recovery actions that work for a variety of circumstances. An instance may crash due to an out-of-memory condition, a bug, a deleted file, or an isolated network failure, but simply terminating and replacing the instance using Auto Scaling resolves all these cases. There's no need to come up with a separate recovery action for each cause when simply re-creating the instance solves all of them.

Auto Scaling Options

After you create an Auto Scaling group, you can leave it be, and it will continue to maintain the minimum or desired number of instances indefinitely. However, maintaining the current number of instances is just one option. Auto Scaling provides several other options to scale out the number of instances to meet demand.

Manual Scaling

If you change the minimum, desired, or maximum values at any time after creating the group, Auto Scaling will immediately adjust. For example, if you have the desired capacity set to 2 and change it to 4, Auto Scaling will launch two more instances. If you have four instances and set the desired capacity to 2, Auto Scaling will terminate two instances. Think of the desired capacity as a thermostat!

Dynamic Scaling Policies

Most AWS-managed resources are elastic; that is, they automatically scale to accommodate increased load. Some examples include S3, load balancers, Internet gateways, and network address translation (NAT) gateways. Regardless of how much traffic you throw at them, AWS is responsible for ensuring they remain available while continuing to perform well. But when it comes to your EC2 instances, you're responsible for ensuring that they're powerful and plentiful enough to meet demand.

Running out of instance resources—be it CPU utilization, memory, or disk space—will almost always result in the failure of whatever you're running on it. To ensure that your instances never become overburdened, dynamic scaling policies automatically provision more instances *before* they hit that point. Auto Scaling generates the following aggregate metrics for all instances within the group:

- Aggregate CPU utilization
- Average request count per target
- Average network bytes in
- Average network bytes out

You're not limited to using just these native metrics. You can also use metric filters to extract metrics from CloudWatch logs and use those. As an example, your application may generate logs that indicate how long it takes to complete a process. If the process takes too long, you could have Auto Scaling spin up new instances.

Dynamic scaling policies work by monitoring a CloudWatch alarm and scaling out—by increasing the desired capacity—when the alarm is breaching. You can choose from three dynamic scaling policies: simple, step, and target tracking.

Simple Scaling Policies

With a *simple scaling policy*, whenever the metric rises above the threshold, Auto Scaling simply increases the desired capacity. How much it increases the desired capacity, however, depends on which of the following *adjustment types* you choose:

ChangeInCapacity   Increases the capacity by a specified amount. For instance, you start with a desired capacity of 4 and then have an Auto Scaling add 2 when the load increases.

ExactCapacity   Sets the capacity to a specific value, regardless of the current value. For example, suppose the desired capacity is 4. You create a policy to change the value to 6 when the load increases.

PercentChangeInCapacity   Increases the capacity by a percentage of the current amount. If the current desired capacity is 4 and you specify the percent change in capacity as 50 percent, then Auto Scaling will bump the desired capacity to 6.

For example, suppose you have four instances and create a simple scaling policy that specifies a PercentChangeInCapacity adjustment of 50 percent. When the monitored alarm triggers, Auto Scaling will increment the desired capacity by 2, which will in turn add two instances to the Auto Scaling group, resulting in a total of 6.

After Auto Scaling completes the adjustment, it waits a *cooldown period* before executing the policy again, even if the alarm is still breaching. The default cooldown period is 300 seconds, but you can set it as high as you want or as low as 0—effectively disabling it. Note that if an instance is unhealthy, Auto Scaling will not wait for the cooldown period before replacing the unhealthy instance.

Referring to the preceding example, suppose after the scaling adjustment completes and the cooldown period expires the monitored alarm drops below the threshold. At this point, the desired capacity is 6. If the alarm triggers again, the simple scaling action will execute again and add three more instances. Keep in mind that Auto Scaling will never increase the desired capacity beyond the group's maximum setting.

Step Scaling Policies

If the demand on your application is rapidly increasing, a simple scaling policy may not add enough instances quickly enough. Using a *step scaling policy*, you can instead add instances based on how much the aggregate metric exceeds the threshold.

To illustrate, suppose your group starts out with four instances. You want to add more instances to the group as the average CPU utilization of the group increases. When the utilization hits 50 percent, you want to add two more instances. When it goes above 60 percent, you want to add four more instances.

You'd first create a CloudWatch Alarm to monitor the average CPU utilization and set the alarm threshold to 50 percent, since this is the utilization level at which you want to start increasing the desired capacity.

You'd must then specify at least one step adjustment. Each step adjustment consists of the following:

- A lower bound
- An upper bound
- The adjustment type
- The amount by which to increase the desired capacity

The upper and lower bound define a range that the metric has to fall within for the step adjustment to execute. Suppose for the first step you set a lower bound of 50 and an upper bound of 60, with a ChangeInCapacity adjustment of 2. When the alarm triggers, Auto Scaling will consider the metric value of the group's average CPU utilization. Suppose it's 55 percent. Because 55 is between 50 and 60, Auto Scaling will execute the action specified in this step, which is to add two instances to the desired capacity.

Suppose now that you create another step with a lower bound of 60 and an upper bound of infinity. You also set a ChangeInCapacity adjustment of 4. If the average CPU utilization increases to 62 percent, Auto Scaling will note that 60 <= 62 < infinity and will execute the action for this step, adding four instances to the desired capacity.

You might be wondering what would happen if the utilization were 60 percent. Step ranges can't overlap. A metric of 60 percent would fall within the lower bound of the second step.

With a step scaling policy, you can optionally specify a *warm-up time*, which is how long Auto Scaling will wait until considering the metrics of newly added instances. The default warm up time is 300 seconds. Note that there are no cooldown periods in step scaling policies.

Target Tracking Policies

If step scaling policies are too involved for your taste, you can instead use *target tracking policies*. All you do is select a metric and target value, and Auto Scaling will create a CloudWatch Alarm and a scaling policy to adjust the number of instances to keep the metric near that target.

The metric you choose must change proportionally to the instance load. Metrics like this include average CPU utilization for the group, the number of SQS messages in a queue, and request count per target. Aggregate metrics like the total request count for the ALB don't change proportionally to the load on an individual instance and aren't appropriate for use in a target tracking policy.

Note that in addition to scaling out, target tracking will scale in by deleting instances to maintain the target metric value. If you don't want this behavior, you can disable scaling in. Also, just as with a step scaling policy, you can optionally specify a warm-up time.

Scheduled Actions

Scheduled actions are useful if you have a predictable load pattern and want to adjust your capacity proactively, ensuring you have enough instances *before* demand hits.

When you create a scheduled action, you must specify the following:

- A minimum, maximum, or desired capacity value
- A start date and time

You may optionally set the policy to recur at regular intervals, which is useful if you have a repeating load pattern. You can also set an end time, after which the scheduled policy gets deleted.

To illustrate how you might use a scheduled action, suppose you normally run only two instances in your Auto Scaling group during the week. But on Friday, things get busy, and you know you'll need four instances to keep up. You'd start by creating a scheduled action that sets the desired capacity to 2 and recurs every Saturday, as shown in Figure 10.1.

The start date is January 5, which is a Saturday. To handle the expected Friday spike, you'd create another weekly recurring policy to set the desired capacity to 4, as shown in Figure 10.2.

FIGURE 10.1 Scheduled action setting the desired capacity to 2 every Saturday

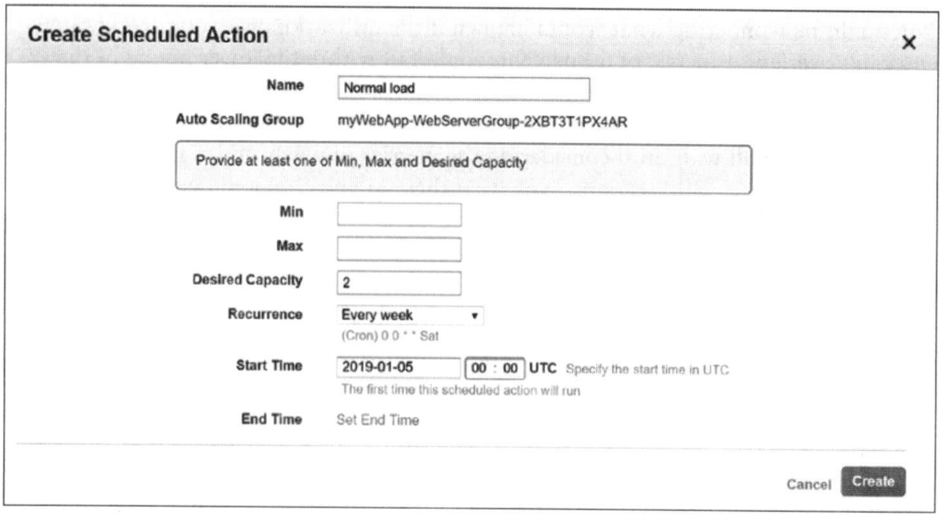

FIGURE 10.2 Scheduled action setting the desired capacity to 4 every Friday

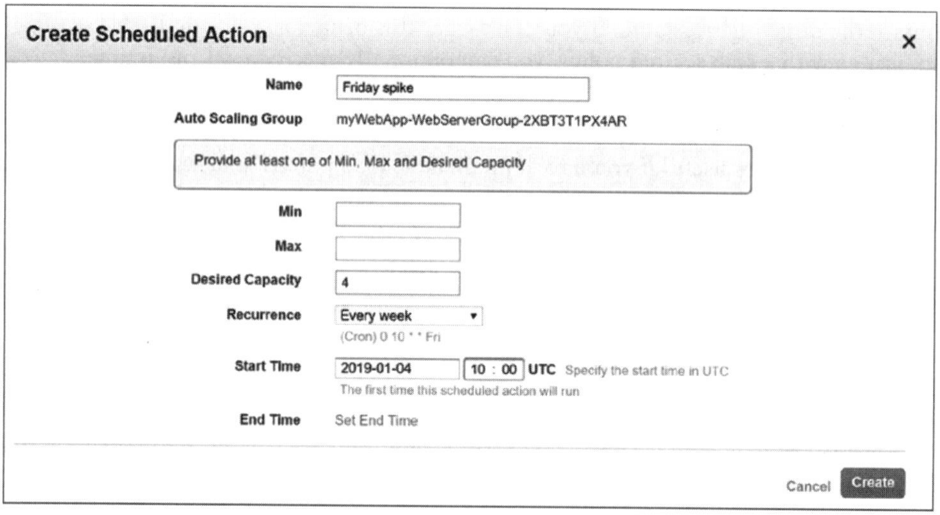

This action will run every Friday, setting the desired capacity to 4, prior to the anticipated increased load.

Note that you can combine scheduled actions with dynamic scaling policies. For example, if you're running an e-commerce site, you can use a scheduled action to increase the maximum group size during busy shopping seasons and then rely on dynamic scaling policies to increase the desired capacity as needed.

Data Backup and Recovery

As you learned in Chapter 2, "Amazon Elastic Compute Cloud and Amazon Elastic Block Store," and Chapter 3, "AWS Storage," AWS offers several different data storage options. Although Amazon takes great pains to ensure the durability of your data stored on these services, data loss is always a possibility. AWS offers several options to help you store your data in a resilient manner as well as maintain backups to recover from when the inevitable data loss occurs.

S3

All S3 storage classes except One Zone-Infrequent Access distribute objects across multiple availability zones. To avoid data loss from the failure of a single availability zone, be sure to use one of the other storage classes.

To guard against deletion and data corruption, enable S3 versioning. With versioning enabled, S3 never overwrites or deletes an object. Instead, modifying an object creates a new version of it that you can revert to if needed. Also, instead of deleting an object, S3 inserts a delete marker and removes the object from view. But the object and all of its versions still exist.

To protect your data against multiple availability zone failures—or the failure of an entire region—you can enable cross-region replication between a source bucket in one region and destination bucket in another. After you enable it, S3 will synchronously copy every object from the source bucket to the destination. Note that cross-region replication requires versioning to be enabled on both buckets. Also, deletes on the source bucket don't get replicated.

Elastic File System

AWS Elastic File System (EFS) provides a managed Network File System (NFS) that can be shared among EC2 instances or on-premises servers. EFS filesystems are stored across multiple zones in a region, allowing you to withstand an availability zone failure.

To protect against data loss and corruption, you can back up individual files to an S3 bucket or another EFS filesystem in the same region. You can also use the AWS Backup Service to schedule incremental backups of your EFS filesystem.

Elastic Block Storage

EBS automatically replicates volumes across multiple availability zones in a region so that they're resilient to a single availability zone failure. Even so, data corruption is always a possibility.

One of the easiest ways to back up an EBS volume is to take a snapshot of it. AWS stores EBS snapshots across multiple availability zones in S3. You can either create a snapshot manually or use the Amazon Data Lifecycle Manager to automatically create a snapshot for you at regular intervals. To use the Amazon Data Lifecycle Manager, you create a Snapshot Lifecycle Policy and specify an interval of 12 or 24 hours, as well as a snapshot creation time. You also must specify the number of automatic snapshots to retain, up to 1,000 snapshots.

One important but often overlooked piece of data that gets stored on EBS is application logs. An application may create log files on the instance it's running on, but if the instance crashes or loses connectivity, those files won't be able to tell you what's wrong. And if EC2 Auto Scaling terminates an instance and its EBS volume along with it, any log files stored locally are gone permanently. CloudWatch Logs can collect and store application logs from an instance in real time, letting you search and analyze them even after the instance is gone.

Database Resiliency

When it comes to protecting your databases, your options vary depending on whether you're using a relational database or DynamoDB.

If you're running your own relational database server, you can use the database engine's native capability to back up the database to a file. You can then store the file in S3 or Glacier for safekeeping. Restoring a database depends on the particulars of the database engine you're running but generally involves copying the backup file back to the destination instance and running an import process to copy the contents of the backup file into the database.

If you're using Amazon Relational Database Service (RDS), you always have the option of taking a simple database instance snapshot. Restoring from a database snapshot can take several minutes and always creates a new database instance.

For additional resiliency, you can use a multi-AZ RDS deployment that maintains a primary instance in one availability zone and a standby instance in another. RDS replicates data synchronously from the primary to the standby. If the primary instance fails, either because of an instance or because of availability zone failure, RDS automatically fails over to the secondary.

For maximum resiliency, use Amazon Aurora with multiple Aurora replicas. Aurora stores your database across three different availability zones. If the primary instance fails, it will fail over to one of the Aurora replicas. Aurora is the only RDS offering that lets you provision up to 15 replicas, which is more than enough to have one in each availability zone!

DynamoDB stores tables across multiple availability zones, which along with giving you low-latency performance provides additional protection against an availability zone failure. For additional resiliency, you can use DynamoDB global tables to replicate your tables to different regions. You can also configure point-in-time recovery to automatically take backups of your DynamoDB tables. Point-in-time recovery lets you restore your DynamoDB tables to any point in time from 35 days until 5 minutes before the current time.

Creating a Resilient Network

All of your AWS resources depend on the network, so designing your network properly is critical. When it comes to network design, you must consider two elements: your virtual private cloud (VPC) design and how users will connect to your resources within your VPC.

VPC Design Considerations

When creating a VPC, make sure you choose a sufficiently large classless interdomain routing (CIDR) block to provide enough IP addresses to assign to your resources. The more redundancy you build into your AWS deployment, the more IP addresses you'll need. Also, when you scale out by adding more resources, you must have enough IP addresses to assign to those additional resources.

When creating your subnets, leave enough unused address space within the CIDR to add additional subnets later. This is important for a couple of reasons. First, AWS occasionally adds additional availability zones to a region. By leaving additional space, you can take advantage of those new availability zones. Your availability requirements may call for only one availability zone today, but it doesn't hurt to leave room in case your needs change. Second, even if you don't ever need an additional availability zone, you may still need a separate subnet to segment your resources for security or ease of management. When you separate an application's components into multiple subnets, it's called a *multi-tier architecture*.

Leave plenty of space in each subnet. Naturally, having more instances will consume more IP addresses. Leave enough free address space in each subnet for resources to use as demand grows. If any of your instances will have multiple private IP addresses, take that into account. Also, remember that EC2 instances aren't the only resources that consume IP address space. Consider Elastic Load Balancing interfaces, database instances, and VPC interface endpoints.

External Connectivity

Your application's availability depends on the availability of the network users use to access it. Remember that the definition of availability is that your application is working as expected. If users can't connect to it because of a network issue, your application is unavailable to them, even if it's healthy and operating normally. Therefore, the method your users use to connect to the AWS cloud needs to meet your availability requirements.

The most common way to connect to AWS is via the Internet, which is generally reliable, but speed and latency vary and can be unpredictable. If getting to your application is painfully slow, you may consider it to be unavailable.

To get around a slow Internet pipe or for mission-critical applications, you may choose to use Direct Connect, either as a primary connection to AWS or as a backup. If you need a fast and reliable connection to AWS—such as if you're pushing a lot of data—Direct Connect offers access speeds of 1 or 10 Gbps with consistent latency. As a secondary connection, if you can't use the Internet to reach your AWS resources, you can get through using Direct Connect.

If you're going to use Direct Connect, VPC peering, or a VPN connection to an external network, make sure your VPC addresses don't overlap with those used in the external network. Also, be sure to leave enough room to assign IP addresses to virtual private gateways and Direct Connect private virtual interfaces.

Designing for Availability

Understanding all the different factors that go into architecting for reliability is important. But as an AWS architect, you need to be able to start with an availability requirement and design a solution that meets it. In this section, you'll learn how to start with a set of application requirements and design different AWS infrastructures to meet three common availability percentages: 99 percent, 99.9 percent, and 99.99 percent.

As you might expect, there's a proportional relationship between availability, complexity, and cost. The higher the availability, the more resources you're going to need to provision, which adds more complexity and, of course, more cost. Although most organizations *want* as much availability as possible, you as an AWS architect must strike a balance between availability and cost. Designs with lower availability may appear, from a technical standpoint, to be inferior. But at the end of the day, the needs and priorities of the organization are what matter.

In the following scenarios, you'll architect a solution for a web-based payroll application that runs as a single executable on a Linux EC2 instance. The web application connects to a backend SQL database for storing data. Static web assets are stored in a public S3 bucket. Users connect to the application by browsing to a public domain name that is hosted on Route 53.

Designing for 99 Percent Availability

In the first scenario, you're aiming for 99 percent availability, or about 3.5 days of downtime per year. Downtime is an inconvenience but certainly not the end of the world.

In this case, you'll create a single EC2 instance to run both the application and a self-hosted SQL database. For backing up data, you'll run a script to automatically back up the database to an S3 bucket with versioning enabled. To keep data sprawl under control, you'll configure S3 Lifecycle policies to move older backups to the Glacier storage class.

To alert users during outages, configure a Route 53 health check to monitor the health of the application by checking for an HTTP 200 OK status code. Then create a Route 53 failover record set to direct users to the application when the health check passes and to a static informational web page when it doesn't. The static web page can be stored in either a public S3 bucket that's configured for website hosting or a CloudFront distribution.

Recovery Process

It's important to test your recovery processes regularly so that you know they work and how long they take. Don't just create a SQL dump and upload it to an S3 bucket. Regularly bring up a new instance and practice restoring a database backup to it.

To quickly build or rebuild your combination application/database instance, use CloudFormation to do the heavy lifting for you. Create a CloudFormation template to create a new instance, install and configure the web and database servers, copy application files from a repository, and set up security groups. The CloudFormation template serves two purposes: rebuilding quickly in case of an instance failure and creating a throwaway instance for practicing database restores and testing application updates.

Availability Calculation

When it comes to estimating availability, you're going to have to make some assumptions. Each failure will require 30 minutes to analyze and decide whether to initiate your recovery process. When you decide to recover, you'll use your CloudFormation template to build a new instance, which takes about 10 minutes. Lastly, you'll restore the database, which takes about 30 minutes. The total downtime for each failure—or your recovery time objective (RTO)—is 70 minutes. Note that the recovery point objective (RPO) you can expect depends on how frequently you're backing up the database. In this case, assume that your application will suffer one failure every quarter. That's a total annual downtime of 280 minutes, or about 4.6 hours.

However, that's just for failures. You also need to consider application and operating system updates. Assuming that updates occur six times a year and require 4 hours of downtime each, that's additional downtime of 24 hours per year. Adding all the numbers together, between failures and updates, the total downtime per year is about 28.6 hours, or 1.19 days. That's 99.67 percent availability!

Although it's tempting to round 99.67 percent up to 99.9 percent, keep in mind that the difference between those two numbers is more than seven hours per year. Providing inflated availability numbers, even if you're honestly just rounding up, can come back to bite you if those numbers turn out later to be out of line with the application's actual availability. Remember that your availability calculations are only as good as the assumptions you make about the frequency and duration of failures and scheduled outages. When it comes to availability, err on the side of caution and round down! Following that advice, the expected availability of the application is 99 percent.

Designing for 99.9 Percent Availability

When designing for 99.9 percent availability, downtime isn't the end of the world, but it's pretty close. You can tolerate about 9 hours a year, but anything beyond that is going to be a problem. To achieve 99.9 percent availability, you'll use a distributed application design.

The application itself will run on multiple instances in multiple availability zones in a single region. You'll configure an ALB to proxy traffic from users to individual instances and perform health checks against those instances. You'll create a launch template to install the application web server and copy over application files from a centralized repository. You'll then create an Auto Scaling group to ensure that you always have at least six running instances (two in each zone).

Why six instances? Suppose that you need four application instances to handle peak demand. To withstand the loss of one zone, having two instances in each zone ensures they can handle 50 percent of the peak demand. That way, if one zone fails—taking two instances with it—the remaining two zones have four instances, which is enough to handle 100 percent of the load.

Recovery Process

For most failures, recovery will be automatic. You'll configure the Auto Scaling group to use the ALB's target health checks to determine the health of each instance. If an instance fails, the ALB will route traffic away from it, and then Auto Scaling will terminate the failed instance and create a new one.

For the database, you'll use a multi-AZ RDS instance. If the primary database instance fails (perhaps because of an entire availability zone failure), RDS will fail over to the secondary instance in a different zone. You'll also configure automated database instance snapshots to guard against database corruption. Enabling automated snapshots also enables point-in-time recovery, reducing your RPO to as little as five minutes.

Availability Calculation

Thanks to the resilience of the redundant, distributed architecture, you expect only two unplanned outage events per year, lasting 60 minutes each. Software and operating system updates occur 10 times a year and last 15 minutes each. Adding up the numbers, you get a total downtime of 270 minutes, or 4.5 hours per year. That's 99.95 percent availability! Again, to be conservative, you want to round that down to 99.9 percent.

Designing for 99.99 Percent Availability

Mission-critical applications, often those that generate a lot of revenue or have a significant impact on human life, must be able to tolerate multiple component failures without having to allocate additional resources.

AWS takes great pains to isolate availability zone failures. However, there's always the possibility that multiple zones in a single region can fail simultaneously. To protect against these regional disruptions, you'll simultaneously run application instances in two different regions, an active region and a passive region. Just as in the preceding scenario, you'll use Auto Scaling to provision the instances into multiple availability zones in each region and ensure that each zone can handle 50 percent of the peak load.

You'll also use multi-AZ RDS; however, there's a catch. The primary and secondary database instances both must run in the same region. RDS doesn't currently allow you to have a primary database instance in one region and secondary in another. Therefore, there's no way to fail over from one region to the other without the potential for data loss.

But you do have another option. If you use the MySQL or MariaDB database engine, you can create a multi-AZ read replica in a different region. You can then asynchronously replicate data from one region to the other. With this option, there's a delay between the time data is written to the primary and the time it's replicated to the read replica. Therefore, if the primary database instance fails before the replication has completed, you may lose data. For example, given a replication lag of 10 minutes, you may lose 10 minutes of data.

Recovery Process

As in the preceding example, for an application or database instance failure isolated to the active region, or even for an entire availability zone failure, recovery is automatic. But regional failures require some manual intervention.

Users will access application instances only in the active region until that region fails. You'll create a CloudWatch Alarm to monitor the health of the ALB in the active region. If the ALB in the active region fails its health check, you must then decide whether to fail over to the passive region.

If you do choose to fail over, you'll need to do two things. First, you'll change the Route 53 resource record to redirect users to the ALB in the standby region. Second, you'll need to promote the read replica in the passive region to be the new primary database instance.

Availability Calculation

Failures requiring manual intervention should be rare and would probably involve either a failure of the active region that requires failover to the passive region or database corruption that requires restoring from a snapshot. Assuming two failures a year, each failure lasting 20 minutes, that amounts to 40 minutes of downtime per year. That equates to just over 99.99 percent availability.

Software and operating system updates require no downtime. When you need to do an update, you simply provision more instances alongside the existing ones. You upgrade the new instances and then gracefully migrate users to the new instances. In the event that an upgrade breaks the application, directing users back to the original instances is trivial and quick.

Summary

When it comes to maximizing availability, avoiding failure in the first place is job one. To avoid catastrophic application failures, run your application on multiple EC2 instances in different availability zones and distribute the load across those instances as much as possible using an application load balancer. If one instance or even an entire availability zone fails, ALB will route users away from the failed instances, and your application will continue to be available.

Another way to avoid application failures is to not overload individual instances. This is where EC2 Auto Scaling comes in. By implementing dynamic scaling policies, you can ensure you always have enough instances to handle increased demand.

But failures are inevitable, so the next highest priority is to recover from them quickly when they occur. When you create an Auto Scaling group, Auto Scaling will ensure you always have a minimum number of healthy instances. When an instance becomes unhealthy, Auto Scaling will terminate and replace it.

One potential cause of application failures is lost or corrupted data. Take advantage of automatic S3 versioning, cross-region replication, and EBS snapshots to keep backups of your application's important data. For the database, use multi-AZ RDS and automated database instance snapshots to keep up-to-date copies of your database stored safely across multiple availability zones.

Exam Essentials

Know how to calculate total availability using both hard dependencies and redundant components. When one resource depends on another, that's a hard dependency. An example would be an application dependent upon a database. When you have resources that don't depend on one another, such as a collection of identical application instances, those components are redundant.

Understand the difference between traditional and cloud-native applications. Traditional applications are written to run on Linux or Windows servers and often use a standard database component such as a SQL server or a nonrelational database like Redis or MongoDB. Cloud-native applications, on the other hand, use compute, database, or networking resources available only in the cloud, such as Lambda and DynamoDB.

Be able to configure EC2 Auto Scaling. Auto Scaling can help you avoid application failures by automatically provisioning new instances when you need them, avoiding instance failures caused by resource exhaustion. When an instance failure does occur, Auto Scaling steps in and creates a replacement.

Know how application load balancers and EC2 Auto Scaling work together to increase application availability. When Auto Scaling creates a new instance, it automatically adds it to the associated Elastic Load Balancing target group. When it removes an instance, Auto Scaling removes the instance from the associated Elastic Load Balancing target group.

Understand the different backup and recovery options for S3, EBS, and EFS. S3 offers versioning and cross-region replication. To back up data stored on EBS volumes, you can take manual or automated EBS snapshots. To back up your EFS filesystem, you can use AWS Backup or you can copy files from one EFS filesystem to another.

Review Questions

1. What's the minimum level of availability you need to stay under 30 minutes of downtime per month?
   A. 99 percent
   B. 99.9 percent
   C. 99.95 percent
   D. 99.999 percent

2. Your application runs on two EC2 instances in one availability zone. An elastic load balancer distributes user traffic evenly across the healthy instances. The application on each instance connects to a single RDS database instance. Assuming each EC2 instance has an availability of 90 percent and the RDS instance has an availability of 95 percent, what is the total application availability?
   A. 94.05 percent
   B. 99 percent
   C. 99.9 percent
   D. 99.95 percent

3. Your organization is designing a new application to run on AWS. The developers have asked you to recommend a database that will perform well in all regions. Which database should you recommend for maximum availability?
   A. Multi-AZ RDS using MySQL
   B. DynamoDB
   C. Multi-AZ RDS using Aurora
   D. A self-hosted SQL database

4. Which of the following can help you increase the availability of a web application? (Choose all that apply.)
   A. Store web assets in an S3 bucket instead of on the application instance.
   B. Use instance classes large enough to handle your application's peak load.
   C. Scale your instances in.
   D. Scale your instances out.

5. You've configured an EC2 Auto Scaling group to use a launch configuration to provision and install an application on several instances. You now need to reconfigure Auto Scaling to install an additional application on new instances. Which of the following should you do?
   A. Modify the launch configuration.
   B. Create a launch template and configure the Auto Scaling group to use it.

C. Modify the launch template.
D. Modify the CloudFormation template.

6. You create an Auto Scaling group with a minimum group size of 3, a maximum group size of 10, and a desired capacity of 5. You then manually terminate two instances in the group. Which of the following will Auto Scaling do?
   A. Create two new instances
   B. Reduce the desired capacity to 3
   C. Nothing
   D. Increment the minimum group size to 5

7. Which of the following can Auto Scaling use for instance health checks? (Choose all that apply.)
   A. ELB health checks
   B. CloudWatch Alarms
   C. Route 53 health checks
   D. EC2 system checks
   E. EC2 instance checks

8. You're running an application that receives a spike in traffic on the first day of every month. You want to configure Auto Scaling to add more instances before the spike begins and then add additional instances in proportion to the CPU utilization of each instance. Which of the following should you implement? (Choose all that apply.)
   A. Target tracking policies
   B. Scheduled actions
   C. Step scaling policies
   D. Simple scaling policies

9. Which of the following provide the most protection against data corruption and accidental deletion for existing objects stored in S3? (Choose two.)
   A. Versioning
   B. Bucket policies
   C. Cross-region replication
   D. Using the Standard storage class

10. You need to maintain three days of backups for binary files stored across several EC2 instances in a spot fleet. What's the best way to achieve this?
    A. Stream the files to CloudWatch Logs.
    B. Create an Elastic File System and back up the files to it using a `cron` job.
    C. Create a Snapshot Lifecycle Policy to snapshot each instance every 24 hours and retain the latest three snapshots.
    D. Create a Snapshot Lifecycle Policy to snapshot each instance every 4 hours and retain the latest 18 snapshots.

11. You plan to run multi-AZ RDS across three availability zones in a region. You want to have two read replicas per zone. Which database engine should you choose?

   A. MySQL

   B. PostgreSQL

   C. MySQL

   D. Aurora

12. You're running an RDS instance in one availability zone. What should you implement to be able to achieve a recovery point objective (RPO) of five minutes?

   A. Configure multi-AZ.

   B. Enable automated snapshots.

   C. Add a read replica in the same region.

   D. Add a read replica in a different region.

13. When creating subnets in a VPC, what are two reasons to leave sufficient space in the VPC for more subnets later? (Choose two.)

   A. You may need to add another tier for your application.

   B. You may need to implement RDS.

   C. AWS occasionally adds more availability zones to a region.

   D. You may need to add a secondary CIDR to the VPC.

14. You plan to deploy 50 EC2 instances, each with two private IP addresses. To put all of these instances in a single subnet, which subnet CIDRs could you use? (Choose all that apply.)

   A. 172.21.0.0/25

   B. 172.21.0.0/26

   C. 10.0.0.0/8

   D. 10.0.0.0/21

15. You're currently connecting to your AWS resources using a 10 Gbps Internet connection at your office. You also have end users around the world who access the same AWS resources. What are two reasons you may consider using Direct Connect in addition to your Internet connection? (Choose two.)

   A. Lower latency

   B. Higher bandwidth

   C. Better end-user experience

   D. Increased security

16. Before connecting a VPC to your data center, what must you do to ensure proper connectivity?

   A. Use IAM policies to restrict access to AWS resources.

   B. Ensure the IP address ranges in the networks don't overlap.

C. Ensure security groups on your data center firewalls are properly configured.

D. Use in-transit encryption.

17. You plan to run a stand-alone Linux application on AWS and need 99 percent availability. The application doesn't require a database, and only a few users will access it. You will occasionally need to terminate and re-create the instance using a different AMI. Which of the following should you use? (Choose all that apply.)

   A. CloudFormation
   B. Auto Scaling
   C. User data
   D. Dynamic scaling policies

18. You need eight instances running simultaneously in a single region. Assuming three availability zones are available, what's the minimum number of instances you must run in each zone to be able to withstand a single zone failure?

   A. 3
   B. 16
   C. 8
   D. 4

19. If your application is down for 45 minutes a year, what is its approximate availability?

   A. 99 percent
   B. 99.9 percent
   C. 99.99 percent
   D. 99.95 percent

20. You're running an application in two regions and using multi-AZ RDS with read replicas in both regions. Users normally access the application in only one region by browsing to a public domain name that resolves to an elastic load balancer. If that region fails, which of the following should you do to fail over to the other region? (Choose all that apply.)

   A. Update the DNS record to point to the load balancer in the other region.
   B. Point the load balancer to the other region.
   C. Failover to the database in the other region.
   D. Restore the database from a snapshot.

Chapter 11

The Performance Efficiency Pillar

THE AWS CERTIFIED SOLUTIONS ARCHITECT ASSOCIATE EXAM OBJECTIVES COVERED IN THIS CHAPTER MAY INCLUDE, BUT ARE NOT LIMITED TO, THE FOLLOWING:

✓ **Domain 1: Design Resilient Architectures**

- 1.1 Design a multi-tier architecture solution
- 1.2 Design highly available and/or fault-tolerant architectures
- 1.3 Design decoupling mechanisms using AWS services
- 1.4 Choose appropriate resilient storage

✓ **Domain 2: Design High-Performing Architectures**

- 2.1 Identify elastic and scalable compute solutions for a workload
- 2.2 Select high-performing and scalable storage solutions for a workload
- 2.3 Select high-performing networking solutions for a workload
- 2.4 Choose high-performing database solutions for a workload

Introduction

By this point in the book, you should have a pretty good feel for the way the basic AWS services are used to deploy and manage your applications. And, from reading Chapter 10, "The Reliability Pillar," you should also now understand how to respond to failures.

This chapter will teach you the design principles that'll help you squeeze the most performance possible out of your cloud resources. It's not enough to just make sure everything is running: getting full value from AWS requires that you leverage the features that set the cloud apart from legacy environments. For the purposes of this chapter, those will include the ability to:

- Quickly launch test environments to experiment and measure performance.
- Quickly and automatically scale operations up and down to meet changing demand.
- Abstract your operations by using managed services to automate infrastructure provisioning.
- Automate the use of container and serverless technologies in response to external events.
- Automate the deployment of complex, full-stack infrastructure through scripts and templates.
- Understand and optimize operations based on the smart consumption of system data.

Optimizing Performance for the Core AWS Services

Through the book's earlier chapters, you experienced the power and flexibility of Amazon's compute, storage, database, and networking services. You've seen how to tune, launch, and administer those tools to support your applications, and you're aware of some of the many configuration options that come built-in with the platform.

In this section, you'll take all that knowledge to the next level. You'll explore the ways you can apply some key design principles to solve—and prevent—real-world performance problems.

Compute

What, ultimately, is the goal of all "compute" services? To quickly and efficiently apply *some kind of* compute resource to a workload demand. The key there is "some kind of," as it acknowledges that not every problem is best addressed by the same kind of compute resource. To get it right, you'll need to put together just the right blend of functionality and power.

EC2 Instance Types

As you saw in Chapter 2, "Amazon Elastic Compute Cloud and Amazon Elastic Block Store," the performance you'll get from a particular EC2 instance type is defined by a number of configuration variables. Your job, when building your deployment, is to match the instance type you choose as closely as possible to the workload you expect to face. Of course, you can always change the type of a running instance if necessary or just add more of the same.

Table 11.1 lists the configuration parameter categories available to help you select the instance type that's most compatible with your workload. Parameters define the resources allocated to an instance built from a particular type and the hardware capabilities of the underlying hardware.

TABLE 11.1 Instance type parameter descriptions

Instance parameter	Description
ECUs	EC2 compute units—useful for comparing the compute power of one instance type to another
vCPUs	The number of virtual CPUs allotted an instance built from this instance type
Physical Processor	The processor family (e.g., Intel Xeon E52676v3) used by the host server
Clock Speed	The clock speed used by a host server
Memory	The amount of RAM allotted to an instance built from this instance type
Instance Storage	The size of the local (ephemeral) instance store volume used by this instance type

TABLE 11.1 Instance type parameter descriptions *(continued)*

Instance parameter	Description
EBS-Optimized Available	Whether such an instance can be configured to use EBS optimization for dedicated I/O throughput
Network Performance	The data transfer rate allowed to instances built from this instance type
IPv6 Support	Whether IPv6 addressing support is available
Processor Architecture	Whether the hardware server uses a 32- or 64-bit processor
Intel AES-NI	Whether the hardware host uses the Advanced Encryption Standard-New Instructions (AES-NI) encryption instruction set
Intel AVX	Whether the hardware host uses a floating-point intensive instruction set for improved graphics/analytics performance
Intel Turbo	Whether the hardware host can benefit from short-term performance boosts

Auto Scaling

Can't get everything you need from a single EC2 instance? Then feel free to scale out. Scaling out (or scaling *horizontally* as it's sometimes called) involves supporting application demand by adding new resources that will run identical workloads in parallel with existing instances.

A typical scale-out scenario would involve a single EC2 instance hosting an e-commerce web server that's struggling to keep up with rising customer demand. The account administrators could take a snapshot of the EBS volume attached to the existing instance and use it to generate an image for a private EC2 AMI. They could then launch multiple instances using the AMI they've created and use a load balancer to intelligently direct customer traffic among the parallel instances. (You'll learn more about load balancing a bit later in this chapter.)

That's scaling. *Auto Scaling* is an AWS tool that can be configured to automatically launch or shut down instances to meet changing demand. As demand rises, consuming a predefined percentage of your instance's resources, EC2 will automatically launch one or more exact copies of your instance to share the burden. When demand falls, your deployment will be scaled back to ensure there aren't costly, unused resources running.

Auto Scaling groups are built around *launch configurations* that define the kind of instance you want deployed. You could specify a custom EC2 AMI preloaded with your

Optimizing Performance for the Core AWS Services

application (the way you saw in Chapter 2), or you could choose a standard AMI that you provision by passing user data at launch time. Exercise 11.1 demonstrates how you can build your own Auto Scaling group.

EXERCISE 11.1

Configure and Launch an Application Using Auto Scaling

1. From the EC2 Dashboard, create a launch configuration using the Ubuntu Server LTS AMI on the Quick Start tab and a t2.micro instance type.

2. On the Create Launch Configuration page, name your configuration. You don't need to select an IAM role or enable monitoring for this demo.

3. Save the following commands to a file on your local computer, name the file **start .sh**, expand the Advanced Details section, click the As File radio button, and select your start.sh script from your computer. This will install the Apache web server and create a simple index.html web page.

   ```
   #!/bin/bash
   apt-get update
   apt-get install -y apache2
   echo "Welcome to my website" > index.html
   cp index.html /var/www/html
   ```

4. Choose (or create) a security group that will permit all HTTP traffic via port 80 and then create the configuration, making sure to select a valid key pair so that you'll be able to use SSH to log into the instance later if necessary.

5. Create an Auto Scaling group that will use your new launch configuration. Give your group a name and leave the Group Size value at 1 instance.

6. Select a virtual private cloud (VPC) into which your instances will launch, click once inside the Subnet field, and select the subnet where you'd like your instances to live. Assuming you're working with your default VPC, any one of the public subnets you'll be shown should be fine.

7. On the Configure Scaling Policies page, select the "Use scaling policies to adjust the capacity of this group" radio button and edit the value to read **Scale between 1 and 2 instances**.

8. Leave the Metric Type value as Average CPU Utilization and edit its Target Value to 5 percent.

   This unusually low value will make it easier to confirm that the auto scaling is working—normally a value of 70 percent to 80 percent would make sense.

EXERCISE 11.1 *(continued)*

9. For this demo, you don't need to configure notifications or tags. Click Create to finish creating the group.

10. Your group will immediately try to start the first instance. It may take a few minutes for everything to load. When it does, point your browser to the IP address associated with the instance (you can retrieve that from the EC2 Instances Dashboard).

    Considering how small the index.html page is, it might be hard to get the Auto Scaling tool to launch a second instance. You'll probably need to "cheat" by running a busywork command such as the following on the server command line to raise your CPU level high enough:

    ```
    $ while true; do true; done
    ```

> **Note:** Normally, you would select the "Receive traffic from one or more load balancers" option from the Configure Auto Scaling group details page so that your customers would be able to access all the instances launched from the same URL. For this demo, we'll just pretend there's a balancer in place.

Serverless Workloads

You can build elastic compute infrastructure to efficiently respond to changing environments without using an EC2 instance. Container tools like Docker or code-only functions like Lambda can provide ultra-lightweight and "instant-on" compute functionality with far less overhead than you'd need to run a full server.

Container management tools like the Amazon Elastic Container Service—and its front-end abstraction AWS Fargate—can maximize your resource utilization, letting you tightly pack multiple fast-loading workloads on a single instance. AWS container environments—which include the Kubernetes-based EKS service—are also designed to be scriptable so that their administration can easily be automated.

Unlike EC2 instances or ECS containers, AWS Lambda works without the need to provision an ongoing server host. Lambda functions run code for brief spurts (currently up to 15 minutes) in response to network events. Lambda functions can be closely integrated with other AWS services, including by exposing your function to external application programming interface (API) requests using Amazon API Gateway.

Table 11.2 describes some use cases that are good fits with each of the three compute technology categories.

TABLE 11.2 Common use cases for compute categories

Technology	Use cases
EC2 instances	Long-lived, more complex web applications
	Processes requiring in-depth monitoring and tracking
ECS containers	Highly scalable, automated applications
	Applications requiring full control over underlying resources
	Microservices deployments
	Testing environments
Lambda functions	Retrieving data from backend databases
	Parsing data streams
	Processing transactions

Storage

You've already seen how Elastic Block Store (EBS) volumes provide durable storage where IOPS, latency, and throughput settings determine your data access performance (see Chapter 2). You also know about the scalable and multitiered object storage available through S3 and Glacier (see Chapter 3, "AWS Storage"). Chapter 3 is also where you encountered the Amazon Elastic File System (EFS), Amazon FSx for Windows File Server, and Amazon FSx for Lustre, which offer shared file storage for instances within a VPC or over an AWS Direct Connect connection.

Here, I'll discuss some performance optimizations impacting cloud storage and the services that use it.

RAID-Optimized EBS Volumes

You can optimize the performance and reliability of your disk-based data using software versions of a redundant array of independent disks (RAID). This is something admins have been doing in the data center for many years, but some RAID configurations can also make a big difference when applied to EBS disks in the AWS cloud (although on AWS they're managed from within the OS).

RAID works by combining the space of multiple drives into a single logical drive and then spreading and/or replicating your data across the entire array. The non-redundant *RAID 0* standard involves *data striping*, where data is segmented and stored across more than one device, allowing concurrent operations to bypass a single disk's access limitations. This can greatly speed up I/O performance for transaction-heavy operations like busy databases. RAID 0 does not add durability.

RAID 1 creates mirrored sets of data on multiple volumes. Traditional RAID 1 arrays don't use striping or parity, so they won't provide improved performance, but they do make your data more reliable since the failure of a single disk would not be catastrophic. RAID 1 is common for scenarios where data durability is critical and enhanced performance is not necessary.

RAID 5 and RAID 6 configurations can combine both performance and reliability features, but because of IOPS consumption considerations, they're not recommended by AWS. AWS also does not recommend using RAID-configured volumes as instance boot drives since the failure of any single drive can leave the instance unbootable.

Setting up a RAID array involves creating at least two EBS volumes of the same size and IOPS setting and attaching them to a running instance. You then use traditional OS tools on the instance—mdadm on Linux and diskpart on Windows—to configure the volumes for RAID. Make sure that the EC2 instance you're working with has enough available bandwidth to handle the extra load.

S3 Cross-Region Replication

Even though S3 is a global service, the data in your buckets has to exist within a physical AWS region. That's why you need to select a region when you create new buckets. Now, since your data is tied to a single region, there will be times when, for regulatory or performance considerations, you'll want data "boots" on the ground in more than one place. Replication can be used to minimize latency for geographically dispersed clients or to increase reliability and durability.

When configured, S3 Cross-Region Replication (CRR) will automatically (asynchronously) sync contents of a bucket in one region with a bucket in a second region. You set up CRR by creating a *replication* rule for your source bucket (meaning the bucket whose contents will be replicated). You can choose to have all objects in your source replicated or only those whose names match a specified prefix.

The replication rule must also define a destination bucket (meaning the bucket to which your replicated objects will be sent). The destination can be a bucket in a different region in your account, or even one from a different account.

Once enabled, copies of objects saved to the source bucket will be created in the destination bucket. When an object is deleted from the source bucket, it will similarly disappear from the destination (unless versioning is enabled on the destination bucket). Exercise 11.2 will guide you through the process of syncing two buckets through CRR.

EXERCISE 11.2

Sync Two S3 Buckets as Cross-Region Replicas

1. Use the S3 Console to create two buckets in two separate regions. One will be your source and the other your destination.

2. Click the name of your source bucket in the S3 Console, then click the Management tab, then the Replication button, and finally click Add Rule. If versioning wasn't already enabled for the bucket, you can click the Enable Versioning button to set it.

3. Define the objects you want replicated from your source bucket. Choose Entire Bucket for this demo.

4. In the Destination definition page, click once in the Select Bucket field and select the second bucket you created as the destination. Note how you can have the storage class and ownership of your objects automatically changed when copies are created in the destination bucket. Leave those options unchecked for this demo.

5. Select Create New Role in the IAM Role dialog box. This will create the permissions necessary to move assets between buckets. You can then review your settings and save the rule.

6. Confirm the replication is working by uploading a file to the source bucket and then checking to see that a copy has been created in the destination bucket.

Amazon S3 Transfer Acceleration

If your team or customers often find themselves transferring larger files between their local PCs and S3 buckets, then you might benefit from faster transfer rates. S3 Transfer Acceleration adds a per-gigabyte cost to transfers (often in the $0.04 range), but by routing data through CloudFront edge locations, they can speed things up considerably.

To find out whether transfer acceleration would work for your use case, enter the following URL in your browser, specifying an existing bucket name and its AWS region:

s3-accelerate-speedtest.s3-accelerate.amazonaws.com/
en/accelerate-speed-comparsion.html?region=us-east-1
&origBucketName=*my-bucket-name*

You'll land on a Speed Comparison page that will, over the course of a few minutes, calculate the estimated difference between speeds you'll get using S3 Direct (the default S3 transfer process) and S3 Transfer Acceleration. If Transfer Acceleration makes sense for you, you'll need to enable it for your bucket. Exercise 11.3 shows you the AWS CLI command that will do that for you.

> **EXERCISE 11.3**
>
> **Upload to an S3 Bucket Using Transfer Acceleration**
>
> 1. Enable S3 Transfer Acceleration on an existing bucket (substitute the name of your bucket for *my-bucket-name*).
>
>    ```
>    $ aws s3api put-bucket-accelerate-configuration \
>        --bucket my-bucket-name \
>        --accelerate-configuration Status=Enabled
>    ```
>
> 2. Transfer a file by specifying the `s3-accelerate` endpoint. Make sure you're using the correct file and bucket names and region.
>
>    ```
>    $ aws s3 cp filename.mp4 s3://my-bucket-name \
>        --region us-east-1 \
>        --endpoint-url http://s3-accelerate.amazonaws.com
>    ```

CloudFront and S3 Origins

Besides the static website hosting you explored in Chapter 3, S3 is also an excellent platform for hosting a wide range of files, media objects, resources (like EC2 AMIs), and data. Access to those objects can be optimized using CloudFront.

One example might involve creating a CloudFront distribution that uses an S3 bucket containing videos and images as an origin rather than a load balancer sitting in front of EC2 instances. Your EC2-based application could point to the distribution to satisfy client requests for media files, avoiding the bottlenecks and extra costs associated with hosting them locally.

Your job as a solutions architect is to assess the kinds of data your deployments will generate and consume and where that data will be most efficiently and affordably kept. More often than you might at first imagine, the answer to that second question will be S3. The trick is to keep an open mind and be creative.

Database

When designing your data management environment, you'll have some critical decisions to make. For instance, should you build your own platform on an EC2 instance or leave the heavy infrastructure lifting to AWS by opting for Amazon's Relational Database Service (RDS)?

Doing it yourself can be cheaper and permits more control and customization. But on the other hand, RDS—being a fully managed service—frees you from significant administration worries such as software updates and patches, replication, and network accessibility. You're responsible only for making basic configuration choices, like these:

- Choosing the right RDS instance type
- Optimizing the database through intelligently designing schemas, indexes, and views

- Optionally configuring option groups to define database functionality that can be applied to RDS instances
- Optionally configuring parameter groups to define finely tuned database behavior controls that can be applied to RDS instances

Building your own database solutions using EC2 resources gives you control over a much wider range of options, including the following:

Consistency, Availability, and Partition Tolerance   These are the elements that make up the CAP theorem of distributed data. To prevent data corruption, distributed data store administrators must choose between priorities, because it's impossible to maintain perfect consistency, availability, and reliability concurrently.

Latency   The storage volume type you choose—with its IOPS class and architecture (SSD or magnetic)—will go a long way to determine the read and write performance of your database.

Durability   How will you protect your data from hardware failure? Will you provision replicated and geographically diverse instances? This was discussed at length in Chapter 10, "The Reliability Pillar."

Scalability   Does your design allow for automated resource growth?

Nondatabase Hosting   Data can sometimes be most effectively hosted on nondatabase platforms such as AWS S3, where tools like the Amazon Redshift Spectrum, Athena, and Elastic Map Reduce (EMR) data analysis engines can effectively connect and consume your data.

Amazon Redshift (which we discussed in Chapter 5, "Database Services") isn't the only available data warehouse tool. Various integrations with third-party data management tools are commonly deployed on AWS for highly scalable SQL operations. Table 11.3 describes some popular solutions.

TABLE 11.3   Third-party data warehousing and management tools

Technology	Role	Platform
Pivotal Greenplum Database	Massively parallel data warehouse for big data analytics and data science	Available as an EC2 marketplace AMI
Snowflake	Petabyte-scale data warehouse SaaS designed from the ground up for cloud environments	Available through the AWS marketplace
Presto	Distributed Hadoop-based SQL query engine for weakly structured data analytics	Deploy through Amazon Athena or Amazon EMR

Network Optimization and Load Balancing

Cloud computing workloads live and die with network connectivity. So, you've got a keen interest in ensuring your resources and customers are able to connect with and access your resources as quickly and reliably as possible. The geolocation and latency-based routing provided by Route 53 and CloudFront are important elements of a strong networking policy, as are VPC endpoints and the high-speed connectivity offered by AWS Direct Connect.

High-bandwidth EC2 instance types can also make a significant difference. When manually enabled on a compatible instance type, enhanced networking functionality can give you network speeds of up to 100 Gbps. There are three enhanced networking technologies: Intel 82599 Virtual Function (VF) interface, Elastic Network Adapter (ENA), and Elastic Fabric Adapter (EFA).

Each technology is available on only certain instance types. The following sites provide details and clear instructions for enabling enhanced networking from within various flavors of the Linux server OS:

docs.aws.amazon.com/AWSEC2/latest/UserGuide/enhanced-networking.html

docs.aws.amazon.com/AWSEC2/latest/UserGuide/efa.html

But perhaps the most important network enhancing technology of them all is load balancing. Sooner or later the odds are that you'll find yourself running multiple instances of your application to keep up with demand. Adding resources to help out a popular app is great, but how are your customers going to find all those new servers? Won't they all still end up fighting their way into the poor, overworked server they've always used?

A load balancer is a software service that sits in front of your infrastructure and answers all incoming requests for content. Whether you use a DNS server like Route 53 or CloudFront, the idea is to associate a domain (like example.com) with the address used by your load balancer rather than to any one of your application servers.

When the balancer receives a request, it will, as illustrated in Figure 11.1, route it to any one of your backend application servers and then make sure that the response is sent back to the original client. When you add or remove servers, you only need to update the load balancer software with the addressing and availability status changes, and your customers won't be any the wiser.

A load balancer automates much of the process for you. Because the service is scalable, you don't need to worry about unexpected and violent changes in traffic patterns, and you can design complex relationships that leverage the unique features of your particular application infrastructure.

Originally, a single EC2 load balancer type would handle HTTP, HTTPS, and TCP workloads. That Classic load balancer is still available, but it's been deprecated. Besides the Classic, the Elastic Load Balancing service now offers two separate balancer types: the *application load balancer* for HTTP and HTTPS and the *network load balancer* for TCP traffic. The new balancer types come with a number of functions unavailable to the Classic version, including the ability to register targets outside the VPC and support for containerized applications.

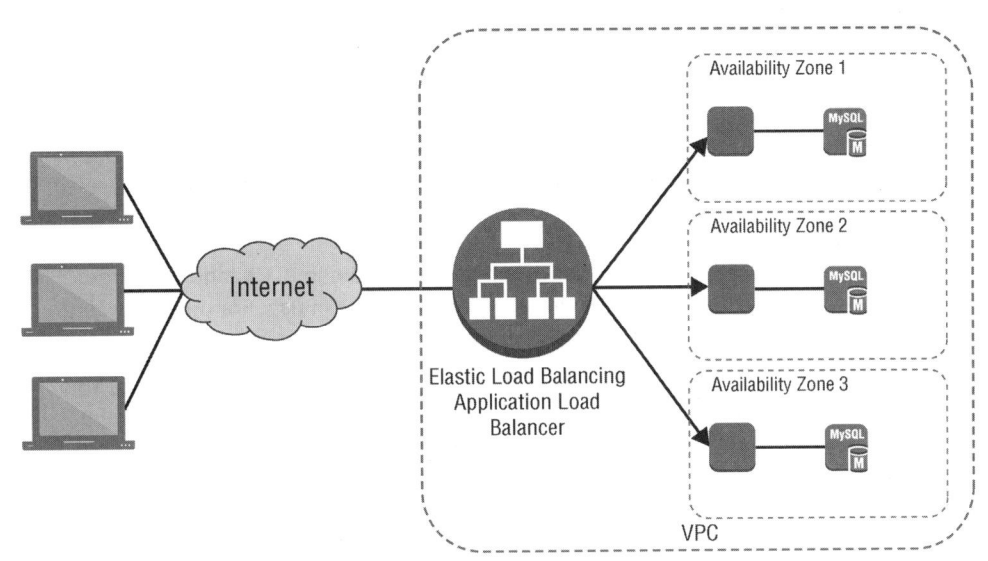

FIGURE 11.1 The data flow of a typical load balancing operation

The application load balancer operates at the application layer (layer 7 of the Open Systems Interconnection [OSI] model). Layer 7 protocols permit host and path-based routing, which means you can use this load balancer to route traffic among the multiple services of microservice or other tiered architectures.

The network load balancer functions at layer 4 (the transport layer) and manages traffic based on TCP port numbers. Network load balancers are built to manage high volumes and traffic spikes and are tightly integrated with Auto Scaling, the Elastic Container Service, and CloudFormation. Network load balancers should also be used for applications that don't use HTTP or HTTPS for their internal communications and for AWS-managed virtual private network (VPN) connections.

Try it for yourself. Exercise 11.4 will guide you through configuring and deploying a simple load balancer using EC2.

EXERCISE 11.4

Create and Deploy an EC2 Load Balancer

1. Create one instance in each of two subnets. You select the subnets on the Configure Instance Details page. Make a note of the subnet designations you choose.

2. Use SSH to access your instances, install the Apache web server, and create unique index.html files containing a short note describing the server—perhaps including the server's IP address. This will make it easier to know which instance your load balancer has loaded when you test everything out later.

EXERCISE 11.4 *(continued)*

3. From the EC2 Dashboard, create a load balancer. Select an application load balancer, give it a name, and specify the Internet-facing scheme, IPv4 for the IP address type, and the default HTTP and port 80 values for the listener. Make sure you're using the right VPC (the default should be fine) and select (and note) two availability zones. For this demo, you can ignore the warning about using an unencrypted listener.

4. Select (or create) a security group that allows access to port 80.

5. When prompted, create a new target group and give it a name. Change the Target type value to IP and add a path to a file the health check can use. For this demo, enter /index.html.

6. Register your two EC2 instances as targets to the group using their private IPs. You can retrieve those private IP addresses from the EC2 Instances console or from the command prompt of the instance itself using the ip addr command.

7. Once your balancer is active (which can take a couple of minutes), get the DNS name of your balancer from the Load Balancers console and paste it into your browser. The index.html page of one of your instances should load. Refresh the browser a few times so that you can confirm that both instances are being accessed.

8. If you like, you can shut down one of the instances without causing any interruption to the load balancer service.

Infrastructure Automation

One of the biggest benefits of virtualization is the ability to define your resources within a script. Every single service, object, and process within the Amazon cloud can be managed using textual abstractions, allowing for an infrastructure as code (IaC) environment.

You've already seen this kind of administration through the many AWS CLI examples used in this book. But building on the CLI and SDK foundations, sophisticated operations can be heavily automated using both native AWS and third-party tools.

CloudFormation

You can represent infrastructure resource stacks using AWS CloudFormation. CloudFormation templates are JSON or YAML-formatted text files that define the complete inventory of AWS resources you would like to run as part of a particular project.

CloudFormation templates are easy to copy or export to reliably re-create resource stacks elsewhere. This lets you design an environment to test a beta version of your application and then easily create identical environments for staging and production. Dynamic change can be incorporated into templates by parameterizing key values. This way, you could specify, say, one VPC or instance type for your test environment and a different profile for production.

Templates can be created in multiple ways:

- Using the browser-based drag-and-drop interface
- Working with prebuilt sample templates defining popular environments such as a LAMP web server or WordPress instance
- Manually writing and uploading a template document

A CloudFormation *stack* is the group of resources defined by your template. You can launch your stack resources by loading a template and creating the stack. AWS will report when the stack has successfully loaded or, in the event of a failure, will roll back changes.

You can update a running stack by submitting a modified version of a template. Deleting a stack will shut down and remove its resources. Try a CloudFormation on your own with Exercise 11.5.

EXERCISE 11.5

Launch a Simple CloudFormation Template

1. From the CloudFormation page, click Create Stack. (Make sure there's a valid key pair available in your selected region.)
2. Click the Select A Sample template radio button, click once inside the field, and select the LAMP Stack option.
3. Click the View In Designer link and spend a few minutes reading through the template text in the bottom panel to get a feel for how all the elements are organized. Values such as the password fields will be automatically populated in later steps.
4. When you're done exploring, click the Create Stack icon at the top of the page, which will take you back to the Select Template page. Click Next in the Create stack page you'll be sent to.
5. Provide a stack name, database name, two passwords, and a DBUser name.
6. Select an instance type and provide the name of an existing, valid EC2 key pair so that you'll have SSH access. Then click Next.
7. The defaults from the Options page are all fine for this demo. Click Next, review your settings, and click Create.
8. It will take a minute or two to launch the resources. While you're waiting, you can view progress in the CloudFormation dashboard. Assuming your stack is selected, the Outputs tab in the dashboard will include a website URL through which you can access the site's public page. The first time you follow the link, you'll probably see the PHP site configuration page. If you plan to leave this site running, make sure that page does not remain publicly available.

EXERCISE 11.5 (continued)

9. Click the other tabs in the dashboard, including Template and Parameters (where you'll see a list of the current parameter values).

10. When you're done, make sure you delete your stack (from the Actions pull-down menu) so that you aren't billed for any services you're not using.

Third-Party Automation Solutions

Of course, you can also create your own infrastructure automation scripts using Bash or Windows PowerShell. Such scripts can connect directly to your AWS account and launch, manipulate, and delete resource stacks.

Besides Bash and PowerShell scripts, third-party configuration management tools such as Puppet, Chef, and Ansible can be used to closely manage AWS infrastructure. Beyond simply defining and launching resource stacks, these managers can also actively engage with your deployments through operations such as configuration enforcement (to continuously ensure that your resources aren't drifting away from their configuration goals), version control (to ensure that your code and software packages are kept up-to-date), and change management.

The potential value of configuration management tools for cloud deployments is so obvious that AWS has a managed service dedicated entirely to integrating Chef and Puppet tools: AWS OpsWorks.

AWS OpsWorks: Chef

Using OpsWorks, you can build an infrastructure controlled by Chef by adding a *stack* (a container for EC2 instances and related resources) and one or more *layers*. Layers are the definitions that add functionality to your stack. A stack might include a Node.js app server layer along with a load balancer, an Amazon EC2 Container Service (ECS) cluster, or an RDS database instance. Chef recipes for managing your infrastructure can also be added as layers.

OpsWorks Chef deployments also include one or more *apps*. An app points to the application code you want to install on the instances running in your stack along with some metadata providing accessibility to a remote code repository.

AWS OpsWorks: Puppet

Firing up OpsWorks for Puppet Enterprise will allow you to launch an EC2 instance as a Puppet Master server, configure the R10k environment and module toolset, and give it access to your remote code repository. Once a server is running and you've logged in using the provided credentials, you'll be ready to begin deploying your application to Puppet nodes.

Your OpsWorks experience will probably be more successful if you're already familiar with administering either Chef or Puppet locally.

Reviewing and Optimizing Infrastructure Configurations

If you want to achieve and maintain high-performance standards for your applications, you'll need solid insights into how they're doing. To get those insights, you'll establish a methodology for monitoring change. Your monitoring should have these four goals:

- Watch for changes to your resource configurations. Whether a change occurred through the intervention of a member of your admin team or through an unauthorized hacker, it's important to find out fast and assess the implications. The AWS Config service—discussed in Chapter 7, "CloudTrail, CloudWatch, and AWS Config"—will give you all the information you need to get started.

- Watch AWS for announcements of changes to its services. Amazon regularly releases updates to existing services and entirely new services. To be sure your current configuration is the best one for your needs, keep up with the latest available features and functionality. One way to do that is by watching the AWS Release Notes page (aws.amazon.com/releasenotes). You can subscribe to update alerts through the AWS Email Preference Center (pages.awscloud.com/communication-preferences.html).

- Passively monitor the performance and availability of your application. Passive monitoring is primarily concerned with analyzing historical events to understand what—if anything—went wrong and how such failures can be avoided in the future. The most important tool for this kind of monitoring is system logs. On AWS, the richest source of log data is CloudWatch, as you learned in Chapter 7.

- Actively test the performance and availability of your application to proactively identify—and fix—potential problems before they can cause any damage. This will usually involve load testing.

Load Testing

One particularly useful form of active infrastructure monitoring is load—or stress—testing. The idea is to subject your infrastructure to simulated workloads and carefully observe how it handles the stress. You can try to devise some method of your own that will generate enough traffic of the right kind to create a useful simulation, but plenty of third-party tools are available to do it for you. Search the AWS Marketplace site (aws.amazon.com/marketplace) for load testing.

In most cases—so you don't prevent your actual users from accessing your application as normal—you should run your stress tests against some kind of test environment that's configured to run at production scale. Orchestration tools like CloudFront or, for containers, Amazon Elastic Container Service (ECS), can be useful for quickly building such an environment. Your application might not respond quite the same way to requests from different parts of the world. So, you should try to simulate traffic from multiple geographical origins into your testing.

Make sure that your testing doesn't break any AWS usage rules. Before launching some kinds of vulnerability or penetration tests against your infrastructure, for instance, you need to request explicit permission from AWS. You can find out more on Amazon's Vulnerability and Penetration Testing page: aws.amazon.com/security/penetration-testing.

Of course, don't forget to set CloudWatch up in advance so you can capture the real-time results of your tests. To get the most out of your testing, it's important to establish performance baselines and consistent key performance indicators (KPIs) so that you'll have an objective set of standards against which to compare your results. Knowing your normal *aggregate cumulative cost per request* or *time to first byte*, for instance, will give important context to what your testing returns.

Visualization

All the active and passive testing in the world won't do you any good if you don't have a user-friendly way to consume the data it produces. As you saw in Chapter 7, you can trigger notifications when preset performance thresholds are exceeded. Having key team members receive Simple Notification Service (SNS) messages or email alerts warning them about abnormal events can be an effective way to make your monitoring data count.

In many cases, however, there's no replacement for good old-fashioned charts and graphs. Some AWS services have their own built-in visualizations. With an instance selected in the EC2 Instances console, as an example, you can click the Monitoring tab in the lower half of the screen and view a dozen or so metrics rendered as data charts.

But the place where you'll get the greatest visualization bang for your buck is in CloudWatch dashboards. CloudWatch lets you create multiple dashboards. As you'll see when you work through Exercise 11.6, a dashboard can contain multiple *widgets*, each of which can be defined to display a customized combination of metrics pulled from individual AWS services. Keeping a well-designed dashboard open in a browser tab where you can check on it from time to time is potentially a useful monitoring strategy.

EXERCISE 11.6

Create a CloudWatch Dashboard

1. So that you'll get a better idea of how things will look, make sure you have at least one EC2 instance running and at least one S3 bucket with some files in it.

2. From the CloudWatch console, click Dashboards, click Create Dashboard, and give your dashboard a name. Select the Line widget type and click Configure.

3. With the All Metrics tab in the bottom half of the screen selected, click EC2 and then click Per-Instance Metrics. Check the boxes for a few metrics associated with your running instance. StatusCheckFailed, NetworkOut, NetworkIn, and CPUUtilization would make for a nice sample. Note how those metrics are now represented in the graph at the top of the screen. When you're satisfied with your selections, click Create Widget.

4. Your first widget will now appear in your dashboard. You can click the widget to display it full-screen. Selecting a particular metric from the legend at the bottom of the graph will show you only that metric, making it easier to understand what the values represent.

5. If your widget is still full-screen, click Close to return to the dashboard.

6. Click Add Widget to create a second widget. This time, select Stacked Area and then Storage Metrics. Select S3 metrics and then the NumberOfObjects and BucketSizeBytes metrics for your bucket. Create the widget.

7. Once you've added all the widgets and metrics you'll need to monitor, click Save Dashboard.

From now on, whenever you browse to CloudWatch, you'll be able to retrieve and view this dashboard with all of its metrics. Bear in mind that, outside of the Free Tier, a single dashboard will cost around $3 per month.

Optimizing Data Operations

Moving data efficiently is pretty much the whole story when it comes to successful applications. Your code may be a thing of great beauty, but if everything slows down while you're waiting for a large data transfer over a weak connection, then it's mostly wasted effort.

Sometimes you can't control the network connection you're working with, and sometimes you can't reduce the volume of data in play. But it's usually possible to apply some clever data management tricks to speed things up noticeably. Data caching, partitioning, compression, and decoupling are good examples of the kinds of tricks you can use, so let's talk about how they work on AWS.

Caching

Does your application provide lots of frequently accessed data like a product catalog or other media that's kept in your S3 buckets or a relational database? So, why not move copies of the data closer to your clients?

A caching solution does just that. If an object is likely to be requested repeatedly, a copy can be written to system memory on a server—or a purpose-built cache database—from which it can be retrieved much more quickly. Subsequent requests for that data will see the cached copy rather than having to wait for the time it would take to read from the source. The cached copy will remain the authoritative version until its maximum *time to live* (TTL) is reached, at which point it will be flushed and replaced by an updated version that's read from the data source.

Amazon ElastiCache

An ElastiCache *cluster* consists of one or more *nodes*. As Figure 11.2 illustrates, a node is a compute instance—built from an EC2 instance type you select—that does the actual work of processing and serving data to your clients. The number of nodes and the particular instance type you choose will depend on the demands your clients are likely to place on the application. Finding the ideal configuration combination that's best for your needs will require some of the same considerations you'd use to provision for any EC2 workload.

FIGURE 11.2 A typical ElastiCache configuration with an ElastiCache cluster containing two cache nodes serving data for frontend EC2 instances

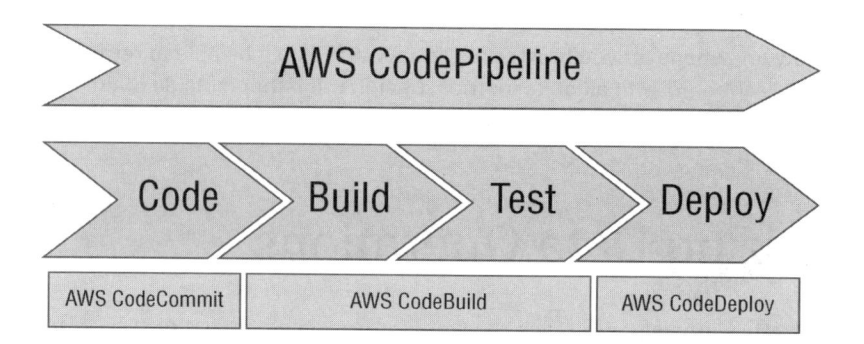

ElastiCache clusters can be created using either one of two engines: Memcached or Redis. Memcached can be simpler to configure and deploy, is easily scalable, and because it works with multiple threads, can run faster. However, because it reads and writes only objects (BLOBs) as in-memory key/value data stores, Memcached isn't always flexible enough to meet the needs of all projects.

Redis, on the other hand, supports more complex data types such as strings, lists, and sets. Data can also be persisted to a disk, making it possible to snapshot data for later recovery. Redis lets you sort and rank data, allowing you to support, say, an online gaming application that maintains a leaderboard ranking the top users. The fact that Redis can persist data makes it possible to use Redis to maintain session caching, potentially improving performance.

Typically, ElastiCache will make the current version of both Redis and Memcached available, along with a few previous versions. This makes it possible to support as wide a range of application needs as possible.

Once your ElastiCache cluster is running, you can retrieve its endpoint from the ElastiCache dashboard and connect to the cluster from your application. For a WordPress instance, for example, this can be as simple as adding a single line to the `wp-config.php` file containing something like the following syntax:

```
define('WP_REDIS_HOST', 'your_cluster_name.amazonaws.com');
```

You can enjoy the benefits of caching without using ElastiCache. For some use cases, you might want to save some money and simply run a reverse proxy like Varnish on your actual EC2 instance.

Other Caching Solutions

Besides caching at the application level, you can also achieve performance gains using broadly similar techniques on the data in your database.

As you saw in Chapter 5, you can add as many as five read replicas to a running database instance in RDS (RDS Aurora instances can have as many as 15). Read replicas are available for RDS MySQL, MariaDB, PostgreSQL, and Aurora. The replica will be an exact copy of the original but will be read-only. By routing some of your traffic to the replica, you lighten the load on the source database, improving response times for both copies. In addition to performance benefits, you can, if necessary, promote a read replica to your primary database should the original fail.

Effectively, a read replica is a kind of cache that leverages relatively cheap storage overheads to provide quick and highly available copies of your data.

Another type of caching—which you saw in Chapter 8—is CloudFront. As you'll remember, a CloudFront distribution can cache copies of S3-based media on edge locations spread around the world and serve requests from the copy located closest to the client.

Partitioning/Sharding

Traffic volume on your RDS database can increase to the point that a single instance simply can't handle it. Although vertical scaling may not be an option, horizontal scaling through partitioning (or sharding) can make a big difference. It's not as simple a solution as the read-only read replicas just discussed, but partitioning can be configured by adding appropriate logic to the application layer so that client requests are directed to the right endpoints.

Sending data in real time through Kinesis streams—using the Amazon DynamoDB Streams Kinesis Adapter—lets you process streams of data records in DynamoDB databases. Kinesis organizes records into shards that contain data along with critical meta information. If one shard hits a limit, the workload can be further distributed by subdividing it into multiple shards.

As it's highly scalable and fully managed, Kinesis streams maintain near-real-time performance and, by extension, can have a significant impact on overall data processing performance.

Compression

If bandwidth is a finite resource and it's not practical to reduce the *volume* of data you're regularly transferring, there's still hope. You can either reduce the *size* of your data or bypass the network with its built-in limits altogether.

How do you reduce data size? By borrowing a solution that, once upon a time, was a necessary part of working with expensive and limited storage space. Are you old enough to remember how we used to fit a full operating system along with our program and work files onto a 10 MB drive? Disk compression.

Storage is no longer a serious problem, but network transfers are. You can, therefore, stretch the value you get from your network connections by compressing your data before sending it. This is something that can be incorporated into your application code. But it's also used directly within some AWS services. CloudFront, for instance, can be configured to automatically compress the files sent in response to client requests, greatly reducing download times.

And bypassing the network? You can ask Amazon to send you a physical petabyte-sized storage device called a *Snowball*. You copy your archives onto the encrypted device and then ship it back to Amazon. Within a short time, your data will be uploaded to buckets in S3.

Summary

The performance and availability of EC2 instances can be managed by choosing the right instance type and by provisioning as many instances as necessary through Auto Scaling. Compute workloads can sometimes be more effectively delivered through serverless (Lambda) or container (Docker/ECS) models.

Access to EBS volume-based data and data reliability can be enhanced by configuring RAID arrays. Cross-region replication and S3 transfer acceleration can improve S3 data administration.

Carefully configuring details such as schemas, indexes, views, option groups, and parameter groups can improve the performance of your RDS databases. For manually hosted databases, latency issues can be addressed through volume class (magnetic versus SSD) and IOPS level; durability is improved through replication; and scalability can be manually configured.

Load balancers listen for incoming requests and distribute them intelligently among all available backend resources. This allows for automated health checks and request routing policies that make the best use of your resources and provide the best end-user experience possible.

Scripting resource provisioning using AWS-native tools like CloudFormation or third-party infrastructure automation tools like Chef and Puppet (working through AWS OpsWorks) can make it easy to quickly launch complicated resource stacks. Doing so can make it possible to maintain multiple on-demand environments for testing, staging, and production.

You can monitor the state and performance of your resources using CloudWatch, CloudTrail, and AWS Config. It's important to incorporate load testing and report visualization (through CloudWatch dashboards) into your administration cycles.

You can decrease data response times by replication through caching, improve data processing performance through partitioning (sharding), and transfer large data stores more quickly through compression.

Exam Essentials

Understand the way EC2 instance types are defined.   Each EC2 instance type is designed to service a different function. You should be familiar with the individual parameters that are used to measure the way each type performs.

Understand Auto Scaling.   The best way to efficiently meet changing demands on your application is to automate the process of provisioning and launching new resources as demand grows and shutting unused resources down as demand drops. Auto Scaling is the way AWS manages this functionality for EC2.

Understand the value of serverless and container computing compared with EC2 instances.   By reducing the underlying resource footprint of your service, serverless functions, such as those offered through AWS Lambda, and clusters of Docker containers (offered through AWS ECS) can provide more responsive and cost-effective compute solutions. You should understand which model will work best for a given use case.

Understand how to optimize storage performance and reliability.   Data stored on EBS volumes (using RAID) or in S3 buckets (using cross-region replication) can be replicated for greater durability and/or performance. You can also enable faster transfers in and out from S3 buckets using S3 Transfer Acceleration or CloudFront.

Know how to optimize database performance.   The performance, reliability, and costs you'll experience running a database on AWS will depend on the choices you make. How should you configure your underlying storage? Should you go with the managed RDS, build your database on an EC2 instance, or choose Redshift or a third-party integration for data warehouse management?

Understand when and how to use a load balancer.   Load balancers direct incoming client requests among multiple backend servers. This allows for scalable and reliable application delivery, because failing or overloaded instances can be automatically replaced or supported by others. Load balancing of one sort or another is a critical component of many—if not most—EC2 and RDS workloads.

Know the value and function of infrastructure automation tools. CloudFormation templates can be used to define and then automatically provision and launch complex resource stacks in multiple environments. Similarly, third-party solutions like Chef and Puppet can be deployed through AWS OpsWorks to script environments. Continuous integration and continuous deployment processes can also be managed through AWS developer tools.

Know the importance of infrastructure monitoring and the tools available for the job. It's critical to stay on top of changes and performance trends within your infrastructure. Active load simulation testing is an important ongoing component of your administration process, as is visualizing the data that's produced using tools like CloudWatch dashboards.

Understand how caching can be used to improve performance. ElastiCache and CloudFront are two services that replicate data and store it to make it quickly available to consuming clients. ElastiCache can use either Memcached or Redis—you should know which will work best for a given workload.

Review Questions

1. Which of the following are parameters used to describe the performance of specific EC2 instance types? (Choose three.)
   A. ECUs (EC2 compute units)
   B. vCPUs (virtual CPUs)
   C. ACCpR (Aggregate Cumulative Cost per Request)
   D. Intel AES-NI
   E. Maximum read replicas

2. As the popularity of your EC2-based application grows, you need to improve your infrastructure so it can better handle fluctuations in demand. Which of the following are normally necessary components for successful Auto Scaling? (Choose three.)
   A. Launch configuration
   B. Load balancer
   C. Custom-defined EC2 AMI
   D. A `start.sh` script
   E. An AWS OpsWorks stack

3. Which of the following best describes the role that launch configurations play in Auto Scaling?
   A. Define the capacity metric that will trigger a scaling change.
   B. Define the AMI to be deployed by Auto Scaling operations.
   C. Control the minimum and maximum number of instances to allow.
   D. Define the associated load balancer.

4. You're considering building your new e-commerce application using a microservices architecture where individual servers are tasked with separate but complementary tasks (document server, database, cache, etc.). Which of the following is probably the best platform?
   A. Elastic Container Service
   B. Lambda
   C. ECR
   D. Elastic Beanstalk

5. Your EC2 deployment profile would benefit from a traditional RAID configuration for the EBS volumes you're using. Where are RAID-optimized EBS volume configurations performed?
   A. From the EBS dashboard
   B. From the EC2 Storage Optimization dashboard

C. From the AWS CLI
D. From within the EC2 instance OS

6. Which of the following tools will provide both low-latency access and resilience for your S3-based data?
    A. CloudFront
    B. RAID arrays
    C. Cross-region replication
    D. Transfer Acceleration

7. Which of the following tools uses CloudFront edge locations to speed up data transfers?
    A. Amazon S3 Transfer Acceleration
    B. S3 Cross-Region Replication
    C. EBS Data Transfer Wizard
    D. EC2 Auto Scaling

8. Your multi-tiered application has been experiencing slower than normal data reads and writes. As you work on improving performance, which of the following is *not* a major design consideration for a managed RDS database?
    A. Optimizing indexes
    B. Optimizing scalability
    C. Optimizing schemas
    D. Optimizing views

9. Which of the following are possible advantages of hosting a relational database on an EC2 instance over using the RDS service? (Choose two.)
    A. Automated software patches
    B. Automated OS updates
    C. Out of the box Auto Scaling
    D. Cost savings
    E. Greater host control

10. You've received complaints from users that performance on your EC2-based graphics processing application is slower than normal. Demand has been rising over the past month or two, which could be a factor. Which of the following is the most likely to help? (Choose two.)
    A. Moving your application to Amazon Lightsail
    B. Switching to an EC2 instance that supports enhanced graphics
    C. Deploying Amazon Elasticsearch in front of your instance
    D. Increasing the instance limit on your Auto Scaling group
    E. Putting your application behind a CloudFront distribution

11. Which of the following load balancer types is optimized for TCP-based applications and preserves the source IP address?
    A. Application load balancer
    B. Classic load balancer
    C. Network load balancer
    D. Dynamic load balancer

12. Which of the following can be used to configure a CloudFormation template? (Choose three.)
    A. The CloudFormation drag-and-drop interface
    B. Selecting a prebuilt sample template
    C. Importing a template from AWS CloudDeploy
    D. Creating your own JSON template document
    E. Importing a template from Systems Manager

13. Which of the following details is not a necessary component of a CloudFormation configuration?
    A. Default node name
    B. Stack name
    C. Database name
    D. DBUser name

14. Which of the following can be integrated into your AWS workflow through AWS OpsWorks? (Choose two.)
    A. Ansible
    B. Chef
    C. Terraform
    D. SaltStack
    E. Puppet

15. Which of the following are important elements of a successful resource monitoring protocol? (Choose two.)
    A. CloudWatch dashboards
    B. CloudWatch OneView
    C. SNS alerts
    D. AWS Config dashboards

16. Which of the following will most enhance the value of the CloudWatch data your resources generate? (Choose two.)
    A. Predefined performance baselines
    B. Predefined key performance indicators (KPIs)

C. Advance permission from AWS
D. A complete record of your account's resource configuration changes
E. A running Service Catalog task

17. Which of the following can be used to audit the changes made to your account and resource configurations?
    A. AWS CloudTrail
    B. AWS CloudWatch
    C. AWS CodePipeline
    D. AWS Config

18. Which of the following caching engines can be integrated with Amazon ElastiCache? (Choose two.)
    A. Varnish
    B. Redis
    C. Memcached
    D. Nginx

19. Which of the following use case scenarios are a good fit for caching using Redis and ElastiCache? (Choose two.)
    A. Your online application requires users' session states to be saved and the behavior of all active users to be compared.
    B. Your online application needs the fastest operation available.
    C. Your admin is not familiar with caching and is looking for a relatively simple setup for a straightforward application performance improvement.
    D. You're not sure what your application needs might be in a month or two, so you want to leave open as many options as possible.

20. Which of the following database engines is not a candidate for read replicas within Amazon RDS?
    A. MySQL
    B. Oracle
    C. MariaDB
    D. PostgreSQL

Chapter 12

The Security Pillar

THE AWS CERTIFIED SOLUTIONS
ARCHITECT ASSOCIATE EXAM
OBJECTIVES COVERED IN THIS CHAPTER
MAY INCLUDE, BUT ARE NOT LIMITED TO,
THE FOLLOWING:

✓ **Domain 3: Design Secure Applications and Architectures**
  - 3.1 Design secure access to AWS resources
  - 3.2 Design secure application tiers
  - 3.3 Select appropriate data security options

Introduction

The primary goal of information security is to protect data, which of course entails protecting the resources that store and provide access to that data. To effectively protect data, you need to ensure three elements of the data:

Confidentiality   The only people or systems that can access data are those authorized to access it. Encryption and access control lists (ACLs) are two common mechanisms for enforcing confidentiality.

Integrity   The data has not been maliciously or accidentally changed. Cryptographic hashing and logging provide two ways to help validate the integrity of data.

Availability   The data is available to those who need it when they need it. Although the data itself may be safe and sound, if an attacker carries out a denial-of-service (DoS) attack to prevent access to that data, then the data is not available to those who need it.

In this chapter, you'll learn how to apply security controls to every system that touches your data—storage, compute, and networking—to ensure the confidentiality, integrity, and availability of your data throughout its lifecycle.

Identity and Access Management

Your AWS credentials let you log into the AWS management console, manage services, view and edit resources, and so on. Security in AWS begins with the foundation of identity, which is managed by the Identity and Access Management (IAM) service (Chapter 6, "Authentication and Authorization—AWS Identity and Access Management"). Because your AWS credentials are the keys to the kingdom, the first order of business is to protect them from accidental exposure and unauthorized use. The second step is to ensure that users have only the permissions they need, and no more.

Protecting AWS Credentials

With some exceptions, if anyone is to interact with your AWS resources, they must be authenticated using a set of credentials. An entity that can take an action on an AWS resource is called a *principal*. You'll sometimes see a principal also referred to as an *identity*. A principal can include the following:

- The root user
- An IAM user
- An IAM role

The root user has unrestricted access to your AWS account, so it's a best practice to avoid using it for routine administrative tasks. Instead, create an IAM user and attach the AdministratorAccess AWS managed policy. For a detailed guide on securing the root user and creating an administrative IAM user, refer to Exercise 6.1 in Chapter 6.

Another crucial step in securing the root user is to enable multifactor authentication (MFA). When you enable MFA, in addition to entering a username and password to log into the root account, you must enter a one-time passcode. This one-time passcode is generated by a physical or virtual MFA device, which is just an application you can install on your smartphone. Use caution when enabling MFA on the root account, because losing your MFA device means being unable to log in as the root user! If this happens, you will need to contact AWS Support to get back in.

For IAM users, you can choose to enforce a password policy to ensure that all IAM users with access to the AWS Management Console have a secure password. A password policy can require any or all of the following:

- Password complexity requirements such as minimum length, the use of lower and upper case, numbers, and nonalphanumeric characters (The minimum password length you can set is six characters.)
- Password expiration
- Preventing password reuse
- Requiring an administrator to reset an expired password

When you create an IAM user, you can choose whether to give that user AWS Management Console access, programmatic access, or both. AWS Management Console access requires a username and password. You may optionally assign an MFA token that the user must possess to sign in. Programmatic access via the AWS API, command-line interface (CLI), or an SDK requires an access key ID and secret access key.

Fine-Grained Authorization

A fundamental concept in information security is the principle of least privilege. You should give IAM principals permissions to only the resources they need and no more. By following this principle, you limit the impact that compromised credentials or malicious users can have on your AWS resources. You achieve tight control over your IAM principals through the use of IAM policies.

IAM principals have no permissions by default. In other words, simply creating an IAM user or role doesn't grant that principal any ability to do anything in your AWS account. IAM users and roles derive their permissions only from IAM policies. An IAM policy is effectively an ACL.

You define the permissions a principal has by associating the principal with one or more policies. A policy contains one or more permission statements. Each permission statement consists of, at minimum, four elements:

Effect   The effect can be allow or deny. The effect controls whether to allow or deny an action on a resource.

Action/Operation   Each AWS service has a set of actions or operations that it can perform on its resources. For example, the EC2 service includes a `RunInstances` action that creates a new instance.

Resource   An action may require you to specify one or more resources. For example, the EC2 `RunInstances` action requires an Amazon Machine Image (AMI) ID. By specifying a resource in a permission statement, you can restrict access on a per-resource basis. For example, you can specify an AMI ID so that the `RunInstances` operation can be performed using only the specified image.

Condition   You may specify conditions that must be met for the permission to be granted. For example, you can allow only users from a particular IP address to perform the specified action. You can also require multifactor authentication or impose time-based restrictions on when an action can be performed.

> An IAM group is not a principal. It's not possible to perform actions under the auspices of a group in the way that you would a role.

AWS Managed Policies

AWS provides hundreds of prepackaged policies called AWS *managed policies*. These cover a variety of permissions typically required by common job roles, including network administrators, database administrators (DBAs), and security auditors. AWS updates these policies periodically to include new services.

Customer-Managed Policies

A *customer-managed policy* is a stand-alone policy that you create and can attach to principals in your AWS account. When you update a customer-managed policy, the changes immediately apply to all the principals that the policy is attached to. Also, IAM doesn't overwrite your existing policy but creates a new version and maintains the last five versions so that you can revert if needed.

Inline Policies

An *inline policy* is a set of permissions that's embedded in an IAM principal or group. Unlike AWS and customer-managed policies that exist independently of any principals, an inline policy is a part of the principal itself. Inline policies are useful when you want to ensure a policy is not inadvertently attached to the wrong principal.

Permissions Boundaries

When you attach a policy to a principal to grant that principal access, the policy is called a *permissions policy*. You can also use a policy to define permissions boundaries. Permissions boundaries let you limit the maximum permissions an IAM principal can be assigned. This can prevent you from accidentally giving a user too many permissions by inadvertently attaching the wrong permissions policy. You set permissions boundaries for a principal by selecting a managed policy that defines the maximum permissions for the principal. For example, you may create the following policy that allows all actions for the EC2 service and then attach that policy to an IAM user as a permissions boundary:

```
{
    "Version": "2012-10-17",
    "Statement": [
        {
            "Effect": "Allow",
            "Action": [
                "ec2:*"
            ],
            "Resource": "*"
        }
    ]
}
```

If you then attach the AdministratorAccess policy—which grants full access to all AWS services—to the user, the user will still only be able to perform actions in EC2. The permissions boundary limits the user to performing only those actions laid out in the permissions boundary policy. Complete Exercise 12.1 to create an IAM user whose permissions are limited by a permissions boundary policy.

EXERCISE 12.1

Create a Limited Administrative User

In this exercise, you'll create an IAM user with limited administrative privileges. You'll use policy boundary permissions to allow the user to perform only actions in the EC2 service.

EXERCISE 12.1 *(continued)*

1. Create a customer managed policy called **LimitedAdminPolicyBoundary**. Populate the policy document with the following content:

   ```
   {
       "Version": "2012-10-17",
       "Statement": [
           {
               "Effect": "Allow",
               "Action": "ec2:*",
               "Resource": "*"
           }
       ]
   }
   ```

2. Create an IAM user with the name of your choice, such as LimitedAdmin.

3. In the IAM Dashboard, click the Users option and then click the username you created.

4. Under Permissions Boundary, click the Set Boundary button.

5. Select the LimitedAdminPolicyBoundary option and click the Set Boundary button.

6. Under Permissions Policy, click Add Permissions.

7. Click the Attach Existing Policies Directly button.

8. Select the AdministratorAccess policy and click the Next: Review button.

9. Click the Add Permissions button.

Even though the AdministratorAccess permissions policy allows full access to all AWS services, the permissions boundary policy overrides that policy and only allows the user to perform actions in EC2.

Roles

A *role* is an IAM principal that doesn't have a password or access key. Like any other IAM principal, a role may have permissions policies and a permissions boundary associated with it. An IAM user or AWS resource can assume a role and inherit the permissions associated with that role.

Roles are particularly useful for allowing applications running on EC2 instances to access AWS resources without having an access key. For example, if you're running an application that requires access to DynamoDB, you can create a role that contains a permissions policy granting the appropriate access to DynamoDB. Such a permissions policy may look like this:

```
{
    "Version": "2012-10-17",
    "Statement": [
        {
            "Effect": "Allow",
            "Action": [
                "dynamodb:CreateTable",
                "dynamodb:PutItem",
                "dynamodb:ListTables",
                "dynamodb:DescribeTable",
                "dynamodb:DeleteItem",
                "dynamodb:GetItem",
                "dynamodb:Query",
                "dynamodb:UpdateItem",
                "dynamodb:UpdateTable"
            ],
            "Resource": "*"
        }
    ]
}
```

Instance Profiles

Recall that in addition to containing permissions policies, a role contains a trust policy that specifies which AWS resource may assume the role. Because the application that needs access to DynamoDB is running on an EC2 instance, the trust policy must grant EC2 permission to assume the role. Such a trust policy would look like this:

```
{
  "Version": "2012-10-17",
  "Statement": [
    {
      "Effect": "Allow",
      "Principal": {
        "Service": "ec2.amazonaws.com"
      },
      "Action": "sts:AssumeRole"
    }
  ]
}
```

> **NOTE** A trust policy is an example of a resource-based policy. When you create a role that grants EC2 access to assume it, IAM automatically creates an instance profile with the same name as the role. For example, suppose you used the previous permissions policy and trust policy documents to create a role named MyAppRole. IAM would automatically create an instance profile also named MyAppRole. You could then associate this instance profile with an EC2 instance, thereby granting the instance permission to assume the role.

To view the instance profile tied to this role, use the AWS CLI to issue the following command:

```
aws iam list-instance-profiles-for-role --role-name MyAppRole
```

You should see something like the following:

```
{
    "InstanceProfiles": [
        {
            "InstanceProfileId": "AIPAJGKR6VR52WTHHXOR6",
            "Roles": [
                {
                    "AssumeRolePolicyDocument": {
                        "Version": "2012-10-17",
                        "Statement": [
                            {
                                "Action": "sts:AssumeRole",
                                "Effect": "Allow",
                                "Principal": {
                                    "Service": "ec2.amazonaws.com"
                                }
                            }
                        ]
                    },
                    "RoleId": "AROAIV2GSBS4E4HAEFTJA",
                    "CreateDate": "2018-10-13T03:20:46Z",
                    "RoleName": "MyAppRole",
                    "Path": "/",
                    "Arn": "arn:aws:iam::xxxxxxxxxxxx:role/MyAppRole"
                }
            ],
```

```
            "CreateDate": "2018-10-13T03:20:46Z",
            "InstanceProfileName": "MyAppRole",
            "Path": "/",
            "Arn": "arn:aws:iam::xxxxxxxxxxxx:instance-profile/MyAppRole"
        }
    ]
}
```

By associating an instance with an instance profile, the instance is able to acquire temporary credentials—an access key, a secret key, and a session token—using the Security Token Service. The Security Token Service places the temporary credentials in the instance metadata and updates them about every six hours. To view the temporary credentials granted by a role named MyAppRole, you could theoretically browse to the following URL from within the instance:

```
http://169.254.169.254/latest/meta-data/iam/
security-credentials/MyAppRole
```

And you would see something like the following:

```
{
  "Code" : "Success",
  "LastUpdated" : "2018-10-14T20:47:19Z",
  "Type" : "AWS-HMAC",
  "AccessKeyId" : "ASIASJ6WQJMEE32SJ56C",
  "SecretAccessKey" : "vcAiY5Tps6U3pbr2TFjzwrrRe2ETXbi0T+mDr1Qi",
  "Token" :
"FQoGZXIvYXdzEBYaDLwVhhKEzhxBvNtRMyK3A7RpKrXELCv61rGatSWBi1Ehg3w9gOBww7jjy9m
CAwTK7kA4S
IyhmyEXQR32McB4xWqjxM3/K4Ij9o5+7ALpegD5p5c0cO7BqGIb4Xb3vZcJiA0pMk7jWRb6afB8c
+iAdP1PhRE
R8oJpAOmUHwC2NoT85tpUbAwPtJc4SCW9HmFvI3Hq5pBOo2c1gB75SdcmaYMR/Xl+HxkV/KM5tqyx
64BypS4uA
ByW9BuoQ5GH+WKHBXOzIFuhVDpvKS2XXaoOOfz/dfdLo/t1n13KkhzXf43NFc4Lunqsd4Zo9o7yr
2D+ezXNLPD
phRN3Itc9dxSaCZY7QE51fgdDUCPBPsQ17okukrcT5jI/R+rY3cL/bBxOQ4VUd47bUcASRxCag2xv
DONMAqDpb
PX4j2Kbgs8avLqEFj4q4RkCOM28gETeqWxEE8XNZEfpXCupr1eeyfPul3BzcZmrTMu22ZvvySyzYJ
QVf9Yijpg
Wa9RcGBFQGKbWAgu5aWxdJvKjDCDjkupyhi2tnBRlRuRXgtXXN19NkDEiVus7rAnZLMRuBIIgbeWt
T6BXSMMjt
HqZ6NpaagDwGHtNmvOv6AEondaO3gU=",
  "Expiration" : "2018-10-15T03:21:21Z"
}
```

Notice that the access key ID, secret access key, and session token are in plain text. These credentials are what an application running on the instance would use to authenticate to the AWS API to perform operations. It is possible for a malicious user or malware to take these credentials and use them to gain unauthorized access. For example, a user may use the preceding credentials to attempt to enumerate resources using the AWS CLI:

```
[ec2-user@ip-172-31-88-201 ~]$ export
AWS_ACCESS_KEY_ID=ASIASJ6WQJMEE32SJ56C
[ec2-user@ip-172-31-88-201 ~]$ export
AWS_SECRET_ACCESS_KEY=vcAiY5Tps6U3pbr2TFjzwrrRe2ETXbi0T+mDr1Qi
[ec2-user@ip-172-31-88-201 ~]$ export
AWS_SESSION_TOKEN=FQoGZXIvYXdzEBYaDLwVhhKEzhxBvNtRMyK3A7RpKrXELCv61rGatSWBi1Eh
g3w9g0Bw
w7jjy9mCAwTK7kA4SIyhmyEXQR32McB4xWqjxM3/K4Ij9o5+7ALpegD5p5c0cO7BqGIb4Xb3vZcJiA
0pMk7jWR
b6afB8c+iAdP1PhRER8oJpAOmUHwC2NoT85tpUbAwPtJc4SCW9HmFvI3Hq5pBOo2c1gB75Sdcma
YMR/Xl+HxkV
/KM5tqyx64BypS4uAByW9BuoQ5GH+WKHBXOzIFuhVDpvKS2XXaoOOfz/dfdLo/t1n13KkhzXf43NFc
4Lunqsd4
Zo9o7yr2D+ezXNLPDphRN3Itc9dxSaCZY7QE51fgdDUCPBPsQ17okukrcT5jI/R+rY3cL/bBx0Q4VU
d47bUcAS
RxCag2xvDONMAqDpbPX4j2Kbgs8avLqEFj4q4RkCOM28gETeqWxEE8XNZEfpXCupr1eeyfPul3BzcZ
mrTMu22Z
vvySyzYJQVf9YijpgWa9RcGBFQGKbWAgu5aWxdJvKjDCDjkupyhi2tnBRlRuRXgtXXN19NkDEiVus7
rAnZLMRu
BIIgbeWtT6BXSMMjtHqZ6NpaagDwGHtNmvOv6AEondaO3gU=
```

The AWS CLI fetches the credentials from environment variables set by the export commands. The user may then attempt to list all DynamoDB tables, like so:

```
[ec2-user@ip-172-31-88-201 ~]$ aws dynamodb list-tables
{
    "TableNames": [
        "MySecretTable",
        "Recipes"
    ]
}
```

Because the role grants the ListTables action, the credentials generated by the Security Token Service also grant that permission. However, if the user tries to use the same credentials to enumerate EC2 instances, the action fails.

```
[ec2-user@ip-172-31-88-201 ~]$ aws ec2 describe-instances

An error occurred (UnauthorizedOperation) when calling the
DescribeInstances operation: You are not authorized to perform this operation.
```

Therefore, it's imperative that you follow the principle of least privilege and don't grant a role more permissions than it requires. As an additional precaution, you can disable the Security Token Service on a per-region basis from the Account Settings menu option in the IAM Dashboard. Later in the chapter, you'll see how Amazon GuardDuty can help you identify when credentials such as these have been compromised.

Assuming a Role

You can create a role and allow any IAM user to assume it. The IAM user that assumes the role can be in the same account as the role or in a different account. When you assume a role, you have only the permissions granted to that role. The permissions of the IAM user you're logged in as are not added to the permissions of the role you assume. For instance, if you are logged in as an IAM user with full access to all AWS services and then assume a role that has read-only access to the EC2 service, you will not have access to any other AWS services. To see how this works for yourself, complete Exercise 12.2.

EXERCISE 12.2

Create and Assume a Role as an IAM User

In this exercise, you'll create a role and assume it.

1. While logged in as an administrative IAM user, click Roles on the menu on the left side of the IAM Dashboard screen and then click the Create Role button.

2. Under Select Type Of Trusted Entity, click the Another AWS Account button.

3. Under Account ID, enter your AWS account number. Click the Next: Permissions button.

4. Select the AmazonEC2ReadOnlyAccess AWS managed policy and click the Next: Tags button.

5. Click the Next: Review button.

6. Enter a name for the role, such as **EC2ReadOnlyRole**.

7. Click the Create Role button.

8. On the navigation bar at the top of the AWS Management Console, click your IAM account name. A drop-down menu will appear.

9. Click the Switch Role link in the drop-down menu, as shown here.

EXERCISE 12.2 *(continued)*

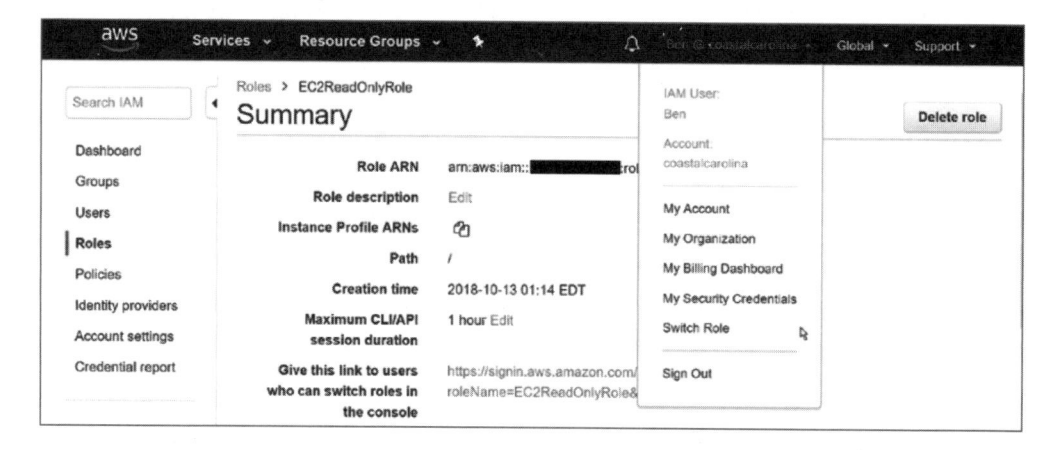

[Screenshot of IAM Roles Summary page for EC2ReadOnlyRole, showing a dropdown menu with options including My Account, My Organization, My Billing Dashboard, My Security Credentials, Switch Role, and Sign Out.]

10. Click the Switch Role button.

11. Enter your 12-digit AWS account number or alias and the name of the role you created in step 6. The following shows an example of a Switch Role screen with the account and role specified.

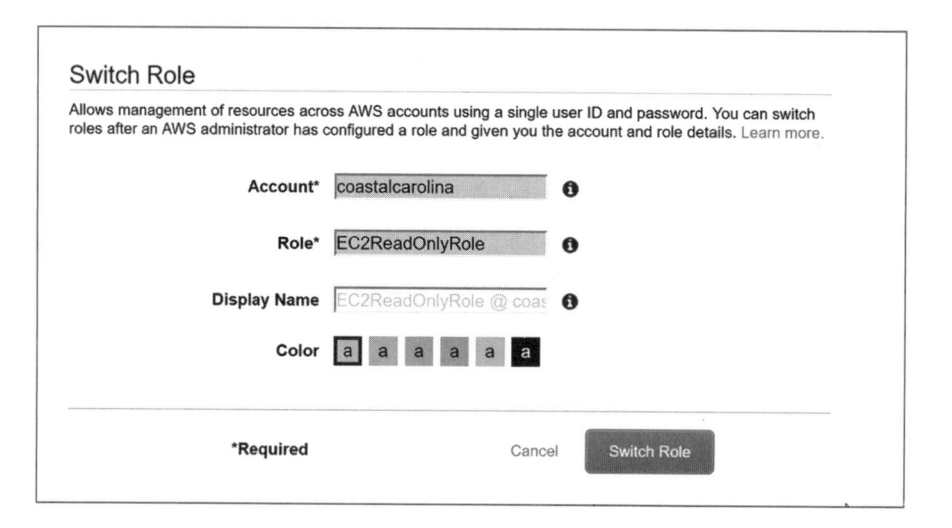

[Screenshot of Switch Role screen with Account: coastalcarolina, Role: EC2ReadOnlyRole, Display Name: EC2ReadOnlyRole @ coas, and color options.]

12. Click the Switch Role button. The name of the role followed will appear in the top navigation bar. You will now be operating under the role you created.

13. Try to launch an EC2 instance. It will fail because the assumed role doesn't have the `RunInstances` permission.
14. To switch back to your IAM user, click the upper-right menu bar where the name of the role followed by your AWS account number is. Then click the link that reads Back To *[your IAM username]*.

Enforcing Service-Level Protection

In addition to defining identity-based IAM policies to control access to resources, some AWS services allow you to define resource-based policies. For example, S3 offers optional bucket policies that control access to objects or entire buckets. The Key Management Service (KMS) requires you to define a key policy to specify the administrators and users of a key. SNS topics have resource policies to control who can publish messages or subscribe to a topic, as well as which delivery protocols they can use. Simple Queue Service (SQS) queues also use resource-based SQS access policies to control who can send to and receive messages from a queue.

Notice that users *without* AWS credentials tend to consume the AWS services that offer resource-based policies. For example, an end user of an application may receive an SNS notification via email, even though they don't have an AWS account. An anonymous user on the Internet may download a publicly accessible object stored in an S3 bucket. Resource-based policies give you control that you wouldn't have with identity-based policies alone. As with identity-based policies, use the principle of least privilege when creating resource-based policies.

Detective Controls

AWS offers a number of detective controls that can keep a record of the events that occur in your AWS environment, as well as alert you to security incidents or potential threats.

CloudTrail

It's important to carefully think about what you want to log and how you'll use the information contained in those logs. CloudTrail (Chapter 7, "CloudTrail, CloudWatch, and AWS Config") can log activities on your AWS account. But before configuring CloudTrail, you need to decide the scope of the logging, including whether to log any of the following:

- Management events, data events, or both
- Read-only events, write-only events, or both

- All resources or just specific ones
- All regions or just specific regions
- Global services

You'll also need to decide how many trails to create. You may decide to have one trail for read-only events and another for write-only events. Or you may choose to have a separate trail for each region.

CloudTrail stores logs in S3, so you'll want to set up bucket policies to control who can read and delete those logs. Also consider setting up an additional trail just to log data events on the S3 bucket containing your CloudTrail management event logs. For additional security, when creating a trail, enable SSE-KMS encryption and log file integrity validation. You can optionally have CloudTrail generate a Simple Notification Service (SNS) notification when it delivers a log file to the S3 bucket. Note that it can take up to 15 minutes between the time an event occurs and when the CloudTrail creates the log file containing the event.

CloudWatch Logs

CloudWatch Logs can aggregate logs from multiple sources for easy storage and searching. The logs that CloudWatch can ingest from various AWS services include the following:

CloudTrail Logs   To easily search and view CloudTrail logs, stream them to CloudWatch Logs. To set this up, refer to the steps in Exercise 7.3 in Chapter 7.

VPC Flow Logs   VPC flow logs include information about traffic ingressing or egressing a VPC. Flow logs include network interface, source and destination IP addresses, ports, protocols, and packet and byte counts. Flow logs do not include DHCP traffic or traffic to the Amazon DNS server. Follow the steps in Exercise 12.3 to configure VPC flow logging and deliver the logs to CloudWatch Logs.

RDS Logs   RDS can stream logs from database engines, including MariaDB, MySQL, Aurora with MySQL compatibility, and Oracle.

Route 53 DNS Queries   You can configure Route 53 to log DNS queries for a hosted zone and stream them to CloudWatch Logs.

Lambda   You can add logging statements to Lambda code. Lambda automatically streams these log events to a log group derived from the name of the function, using the format /aws/lambda/<function name>.

EXERCISE 12.3

Configure VPC Flow Logging

In this exercise, you'll configure VPC flow logging to deliver flow logs to a CloudWatch Logs log group.

1. In the VPC Dashboard, select a VPC you want to log traffic for. Click the Flow Logs tab.
2. Click the Create Flow Log button.
3. In the Destination Log Group field, enter a name of your choice, such as **FlowLogs**.
4. You'll need to specify a role, which AWS can create for you. Click the Set Up Permissions link.
5. A new browser tab will open. Click the Allow button. AWS will create a role called FlowlogsRole.
6. Return to the previous tab with the Create Flow Log wizard. Select the FlowlogsRole.
7. Click the Create button.

It will take up to 10 minutes for logs to begin to flow into the new log group.

Searching Logs with Athena

AWS uses S3 to store various logs, including CloudTrail logs, VPC flow logs, DNS query logs, and S3 server access logs. Athena lets you use the Structured Query Language (SQL) to search data stored in S3. Although you can use CloudWatch Logs to store and search logs, you can't format the output to show you only specific data. For example, suppose you use the following filter pattern to search for all DetachVolume, AttachVolume, and DeleteVolume events in a CloudWatch log stream containing CloudTrail logs:

{ $.eventName = "*tachVolume" || $.eventName = "DeleteVolume" }

CloudWatch Logs will display each matching event in its native JSON format, which as you can see in Figure 12.1 is inherently difficult to read.

Additionally, you may not be interested in every property in the JSON document and want to see only a few key elements. Amazon Athena makes it easy to achieve this. Once you bring data into Athena from S3, you can query it using SQL, sort it, and have Athena display only specific columns, as shown in Figure 12.2.

FIGURE 12.1 CloudWatch Logs showing `AttachVolume`, `DetachVolume`, and `DeleteVolume` events

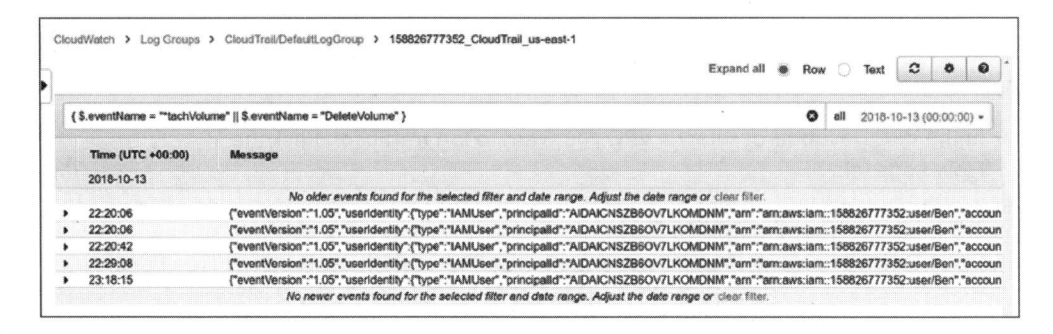

FIGURE 12.2 Athena query results

Because Athena uses SQL, you must define the structure of the data by using a CREATE TABLE Data Definition Language (DDL) statement. The DDL statement maps the values in the source file to columns in a table. AWS provides DDL statements for creating tables to store application load balancer logs, CloudTrail logs, CloudFront logs, and VPC flow logs.

You can import multiple logs into a single Athena database and then use SQL `JOIN` statements to correlate data across those logs. The data formats that Athena supports include the following:

- Comma-separated values (CSV) and tab-separated values (TSV)
- JavaScript Object Notation (JSON)—including CloudTrail logs, VPC flow logs, and DNS query logs
- Apache ORC and Parquet—storage formats for Apache Hadoop, which is a framework for processing large data sets

Auditing Resource Configurations with AWS Config

In addition to monitoring events, your overall security strategy should include monitoring the configuration state of your AWS resources. AWS Config can alert you when a resource configuration in your AWS account changes. It can also compare your resource configurations against a baseline and alert you when the configuration deviates from it, which is useful for validating that you're in compliance with organizational and regulatory requirements.

To compare your resource configurations against a desired baseline, you can implement AWS Config rules. In a busy AWS environment where configuration changes are occurring frequently, it's important to determine which types of changes require further investigation. AWS Config Rules let you define configuration states that are abnormal or suspicious so that you can focus on analyzing and, if necessary, remediating them. AWS offers a variety of managed rules to cover such scenarios. For example, the ec2-volume-inuse-check rule looks for EBS volumes that aren't attached to any instance. With this rule in place, if you create an EBS volume and don't attach it to an instance, AWS Config will report the volume as noncompliant, as shown in Figure 12.3.

FIGURE 12.3 AWS Config showing an EBS volume as noncompliant

Note that the rule gives only the current compliance status of the resource. Suppose that an IT auditor needed to see which EC2 instances this EBS volume was previously attached to. Although you could derive this information from CloudTrail logs, it would be difficult and time-consuming. You'd have to carefully sift through the CloudTrail logs, paying attention to every `AttachVolume` and `DetachVolume` operation. Imagine how much more cumbersome this process would be if you had to do this for hundreds of instances! Instead, you could simply use AWS Config to view the configuration timeline for the resource, as shown in Figure 12.4.

FIGURE 12.4 Configuration timeline for an EBS volume

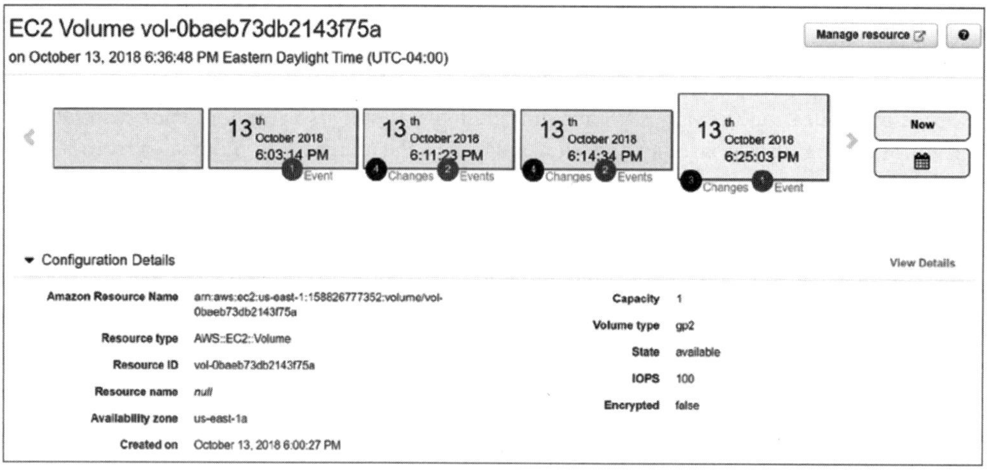

The configuration timeline shows a timestamp for each time the EBS volume was modified. You can click any box in the timeline to view the specific API events that triggered the configuration change, the before-and-after configuration details, and any relationship changes. The latest configuration and relationship changes, shown in Figure 12.5, reveal that the volume was attached to an instance and was then subsequently detached. Note that the act of detaching the volume from an instance placed the volume out of compliance with the ec2-volume-inuse-check rule.

Amazon GuardDuty

Amazon GuardDuty analyzes VPC flow logs, CloudTrail management event logs, and Route 53 DNS query logs, looking for known malicious IP addresses, domain names, and potentially malicious activity.

FIGURE 12.5 EBS volume configuration and relationship changes

Configuration Changes ②		
Field	**From**	**To**
Configuration.State	"in-use"	"available"

Configuration.Attachments.0	▼ Object
	  attachTime: "2018-10-13T22:12:03.000Z"
	  device: "/dev/sdf"
	  instanceId: "i-016cb83c71d221466"
	  state: "attached"
	  volumeId: "vol-0baeb73db2143f75a"
	  deleteOnTermination: false	

Relationship Changes ①		
Field	**From**	**To**
AWS::EC2::Instance	"i-016cb83c71d221466"	

> **NOTE** You do not need to stream any logs to CloudWatch Logs for GuardDuty to be able to analyze them.

When GuardDuty detects a potential security problem, it creates a *finding*, which is a notification that details the questionable activity. GuardDuty displays the finding in the GuardDuty console and also delivers the finding to CloudWatch Events. You can configure an SNS notification to send an alert or take an action in response to such an event. Findings are classified according to the ostensible purpose of the threat, which can be one of the following finding types:

Backdoor   This indicates an EC2 instance has been compromised by malware that can be used to send spam or participate in distributed denial-of-service (DDoS) attacks. This finding may be triggered when the instance communicates on TCP port 25—the standard port for the Simple Mail Transfer Protocol (SMTP)—or when it resolves the domain name of a known command-and-control server used to coordinate DDoS attacks.

Behavior   This indicates an EC2 instance is communicating on a protocol and port that it normally doesn't or is sending an abnormally large amount of traffic to an external host.

Cryptocurrency An EC2 instance is exhibiting network activity indicating that it's operating at a Bitcoin node. It may be sending, receiving, or mining Bitcoin.

Pentest A system running Kali Linux, Pentoo Linux, or Parrot Linux—Linux distributions used for penetration testing—is making API calls against your AWS resources.

Persistence An IAM user with no prior history of doing so has modified user or resource permissions, security groups, routes, or network ACLs.

Policy Root user credentials were used or S3 block public access was disabled.

Recon This indicates a reconnaissance attack may be underway. Behavior that can trigger this finding can include a host from a known malicious IP address probing an EC2 instance on a port that's not blocked by a security group or network ACL. Reconnaissance behavior can also include a malicious IP attempting to invoke an API call against a resource in your AWS account. Another trigger is an IAM user with no history of doing so attempting to enumerate security groups, network ACLs, routes, AWS resources, and IAM user permissions.

ResourceConsumption This indicates that an IAM user has launched an EC2 instance, despite having no history of doing so.

Stealth A password policy was weakened, CloudTrail logging was disabled or modified, or CloudTrail logs were deleted.

Trojan An EC2 instance is exhibiting behavior that indicates a Trojan may be installed. A Trojan is a malicious program that can send data from your AWS environment to an attacker or provide a way for an attacker to collect or store stolen data on your instance.

UnauthorizedAccess Indicates a possible unauthorized attempt to access your AWS resources via an API call or console login. This finding type may also indicate someone attempting to brute-force a Secure Shell (SSH) or Remote Desktop Protocol (RDP) session.

Notice that every finding type relates to either the inappropriate use of AWS credentials or the presence of malware on an EC2 instance. For example, in Figure 12.6, GuardDuty has detected network activity indicative of malware attempting to communicate with a command-and-control server, which is a server run by an attacker used to coordinate various types of attacks.

Remediating this finding might involve identifying and removing the malware or simply terminating the instance and creating a new one. In the case of the potentially inappropriate use of AWS credentials, remediation would include first contacting the authorized owner of the credentials to find out whether the activity was legitimate. If the credentials were used by an unauthorized party, then you would immediately revoke the compromised credentials and issue new ones.

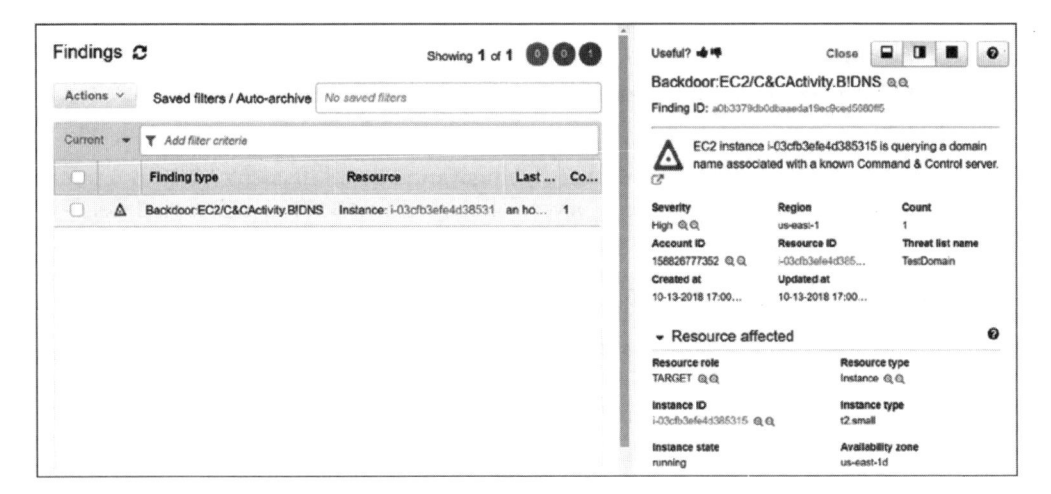

FIGURE 12.6 GuardDuty finding showing a possible malware infection

Amazon Inspector

Amazon Inspector is an agent-based service that looks for vulnerabilities on your EC2 instances. Whereas GuardDuty looks for security threats by inspecting network traffic to and from your instances, the Inspector agent runs an *assessment* on the instance and analyzes its network, filesystem, and process activity. Inspector determines whether any threats or vulnerabilities exist by comparing the collected data against one or more rules packages. Inspector offers five rules packages:

Common Vulnerabilities and Exposures   Common Vulnerabilities and Exposures (CVEs) are common vulnerabilities found in publicly released software, which includes both commercial and open source software for both Linux and Windows.

Center for Internet Security Benchmarks   These include security best practices for Linux and Windows operating system configurations.

Security Best Practices   These rules are a subset of the Center for Internet Security Benchmarks, providing a handful of rules against Linux instances only. Issues that these rules look for include root access via SSH, lack of a secure password policy, and insecure permissions on system directories.

Runtime Behavior Analysis   This rules package detects the use of insecure client and server protocols, unused listening TCP ports, and, on Linux systems, inappropriate file permissions and ownership.

Network Reachability   These rules detect network configurations that make resources in your VPC vulnerable. Some examples include having an instance in a public subnet or running an application listening on a well-known port.

After an assessment runs, Inspector generates a list of findings classified by the following severity levels:

- High—The issue should be resolved immediately.
- Medium—The issue should be resolved at the next possible opportunity.
- Low—The issue should be resolved at your convenience.
- Informational—This indicates a security configuration detail that isn't likely to result in your system being compromised. Nevertheless, you may want to address it depending on your organization's requirements.

Note that the high-, medium-, and low-severity levels indicate that an issue is likely to result in a compromise of the confidentiality, integrity, or availability of information. Figure 12.7 shows a finding with a severity level of medium.

FIGURE 12.7 Inspector finding showing that root users can log in via SSH

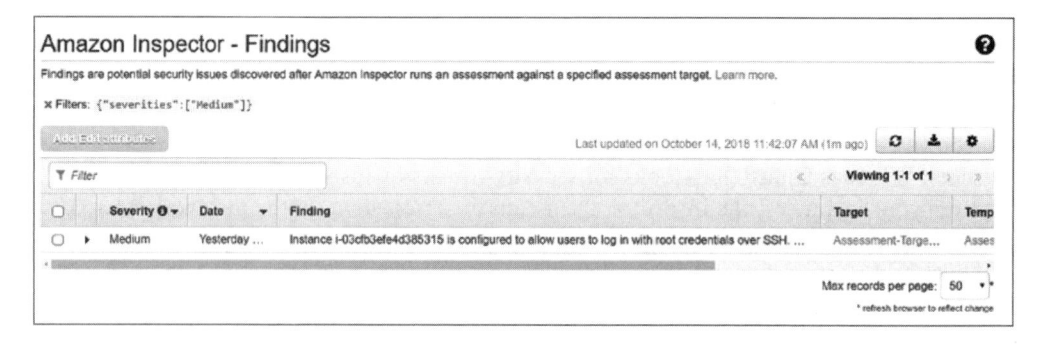

After you resolve any outstanding security vulnerabilities on an instance, you can create a new AMI from that instance and use it going forward when provisioning new instances.

Amazon Detective

Amazon Detective takes information from VPC flow logs, CloudTrail, and GuardDuty and places this information into a graph database. You can visualize the graph model to correlate events to identify and investigate suspicious or interesting activities against your AWS resources. Detective is designed to help you correlate events and see how a given event affects particular resources. This can save you a lot of time that you might otherwise spend digging through CloudTrail logs manually.

> Detective requires GuardDuty to be enabled.

Security Hub

Security Hub is a one-stop shop for the security status of your entire AWS environment. Security Hub collects security information from various AWS services, including Inspector, GuardDuty, and Macie. In addition, Security Hub assesses your account against AWS security best practices and the Payment Card Industry Data Security Standard (PCI DSS). It displays its findings in a user-friendly dashboard that includes charts and tables, making it easier and faster to identify trends concerning security-related issues.

Protecting Network Boundaries

The network can be the first line of defense against attacks. All AWS services depend on the network, so when designing your AWS infrastructure, you should consider how your AWS resources will need to communicate with one another, with external resources, and with users.

Many AWS services have public endpoints that are accessible via the Internet. It's the responsibility of AWS to manage the network security of these endpoints and protect them from attacks. However, you can control network access to and from the resources in your VPCs, such as EC2 instances, RDS instances, and elastic load balancers.

Network Access Control Lists and Security Groups

Each resource within a VPC must reside within a subnet. Network ACLs define what traffic is allowed to and from a subnet. Security groups provide granular control of traffic to and from individual resources, such as instances and elastic load balancer listeners. You should configure your security groups and network ACLs to allow traffic to and from your AWS resources using only the protocols and ports required. If your security requirements call for it, you may also need to restrict communication to specific IP address ranges.

Consider which of your resources need to be accessible from the Internet. A VPC must have an Internet gateway for resources in the VPC to access the Internet. Also, each route table attached to each subnet must have a default route with the Internet gateway as a target. If you're running a multitier application with instances that don't need Internet access—such as database servers—consider placing those in a separate subnet that permits traffic to and from the specific resources that the instance needs to communicate with. As an added precaution, ensure that the route table the subnet is associated with doesn't have a default route.

AWS Web Application Firewall

The Web Application Firewall (WAF) monitors HTTP and HTTPS requests to an application load balancer or CloudFront distribution. WAF protects your applications from common exploits that could result in a denial of service or allow unauthorized access to your application.

Unlike network ACLs and security groups, which allow or deny access based solely on a source IP address or port and protocol, WAF lets you inspect application traffic for signs of malicious activity, including injection of malicious scripts used in cross-site scripting attacks, SQL statements used in SQL injection attacks, and abnormally long query strings. You can block these requests so that they never reach your application.

WAF can also block traffic based on source IP address patterns or geographic location. You can create a Lambda function to check a list of known malicious IP addresses and add them to a WAF block list. You can also create a Lambda function to analyze web server logs and identify IP addresses that generate bad or excessive requests indicative of an HTTP flood attack and add those addresses to a block list.

AWS Shield

Internet-facing applications running on AWS are of necessity vulnerable to DDoS attacks. AWS Shield is a service that helps protect your applications from such an attack. AWS Shield comes in two flavors:

AWS Shield Standard Defends against common Layers 3 and 4 DDoS attacks such as SYN flood and UDP reflection attacks. Shield Standard is automatically activated for all AWS customers.

AWS Shield Advanced Provides the same protection as Shield Standard but also includes protection against Layer 7 attacks, such as HTTP flood attacks that overwhelm an application with HTTP GET or POST requests. For you to obtain Layer 7 protection for an EC2 instance, it must have an elastic IP address. You also get attack notifications, forensic reports, and 24/7 assistance from the AWS DDoS response team. AWS WAF is included at no charge.

Shield mitigates 99 percent of attacks in five minutes or less. It mitigates attacks against CloudFront and Route 53 in less than one second and Elastic Load Balancing in less than five minutes. Shield usually mitigates all other attacks in less than 20 minutes.

Data Encryption

Encrypting data ensures the confidentiality of data by preventing those without the correct key from decrypting and reading it. As an added benefit, encryption also makes it infeasible for someone to modify the original data once it's been encrypted.

Data can exist in one of two states: at rest, sitting on a storage medium such as an EBS volume or S3 bucket, and in transit across the network. How you encrypt data differs according to its state.

Data at Rest

How you encrypt data at rest depends on where it's stored. On AWS, the bulk of your data will probably reside in S3, Elastic Block Store (EBS) volumes, or an Elastic File System (EFS) or a Relational Database Service (RDS) database. Each of these services integrates with KMS and gives you the option of using your own customer-managed customer master key (CMK) or an AWS-managed CMK.

When you use your own CMK, you can configure key policies to control who may use the key to encrypt and decrypt data. You can also rotate, disable, and revoke keys. Using a customer-managed CMK gives you maximum control over your data.

An AWS-managed CMK automatically rotates once a year. You can't disable, rotate, or revoke it. You can view existing CMKs and create new ones by browsing to the Encryption Keys link in the IAM Dashboard.

Most AWS services offer encryption of data at rest using KMS-managed keys, including DynamoDB. Note that enabling encryption for some services such as CloudWatch Logs requires using the AWS CLI.

S3

If your data is stored in an S3 bucket, you have a number of encryption options:

- Server-side encryption with S3-managed Keys (SSE-S3)
- Server-side encryption with KMS-managed keys (SSE-KMS)
- Server-side encryption with customer-provided keys (SSE-C)
- Client-side encryption

Remember that encryption applies per object, so it's possible to have a bucket containing objects using different encryption options or no encryption at all. Applying default encryption at the bucket level does not automatically encrypt existing objects in the bucket but only those created moving forward.

Elastic Block Store

You can encrypt an EBS volume using a KMS-managed key. You can encrypt a volume when you initially create it. However, you cannot directly encrypt a volume created from an unencrypted snapshot or unencrypted AMI. You must first create a snapshot of the unencrypted volume and then encrypt the snapshot. To encrypt an existing unencrypted volume, follow the steps in Exercise 12.4.

> **EXERCISE 12.4**
>
> **Encrypt an EBS Volume**
>
> For this exercise, you'll encrypt an unencrypted volume attached to a running EC2 instance.

EXERCISE 12.4 *(continued)*

1. Create an EC2 instance from an unencrypted snapshot or AMI. If you already have one you can bear to part with, you may use that.

2. At the EC2 Dashboard, click Volumes in the left-side menu.

3. Select the volume attached to your instance.

4. Under the Actions menu, click Create Snapshot. EBS will begin creating an unencrypted snapshot of the volume.

5. On the left-side menu, click Snapshots. Wait for the snapshot to complete.

6. Select the snapshot and under the Actions menu, click Copy.

7. Check the Encrypt This Snapshot check box.

8. Next to the Master Key drop-down, select the KMS key you'd like to use to encrypt the snapshot. You can use your own customer master key, or the default aws/ebs key.

9. Click the Copy button. EBS will begin creating an encrypted copy of the snapshot. The larger the volume, the longer the encryption process takes. A good rule of thumb is about 3 gigabytes per minute.

10. Select the encrypted snapshot and under the Actions menu, click Create Volume.

11. Choose the volume type, size, and availability zone, which can be different than the original volume.

12. Click the Create Volume button. The new volume will be encrypted using the same key that was used to encrypt the snapshot.

You can now detach the unencrypted volume from your instance and attach the new encrypted volume.

When you copy a snapshot, you have the option of choosing the destination region. The key you use to encrypt the destination snapshot must exist in the destination region.

Elastic File System

You can enable encryption for an EFS filesystem when you create it. EFS encryption uses KMS customer master keys to encrypt files and an EFS-managed key to encrypt filesystem metadata, such as file and directory names.

Data in Transit

Encrypting data in transit is enabled through the use of a Transport Layer Security (TLS) certificate. You can use the Amazon Certificate Manager (ACM) to generate a TLS certificate and then install it on an application load balancer or a CloudFront distribution. Refer to Exercise 8.4 in Chapter 8, "The Domain Name System and Network Routing: Amazon Route 53 and Amazon CloudFront," for instructions on creating a TLS encryption certificate and installing it on a CloudFront distribution.

Keep in mind that you cannot export the private key for a TLS certificate generated by ACM. This means you can't install the certificate directly on an EC2 instance or on-premises server. You can, however, import an existing TLS certificate and private key into ACM. The certificate must be valid and not expired when you import it. To use an imported certificate with CloudFront, you must import it into the `us-east-1` region.

Macie

Macie is a service that automatically locates and classifies your sensitive data stored in S3 buckets and shows you how it's being used. Using machine learning, Macie can help improve your cloud security posture by doing the following:

- Recognizing sensitive data like trade secrets and personally identifiable information
- Alerting you if it finds buckets that have permissions that are too lax
- Tracking changes to bucket policies and ACLs
- Classifying other types of data based on custom data identifiers that you define

Macie classifies its findings into either policy findings or sensitive data findings. Policy findings include changes to a bucket that reduce its security, such as changing a bucket policy or removing encryption. Sensitive data findings classify any sensitive data Macie finds in an S3 bucket.

Macie automatically publishes its findings to EventBridge and AWS Security Hub. Other applications can tap into these findings and take some action based on them. For example, you can send findings to an SNS notification topic to email you when it finds an S3 bucket with public permissions. Macie publishes sensitive data findings immediately and policy findings every 15 minutes.

Summary

Effective security controls necessarily add more complexity to your deployment and will occasionally cause things to break. But by implementing security at every level, you can resolve those issues as you go along. Then by the time you're done, you'll be starting with an effective security posture. From that point on, it will be much easier to mitigate vulnerabilities and threats as they trickle in.

It's fine to be a little lax when initially configuring a service just to make sure everything works as expected, but as soon as you do, be sure to implement your security controls. Follow the principle of least privilege, using IAM and resource-based policies and network-based controls. Configure logging early on to make troubleshooting easier. Set up encryption prior to going live so that your data is never sent or stored in the clear. Implement notifications for key events so that you know immediately when a misconfiguration, attack, or error occurs. Finally, automate as much as possible using Auto Scaling, CloudFormation, and preconfigured AMIs. Automation allows you to recover more quickly from an incident. It also reduces human intervention, thus minimizing the chances of mistakes.

Exam Essentials

Be able to configure permissions policies. Permissions policies are the foundation of identity and resource-based policies. Understand the difference between these policies and be able to understand and create them. Identity-based policies are attached to an IAM principal such as a user or role. Resource-based policies apply to a resource and don't include the `principal` element.

Know how permissions boundaries interact with permissions policies. Permissions boundaries define the maximum permissions an IAM principal can have. You set permissions boundaries on a per-identity basis by attaching a managed policy that defines them.

Understand roles and trust policies. A role is a set of permissions that an IAM user, AWS service, or federated user can acquire by assuming that role. Trust policies are resource-based policies that define who can assume the role and under what conditions.

Know your options for collecting, searching, and analyzing logs. CloudWatch Logs can consolidate logs from multiple sources, including CloudTrail, VPC flow logs, DNS query logs, and logs generated by your application. You can search logs from within CloudWatch Logs using filter patterns. For logs stored in S3, you can use Athena to search them using SQL queries.

Be able to identify use cases for GuardDuty, Inspector, Shield, and WAF. GuardDuty analyzes various logs looking for signs of an attack. Inspector uses an agent to inspect your EC2 instances for vulnerabilities. Shield identifies and blocks well-known network-based attacks as they occur. WAF allows you to configure custom rules to block suspicious or undesirable traffic.

Understand how KMS works. Know how to create customer master keys and set key policies. A key policy defines the key users—IAM principals who can use a key to encrypt and decrypt data. A key policy also defines the administrators of a key—those who can enable and disable a key. Key administrators can also modify the key policy.

Review Questions

1. Which of the following options can you not set in a password policy? (Choose two.)
   A. Maximum length
   B. Require the use of numbers.
   C. Prevent multiple users from using the same password.
   D. Require an administrator to reset an expired password.

2. An IAM user is attached to a customer-managed policy granting them sufficient access to carry out their duties. You want to require multifactor authentication (MFA) for this user to use the AWS CLI. What element should you change in the policy?
   A. Resource
   B. Condition
   C. Action
   D. Principal

3. You created an IAM policy that another administrator subsequently modified. You need to restore the policy to its original state but don't remember how it was configured. What should you do to restore the policy? (Choose two.)
   A. Consult CloudTrail global management event logs.
   B. Restore the policy from a snapshot.
   C. Consult CloudTrail data event logs.
   D. Revert to the previous policy version.

4. An IAM user with full access to all EC2 actions in all regions assumes a role that has access to only the EC2 `RunInstances` operation in the `us-east-1` region. What will the user be able to do under the assumed role?
   A. Create a new instance in any region.
   B. Create a new instance in the `us-east-1` region.
   C. Start an existing instance in the `us-east-1` region.
   D. Start an existing instance in any region.

5. Several objects in a S3 bucket are encrypted using a KMS customer master key. Which of the following will give an IAM user permission to decrypt these objects?
   A. Add the user to the key policy as a key user.
   B. Grant the user access to the key using an IAM policy.
   C. Add the user to the key policy as a key administrator.
   D. Add the user as a principal to the bucket policy.

6. You run a public-facing application on EC2 instances. The application is backed by a database running on RDS. Users access it using multiple domain names that are hosted in Route 53. You want to get an idea of what IP addresses are accessing your application. Which of the following would you stream to CloudWatch Logs to get this information?

   A. RDS logs
   B. DNS query logs
   C. VPC flow logs
   D. CloudTrail logs

7. You're running a web server that keeps a detailed log of web requests. You want to determine which IP address has made the most requests in the last 24 hours. What should you do to accomplish this? (Choose two.)

   A. Create a metric filter.
   B. Stream the web server logs to CloudWatch Logs.
   C. Upload the web server log to S3.
   D. Use Athena to query the data.

8. An application running on an EC2 instance has been updated to send large amounts of data to a server in your data center for backup. Previously, the instance generated very little traffic. Which GuardDuty finding type is this likely to trigger?

   A. Behavior
   B. Backdoor
   C. Stealth
   D. ResourceConsumption

9. You've set up an AWS Config managed rule to check whether a particular security group is attached to every instance in a VPC. You receive an SNS notification that an instance is out of compliance. But when you check the instance a few hours later, the security group is attached. Which of the following may help explain the apparent discrepancy? (Choose two.)

   A. The AWS Config timeline
   B. Lambda logs
   C. CloudTrail management event logs
   D. VPC flow logs

10. You want to use Amazon Inspector to analyze the security posture of your EC2 instances running Windows. Which rules package should you not use in your assessment?

    A. Common Vulnerabilities and Exposures
    B. Center for Internet Security Benchmarks
    C. Runtime Behavior Analysis
    D. Security Best Practices

11. You have a distributed application running in datacenters around the world. The application connects to a public Simple Queue Service (SQS) endpoint to send messages to a queue. How can you prevent an attacker from using this endpoint to gain unauthorized access to the queue? (Choose two.)
    A. Network access control lists
    B. Security groups
    C. IAM policies
    D. SQS access policies

12. You're using a public-facing application load balancer to forward traffic to EC2 instances in an Auto Scaling group. What can you do to ensure users on the Internet can reach the load balancer over HTTPS without reaching your instances directly? (Choose two.)
    A. Create a security group that allows all inbound traffic to TCP port 443.
    B. Attach the security group to the instances.
    C. Attach the security group to the load balancer.
    D. Remove the Internet gateway from the VPC.
    E. Create a security group that allows all inbound traffic to TCP port 80.

13. You're running a UDP-based application on an EC2 instance. How can you protect it against a DDoS attack?
    A. Place the instance behind a network load balancer.
    B. Implement a security group to restrict inbound access to the instance.
    C. Place the instance behind an application load balancer.
    D. Enable AWS Shield Standard.

14. You're running a web application on six EC2 instances behind a network load balancer. The web application uses a MySQL database. How can you protect your application against SQL injection attacks?
    A. Enable WAF.
    B. Assign elastic IP addresses to the instances.
    C. Place the instances behind an application load balancer.
    D. Block TCP port 3306.

15. Which services protect against an HTTP flood attack?
    A. GuardDuty
    B. WAF
    C. Shield Standard
    D. Shield Advanced

16. Your security policy requires that you use a KMS key for encrypting S3 objects. It further requires this key be rotated once a year and revoked when misuse is detected. Which key type should you use? (Choose two.)
    A. Customer-managed CMK
    B. AWS-managed CMK
    C. S3-managed key
    D. Customer-provided key

17. A developer is designing an application to run on AWS and has asked for your input in deciding whether to use a SQL database or DynamoDB for storing highly transactional application data. Your security policy requires all application data to be encrypted and encryption keys to be rotated every 90 days. Which AWS service should you recommend for storing application data? (Choose two.)
    A. KMS
    B. RedShift
    C. DynamoDB
    D. RDS

18. You need to copy the data from an unencrypted EBS volume to another region and encrypt it. How can you accomplish this? (Choose two.)
    A. Create an encrypted snapshot of the unencrypted volume.
    B. Simultaneously encrypt and copy the snapshot to the destination region.
    C. Copy the encrypted snapshot to the destination region.
    D. Create an unencrypted snapshot of the unencrypted volume.

19. An instance with an unencrypted EBS volume has an unencrypted EFS filesystem mounted on it. You need to encrypt the data on an existing EFS filesystem using a KMS key. How can you accomplish this?
    A. Encrypt the EBS volume of the instance.
    B. Create a new encrypted EFS filesystem and copy the data to it.
    C. Enable encryption on the existing EFS filesystem.
    D. Use a third-party encryption program to encrypt the data.

20. On which of the following can you *not* use an ACM-generated TLS certificate? (Choose two.)
    A. An S3 bucket
    B. A CloudFront distribution
    C. An application load balancer
    D. An EC2 instance

21. Which of the following assesses the security posture of your AWS resources against AWS best practices?
    A. Detective
    B. Macie
    C. Security Hub
    D. GuardDuty

Chapter 13

The Cost Optimization Pillar

THE AWS CERTIFIED SOLUTIONS ARCHITECT ASSOCIATE EXAM OBJECTIVES COVERED IN THIS CHAPTER MAY INCLUDE, BUT ARE NOT LIMITED TO, THE FOLLOWING:

✓ **Domain 4: Design Cost-Optimized Architectures**

- 4.1 Identify cost-effective storage solutions
- 4.2 Identify cost-effective compute and database services
- 4.3 Design cost-optimized network architectures

Introduction

Through the previous chapters of this book you've explored many AWS cost planning basics. In Chapter 3, "AWS Storage," for instance, you learned how to leverage different data storage classes to ensure the instant access you need for your newer data using, say, the S3 Standard class while using the lower costs available for high-latency, long-term storage (using Glacier, etc.) of older objects.

You're no doubt also familiar with the Free Tier available on AWS accounts within their first year. The Free Tier makes it possible to experiment with light versions of just about any combination of AWS resources until you find the perfect stack for your application. Used right, the tier can help you build excellent solutions architect skills without costing you anything. Don't be shy; use it.

But you're not quite there yet.

Even if you do everything right and design the perfect, multi-tiered, scalable, and highly available application environment, it'll be a complete waste if it ends up costing too much. No plan is complete without a solid cost assessment. And even if you did make it all happen for the right price, are you sure nothing expensive has accidentally been left running on any of the accounts under your control or you haven't been compromised by hackers?

The best protection from unpleasant surprises is by understanding how you're billed for using AWS services, how you can deploy resources in the most cost-effective way, and how you can automate cost event–alerting protocols.

That's what you'll learn about here.

Planning, Tracking, and Controlling Costs

The place to find account-level financial information is the Billing and Cost Management Dashboard, which is accessed via the My Billing Dashboard link in the account drop-down menu at the top of the AWS management console. That's where you can view past bills, manage credits, and enter your tax registration information. The page also includes a quick visual overview of the current month's cost activities.

Right now, however, you're more interested in how the Billing and Cost Management Dashboard also links you to tools you'll use to monitor and control your AWS spending. The precise function of these tools—and even the names AWS gives them—will probably change over time. But their overall goal is clear: to provide deep insight into what your running resources are costing you and how you can closely track them.

AWS Budgets

The Preferences link takes you to a simple page where you choose the way you'd like to consume billing information. You can tell AWS to send ongoing *billing reports* to an S3 bucket you specify. You can also have AWS email you *billing alerts* created through either Amazon CloudWatch Alarms (as you saw in Chapter 7, "CloudTrail, CloudWatch, and AWS Config") or the newer AWS Budgets.

Budgets are meant to track your ongoing resource-related usage and costs and alert you if projected usage levels fall outside of a predefined threshold. You could, for instance, have an email sent alerting you when the total quarterly costs of transferring S3-based data between regions threatens to rise above a certain dollar amount. Or you could create a budget to tell you when the total volume of outbound data transferred to the Internet from On Demand EC2 instances running in a specified AWS region exceeds, say, 100 GB.

You can create and apply *cost allocation tags* to your resources and use them in your budgets as filters. These tags will let you limit your alerts to only those resources belonging to a particular class. You could, perhaps, use this kind of filtering to track only your staging or development environments, but not production.

You can activate and manage tags from the Cost Allocation Tags page, which you access by clicking the Cost Allocation Tags link in Billing. User-defined tags can be created and managed using the Tag Editor accessed through the Resource Groups drop-down menu in the AWS Console. You use that page to find active resources such as EC2 instances and S3 buckets and then create and/or edit tags for those resources. Note the following:

- Tags can take up to 24 hours before they appear in the Billing and Cost Management Dashboard.
- Tags can't be applied to resources that were launched before the tags were created.

You're allowed two free budgets per account. Subsequent tags will cost around $0.02 each per month.

You configure a budget by defining whether you want to track resource cost, usage, or—to be sure you're getting optimal value from your EC2 reserve instances—reserve instance utilization or reserve instance coverage. You can narrow the scope of a budget by filtering it using parameters such as services, regions, tags, and usage type. Notification triggers are defined when thresholds are met (for instance, "Forecasted usage is greater than 80 percent of the budgeted amount") and are sent to Simple Notification Service (SNS) or email addresses.

Exercise 13.1 shows you how to configure a simple budget to protect an account from unexpected, unauthorized, and expensive activity.

> ### EXERCISE 13.1
>
> **Create an AWS Budget to Send an Alert**
>
> 1. Click the Budgets link in the Billing and Cost Management console and then click Create Budget.
>
> 2. Select the Cost Budget option and click Set Your Budget.
>
> 3. Give your budget a name (something descriptive like **Monthly Limit** might work well). Select Monthly for Period and select the Recurring Budget radio button. That will run the budget each month indefinitely. For this exercise, enter, say, **100** in the Budgeted Amount field. That value will be interpreted as a dollar value. Leave all the Aggregate Costs values as Unblended Costs. When you're done, click Configure Alerts.
>
> 4. Select the Actual Costs radio button for Send Alert Based On, and enter **80** in Alert Threshold, and select % Of Budgeted Amount from the drop-down menu. Type your email address in the Email Contacts field, and click Create.
>
>   After you've set this up, you'll receive an email alert whenever your account resources have generated more than $80 (80 percent of your $100 value for Budgeted Amount).
>
>   If you don't need the budget you just created, select its entry in the Dashboard and delete it to make sure you're not billed for it.

Monitoring Tools

You've seen how to set limits to help you control your AWS spending using budget alerts. But the busier and more complicated your resource stacks get, the more you'll need to dive deeply into spending patterns so that you can understand and, if necessary, adjust developing trends. Cost Explorer and Reports—while they won't actively cap your spending—can help you more completely understand what's going on so that you can make smarter choices.

Cost Explorer

Clicking Cost Explorer from the Billing and Cost Management Dashboard, and then on Launch Cost Explorer, takes you to a detailed visualization of your historical AWS usage and costs. The default view covers the last six months, and you can click the Explore Costs link to filter by a prominent Usage Type Group metric (like EC2: Running Hours).

The more you have running, the more you'll want to drill down into the weeds of the data. You can adjust the view, filtering by things such as grouping or service, or download the data as a CSV file.

Clicking New Report lets you open a report template, including Reserve Instance (RI) Utilization or RI Coverage—relevant, obviously, only if you've purchased reserve instances.

AWS Cost and Usage Reports

At first glance, the content delivered by Cost and Usage Reports seems to overlap a bit with Cost Explorer. After all, they both provide data covering your billing costs, rates, and product and pricing attributes.

However, the unique value proposition of Reports is hinted at in the Delivery options page of the Report creation process. There, you can choose to enable support for Athena, Redshift, and QuickSight. Athena is a tool that lets you efficiently query your data using SQL. Redshift is a serious tool for handling serious amounts of data. And Amazon QuickSight is a pay-per-session tool for squeezing business intelligence out of data stores.

So, AWS Reports are meant for accounts that are so busy you struggle to keep track of many moving parts. Reports are optimized for this kind of analytics through their normalized data model that organizes discrete cost components in individual columns.

Reports are configured to be regularly written to an S3 bucket in your account for which you've previously added appropriate permissions. Sample bucket policies are available from the Create Report dialog.

AWS Organizations

You can consolidate the management of multiple AWS accounts using AWS Organizations. Why bother? Because it's not uncommon for larger organizations to have so many projects and resources running on AWS that they need some degree of resource isolation to keep a handle on them. It can also be hugely convenient to maintain some level of bird's-eye-view supervision over the whole thing—and to channel billing through a single payer.

AWS Organizations lets you create new accounts and invite existing accounts into an organization. Once your accounts are connected, the organization administration can create global access policies much the same way as you might for individual users within a single account.

Account security for an organization is even more critical than for a single account since a breach could potentially impact all the resources in all member accounts. You should therefore be especially careful to enforce all the normal security best practices, including multifactor authorization (MFA), strong passwords, and secured root accounts.

AWS Organizations allows you to:

- Share resources through a unified AWS Single Sign-on configuration that applies global permissions to developers for all your accounts.
- Apply IAM rules globally through service control policies (SCPs).

- Create and manage accounts—along with account users and groups—programmatically.
- Audit, monitor, and secure all your environments for compliance and functional purposes. A single account's CloudTrail service can be configured to watch events from your entire organization.

AWS Organizations has replaced the old and deprecated Consolidated Billing tool.

AWS Trusted Advisor

You can get a quick view of how closely your account configurations are currently following AWS best practices from the Trusted Advisor Dashboard (accessed from the main AWS service menu). The compliance and health data is divided into five categories:

- Cost Optimization checks for underutilized or running but idle resources that might be costing you money.
- Performance checks for configuration mismatches—such as excessive use of EBS magnetic volumes—that might unnecessarily impact performance.
- Security checks for potentially vulnerable configurations, such as wide-open S3 bucket permissions or security groups.
- Fault Tolerance checks for appropriate replication and redundancy configurations to ensure that a single service failure would be unlikely to cause you a catastrophic loss.
- Service Limits checks for your usage of AWS services to make sure you're not approaching default limits.

The results of each check are color-coded to tell you whether you're compliant and are accompanied by a helpful description of the issue. Sometimes, for an S3 bucket-hosted static website that needs open permissions, for instance, you'll want to ignore the warning. But it's nevertheless important to regularly check in with Trusted Advisor to see whether it's caught anything new.

The first thing you'll notice when you visit Trusted Advisor is that most of the metrics are disabled unless you upgrade your support—to either the Business or the Enterprise plan.

Online Calculator Tools

One powerful way to model the cost of multiple resource stacks is by gathering comprehensive pricing information and using it to calculate the price tag for each element of a stack. However, when you consider just how many permutations and combinations of services and price levels can coexist within a single stack—and how often those prices change—then accurate *cost modeling* can feel impossibly complicated.

Fortunately, AWS provides two regularly updated calculators that do all the number crunching for you: the Simple Monthly Calculator and the AWS TCO Calculator.

Simple Monthly Calculator

The Simple Monthly Calculator (`calculator.s3.amazonaws.com/index.html`) lets you select service resources at a fine detail (three EC2 instances built on a Linux m4.large instance type with a 500 GB Provisioned IOPS SSD EBS volume set for 1,000 IOPS in the us-east-1 region, for example). You can also enter the portion of a month you expect your instance to actually be running.

> **NOTE** As of this writing, AWS has plans to deprecate the Simple Monthly Calculator and replace it with the AWS Pricing Calculator (https://calculator.aws/#/). For now, the old calculator is still available (and, in fact, I still find it much more useful than the new version), and there is no specific date set for its retirement.

You should be aware that costs can differ between AWS regions, so make sure you price resources within each of the regions that would, practically, work for your application. The same applies to different EBS and RDS storage types: explore pricing for SSD, Provisioned IOPS, and other storage types.

When you're done entering the stack resources from every relevant service your project will use, the estimated monthly bill will appear in the tab at the top of the page. Clicking that tab will provide a fully itemized breakdown of the monthly cost of your stack, which can be downloaded in CSV format. Saving your estimate will generate a URL that can be shared with colleagues or clients.

The beautiful thing about the monthly calculator—as you'll see for yourself while working through Exercise 13.2—is how closely it mirrors the experience of selecting and then running AWS resources. If you haven't built an accurate cost simulation of a complicated stack you're planning, then you're probably setting yourself up for some surprises later when painful monthly billing statements show up.

EXERCISE 13.2

Build Your Own Stack in Simple Monthly Calculator

1. Start a fresh calculation at `calculator.s3.amazonaws.com/index.html`. Click Reset All if necessary and deselect the Free Usage Tier box at the top.

2. Within the us-west-2 (Oregon) click the Amazon EC2 tab on the left and click the plus sign next to the first Add New Row. Select two instances running **120** hours (5 days) a week of a `Linux c5d.xlarge` instance type. Leave the Billing Option setting as On-Demand.

3. Add two general-purpose EBS volumes of **100** GB each and, in the Data Transfer section, add **200** GB of Data Transfer In and **400** GB of Data Transfer Out. Feel free to click the Estimate tab at the top.

EXERCISE 13.2 *(continued)*

4. From the Services tab once again, click the Amazon S3 link on the left. Enter **200** GB of storage, **250000** PUT Requests, and **1000000** GET Requests for S3 Standard Storage And Requests. Enter **600** GB of storage, and use 10 Lifecycle Transitions for S3 Standard – Infrequent Access (S3 Standard-IA) Storage & Requests.

5. Enter **400** GB/month of Data Transfer Out on the Amazon CloudFront tab.

6. On the Amazon Elastic Load Balancing tab, enter **1** for Number Of Application LBs, **5** for Avg Number Of New Connections/Sec Per ALB, **300** seconds for Avg Connection Duration, and **400** GB for Total Data Processed Per ALB.

7. Now take a good look at the itemized bill in the Estimate tab. Just for comparison, my total came to around $320/month.

AWS Total Cost of Ownership Calculator

The "other" AWS calculator (aws.amazon.com/tco-calculator) estimates the total cost of ownership (TCO) to help you make an apples-to-apples comparison of what a complex workload would cost on-premises as opposed to what it would cost on the AWS cloud. The interface asks you to define the infrastructure you'd need to run your application locally. You'll enter the number and size of the servers, whether they're physical or virtual, which virtualization hypervisor you'll be using (if applicable), and the total storage capacity you'll need.

When you run the calculation, you'll be shown a highly detailed itemized cost sheet calculated over three years. The output will include estimates of on-premises networking and hardware acquisition costs along with AWS support costs, and a detailed explanation of the calculator's methodology and assumptions.

Cost-Optimizing Compute

Perhaps more than anything else, the virtualization revolution makes greatly increased server density possible. That is, you can run more compute workloads sharing the physical assets of a single server than ever before. And of course, getting "more production out of a server" is just another way of saying "getting more value for your investment."

Since cloud computing platforms like AWS live and breathe virtualization, you might assume that following AWS best practices can help you get the most out of your assets. And you'd be correct.

Here we'll spend some time talking about squeezing every drop of digital goodness out of your AWS infrastructure. You'll explore some basic server density tricks such as selecting the EC2 instance type that's the best match for your workload, and, although they were introduced in Chapter 2, "Amazon Elastic Compute Cloud and Amazon Elastic Block Store," we'll dive a bit deeper into effectively using reserved and spot instances.

Maximizing Server Density

You have a keen interest in getting the most out of a server instance—whether it's for an application running on EC2 or a database with AWS Relational Database Service (RDS). Matching your workload to the right instance type can make a big difference not only in performance but also in overall cost.

The M5 instance type family, for instance, is optimized to take advantage of the higher core counts found on Intel Xeon Scalable processors. You can typically pack multiple resource-hungry workloads onto a single, high-end M5 instance (like the m5.24xlarge type), potentially requiring fewer instances overall. Similarly, the high-performing and parallel processing NVIDIA K80 GPU cores, enhanced networking, and default EBS optimization of a single P2 instance can often replace multiple instances of other families for high-demand analytics and machine learning workloads.

The features unique to each EC2 instance type are detailed and compared at aws.amazon.com/ec2/instance-types.

AWS Lambda also provides a kind of server density; it may not be *your* server resources you're conserving by running Lambda functions, but you're certainly getting a lot more value from your compute dollar by following the "serverless" model.

The flexibility potential of virtualization is nowhere more obvious than with container technologies like Docker. Whether you're running Docker clusters on Amazon's Elastic Container Service, running the Amazon Elastic Container Service for Kubernetes (Amazon EKS), or manually provisioning your own EC2 instances as Docker hosts, you'll be able to tightly pack virtual applications or microservices into every corner of your host hardware.

EC2 Reserved Instances

You learned about purchasing reserved instances (RIs) for 24/7 compute needs lasting at least 12 months in Chapter 2. But spending a few extra minutes on some of the finer details will help you make smarter and more effective choices.

Using Traditional Reserved Instances

The basics are straightforward; you can have AWS set aside lower-cost, long-term compute capacity for you by purchasing a reserved instance. (You're not actually "purchasing" something called a *reserved instance [RI]*; instead, you're just buying the right to use a regular EC2 instance at a discounted price.) You can also purchase an instance through the Amazon EC2 Reserved Instance Marketplace, where you effectively take over the time remaining on a reserved instance owned by other AWS users who have changed their usage plans. Just as you would when launching a normal EC2 instance, you select an RI that matches the configuration details you're after. Therefore, you'll search by tenancy (default or dedicated), instance type, and platform (choosing from OSs including Linux, SUSE, Red Hat Enterprise Linux [RHEL], Microsoft Windows Server, and Microsoft SQL Server).

However, exactly how much money you'll save on your instance and how much flexibility it'll give you depends on decisions you make up front. You can, for example, pay a bit more for a *Convertible* RI that will allow you to exchange your instance later as long as the new instance has equal or greater value than the original. Convertible RIs provide a discount of up to 54 percent less than on-demand. A Standard RI can save you up to 75 percent off on-demand, while, obviously, locking you in for the duration of the term to the instance profile you initially selected.

Optionally, you can also specify a single availability zone or, for a cost premium, maintain the flexibility to run your RI in any AZ within your selected region.

You can also schedule an RI to launch within a specified recurring time window. You purchase on a schedule from the Scheduled Instances link in the EC2 Dashboard (rather than the regular Reserved Instances link).

Finally, you can pay for your RI using any one of three payment options: All Upfront (the lowest price), Partial Upfront, and No Upfront (billed hourly).

Using Savings Plans

A recent addition to EC2 payment options is a pair of purchase formats called Savings Plans. Plans are a lot like RIs in that you commit to a one- to three-year period of constant service usage, but they offer significantly more flexibility. There are two flavors of Savings Plans:

Compute Savings Plans Offer up to 66 percent price reductions over On Demand (equivalent to Convertible RIs) for any EMR, ECS, EKS, or Fargate workload, allowing you to switch between any covered resources running in any AWS region.

EC2 Instance Plans Offer price reductions up to 72 percent over On Demand (equivalent to Standard RIs) and limit usage changes to a single AWS region.

In either case, you'll be covered for any usage up to and including your initial hourly commitment ($10/hour, say). Savings Plans will not be used to cover resources already paid for through an active RI.

EC2 Spot Instances

In Chapter 2, you also learned about EC2 spot instances—short-term "rentals" of EC2 instances that offer very low prices in exchange for volatile lifecycles and forced shutdowns. As we just did with RIs, let's look at the details. I'll begin with a few important definitions.

Spot Price The current going rate for spot instances of a given set of launch specifications (type, region, and profile values). Spot prices can rise or fall without warning, sometimes forcing a running instance to shut down.

Spot Instance Interruption   You can choose one of three ways for EC2 to react when your maximum price is exceeded by the spot price: Terminate (permanently delete all associated resources and volumes), Stop (possible only when you're using an EBS-backed AMI), and Hibernate.

Spot Instance Pool   All the unused EC2 instances matching a particular set of launch specifications. Spot fleets are drawn from matching spot instance pools.

Spot Fleet   A group of spot instances (sometimes including on-demand instances) that is launched to meet a spot fleet request. A spot fleet can be made up of instances using multiple launch specifications drawn from multiple spot instance pools.

Request Type   When you configure your request from the Request Spot Instances page, you're offered three request types: Request (a one-time instance request), Request And Maintain (to maintain target capacity using a fleet), and Reserve For Duration (request an uninterrupted instance for between one and six hours).

You can configure your request for spot instances from the EC2 Dashboard. You use the Request Spot Instances page to set details such as your total target capacity (the maximum instances or vCPUs you want running), the AMI and instance type you want launched, whether you want to incorporate load balancing, and any user data you want to pass to your instances at launch time.

You can define your workload by either using a private AMI you built for the task or by passing in user data (the way you did for yourself in Exercise 11.1 in Chapter 11, "The Performance Efficiency Pillar").

The Spot Advisor, which is available from the Spot Instances Dashboard, can helpfully recommend sample configuration profiles for a few common tasks. Once you accept a recommendation, you can either customize it or send it right to the fleet configuration stage with a single click.

Why not work through Exercise 13.3 to see how you can request a simple spot fleet using the AWS CLI?

EXERCISE 13.3

Request a Spot Fleet Using the AWS CLI

1. Before you'll be able to launch your spot request using the AWS CLI, you'll need to gather some information from your AWS account. From the Quick Start tab in the EC2 Launch Instance dialog, copy the AMI ID of, say, the Ubuntu LTS AMI and save it; you'll need it later.

2. From the VPC Dashboard, click the Subnets link on the left and then note the subnet IDs of two subnets in a VPC (your default VPC should work fine). While still in the VPC Dashboard, click Security Groups and note the group ID of a security group that fits the permissions your spot instances will require. Since this is only a simple demo, anything will do. If there aren't any existing groups, create a new one.

EXERCISE 13.3 *(continued)*

3. Now you'll need to create an IAM role that will give the spot manager permission to start and stop EC2 instances. From the IAM Roles page, click Create Role, select AWS Service, select EC2, and then select EC2 – Fleet. Then click Next: Permissions.

4. You'll see the AWSEC2FleetServiceRolePolicy on the next screen. Click Next: Tags, and then Next: Review. You can add a suffix to the role name if you like and then create the role. You'll see your new role in the list of IAM roles. Click it and copy its ARN.

5. With all that done, create a Config.json file that looks something like this (using the correct values from your account):

    ```
    {
        "SpotPrice": "0.04",
        "TargetCapacity": 2,
        "IamFleetRole": "arn:aws:iam::123456789012:role/my-spot-fleet-role",
        "LaunchSpecifications": [
            {
                "ImageId": "ami-1a2b3c4d",
                "SecurityGroups": [
                    {
                        "GroupId": "sg-1a2b3c4d"
                    }
                ],
                "InstanceType": "t2-micro",
                "SubnetId": "subnet-1a2b3c4d, subnet-3c4d5e6f",
            }
        ]
    }
    ```

6. Now run this CLI command pointing to the Config.json file you just created:

    ```
    aws ec2 request-spot-fleet
      --spot-fleet-request-config file://Config.json
    ```

7. The AWS CLI may complain about errors, so carefully read the messages and go back to figure out what went wrong. When the command returns only a spot fleet ID that looks something like the following output example, you'll know you were successful:

    ```
    {
        "SpotFleetRequestId": "sfr-6b8225fe-b6d2-4c58-94d7-d14fdd7877be"
    }
    ```

8. That alone is no proof that everything is running properly. You should head over to the Spot Requests page of the EC2 Dashboard and select the entry for your fleet. Click the History tab in the bottom half of the page and look for messages. Correct any errors that might appear, and if necessary, cancel your request and launch a new one with your corrected configuration.
9. Don't forget to shut down your resources when you're done with them.

Auto Scaling

You've already learned about efficiently increasing your compute resources to meet demand using Auto Scaling. But don't forget that an Auto Scaling configuration should also be set to scale your instances *down* as demand drops. You do this by configuring the scaler to decrease the group size when the average usage of your running instances drops below a set threshold. Doing so can make a big difference to your costs and is one of the greatest benefits of the entire cloud computing model.

Elastic Block Store Lifecycle Manager

The EBS volumes attached to your EC2 instances are well protected by replication, but as with all technology, there's no guarantee that *nothing* bad will ever happen. The best practice has always been to back up your volumes by making snapshots, but adding a new snapshot every few hours can, over time, add up to some intimidating storage costs.

The obvious solution is to delete older snapshots, leaving only fresher copies online. Since, however, you're not likely to get around to regularly doing that on your own, you're better off automating. The EBS Lifecycle Manager (accessed through the EC2 Dashboard) lets you create a rotation policy where you set the frequency at which new snapshots will be created and the maximum number of snapshots to retain.

Summary

AWS budgets can be created to watch all or selected areas of your account usage and alert you when costs approach specified thresholds. You can use cost allocation tags to efficiently focus your budgets on subsets of your running resources.

You can track historical billing costs, rates, and product attributes using both Cost Explorer and Cost and Usage Reports. Reports, however, is designed for the kind of big data analytics needed by large-scale operations.

You can consolidate billing and access management administration of multiple accounts belonging to a single owner using AWS Organizations.

AWS Trusted Advisor monitors and reports on your account configuration status as it relates to best practices and account limits. The Simple Monthly Calculator and AWS Total Cost of Ownership are important tools for helping you find the most cost-effective and efficient way to build your application stack.

Matching the right EC2 instance type and, where appropriate, running container workloads, are effective ways to maximize your use—and minimize the costs—of server resources.

Your use of EC2 reserved instances can be optimized by selecting a convertible RI for greater long-term flexibility and by scheduling reserve availability for recurring blocks of time.

Spot instances can be requested individually or as a fleet and can be set to react to shutdowns in ways that best fit your specific application needs.

Exam Essentials

Know how to effectively monitor costs incurred by your AWS account resource use. AWS offers multiple tools to track your billing and usage activity to help you anticipate and prevent overages, spot and understand usage trends, and deeply analyze usage data. Tracking tools include Budgets, Cost Explorer, Cost and Usage Reports, and Trusted Advisor.

Know how to simulate and anticipate the costs of AWS application stacks. The Simple Monthly Calculator lets you quickly model multiple alternative resource stacks to accurately estimate and compare costs. The AWS Total Cost of Ownership calculator is built to closely compare the cost of deploying an application stack on AWS versus the costs you'd face doing it on-premises.

Understand how to match your application to the right EC2 instance type. An EC2 instance type is built on a hardware profile that makes it particularly efficient for certain kinds of workloads. You should be familiar with at least the general instance family categories and their use cases.

Know how to configure EC2 reserved instances for a wide range of uses. You can reserve RIs as Convertible to allow you to exchange your instance type with another one during the reservation term or to permit use in any availability zone within your region. You can also schedule recurring time blocks for your RI.

Know how to use spot fleets to provide multiple concurrent instances. Complex combinations of multiple customized spot instances can be configured using prebuilt AMIs, durable capacity preferences, and lifecycle behavior.

Review Questions

1. Which of the following best describes the AWS Free Tier?
   A. Free access to AWS services for a new account's first month
   B. Free access to all instance types of AWS EC2 instances for new accounts
   C. Free access to basic levels of AWS services for a new account's first year
   D. Unlimited and open-ended access to the "free tier" of most AWS services

2. Which of the following storage classes provides the least expensive storage *and* transfer rates?
   A. Amazon S3 Glacier
   B. Amazon S3 Standard-Infrequent Access
   C. Amazon S3 Standard
   D. Amazon S3 One Zone-Infrequent Access

3. Which AWS service is best suited to controlling your spending by sending email alerts?
   A. Cost Explorer
   B. Budgets
   C. Organizations
   D. TCO Calculator

4. Your AWS infrastructure is growing and you're beginning to have trouble keeping track of what you're spending. Which AWS service is best suited to analyzing account usage data at scale?
   A. Trusted Advisor
   B. Cost Explorer
   C. Budgets
   D. Cost and Usage Reports

5. Your company wants to more closely coordinate the administration of its multiple AWS accounts, and AWS Organizations can help it do that. How does that change your security profile? (Choose three.)
   A. An organization-level administration account breach is potentially more damaging.
   B. User permissions can be controlled centrally at the organization level.
   C. You should upgrade to use only specially hardened organization-level VPCs.
   D. Standard security best practices such as MFA and strong passwords are even more essential.
   E. You should upgrade all of your existing security groups to account for the changes.

6. Which of the following resource states are monitored by AWS Trusted Advisor? (Choose two.)
   A. Route 53 routing failures
   B. Running but idle EC2 instances
   C. S3 buckets with public read access permissions
   D. EC2 Linux instances that allow root account SSH access
   E. Unencrypted S3 bucket data transfers

7. You're planning a new AWS deployment, and your team is debating whether they'll be better off using an RDS database or one run on an EC2 instance. Which of the following tools will be most helpful?
   A. TCO Calculator
   B. AWS Pricing Calculator
   C. Trusted Advisor
   D. Cost and Usage Reports

8. Which of the following is not a metric you can configure an AWS budget to track?
   A. EBS volume capacity
   B. Resource usage
   C. Reserve instance coverage
   D. Resource cost

9. Which of the following statements are true of cost allocation tags? (Choose two.)
   A. Tags can take up to 24 hours before they appear in the Billing and Cost Management dashboard.
   B. Tags can't be applied to resources that were launched before the tags were created.
   C. You're allowed five free budgets per account.
   D. You can activate and manage cost allocation tags from the Tag Editor page.

10. Your online web store normally requires three EC2 instances to handle traffic but experiences a twofold increase in traffic for the two summer months. Which of the following approaches makes the most sense?
    A. Run three on-demand instances 12 months per year and schedule six reserve instances for the summer months.
    B. Run three spot instances for the summer months and three reserve instances 12 months/year.
    C. Run nine reserve instances for 12 months/year.
    D. Run three reserve instances 12 months/year and purchase three scheduled reserve instances for the summer months.

11. Which of the following settings do you *not* need to provide when configuring a reserved instance?
    A. Payment option
    B. Standard or Convertible RI
    C. Interruption policy
    D. Tenancy

12. Your new web application requires multiple EC2 instances running 24/7 and you're going to purchase reserved instances. Which of the following payment options is the most expensive when configuring a reserve instance?
    A. All Upfront
    B. Partial Upfront
    C. No Upfront
    D. Monthly

13. Which of the following benefits of containers such as Docker can significantly reduce your AWS compute costs? (Choose two.)
    A. Containers can launch quickly.
    B. Containers can deliver increased server density.
    C. Containers make it easy to reliably replicate server environments.
    D. Containers can run using less memory than physical machines.

14. Which of the following is the best usage of an EC2 reserved instance?
    A. An application that will run continuously for six months straight
    B. An application that will run continuously for 36 months straight
    C. An application that runs only during local business hours
    D. An application that runs at unpredictable times and can survive unexpected shutdowns

15. Which of the following describes "unused EC2 instances matching a particular set of launch specifications"?
    A. Request type
    B. Spot instance interruption
    C. Spot fleet
    D. Spot instance pool

16. Which of the following best describes a spot instance interruption?
    A. A spot instance interruption occurs when the spot price rises above your maximum.
    B. A spot instance interruption is the termination of a spot Instance when its workload completes.
    C. A spot instance interruption occurs when a spot request is manually restarted.
    D. A spot instance interruption is the result of a temporary outage in an AWS data center.

17. Which of the following describes the maximum instances or vCPUs you want running?
   A. Spot instance pool
   B. Target capacity
   C. Spot maximum
   D. Spot cap

18. You need to make sure your EBS volumes are regularly backed up, but you're afraid you'll forget to remove older snapshot versions, leading to expensive data bloat. What's the best solution to this problem?
   A. Configure the EBS Lifecycle Manager.
   B. Create a script that will regularly invoke the AWS CLI to prune older snapshots.
   C. Configure an EBS Scheduled Reserved Instance.
   D. Tie a string to your finger.
   E. Configure an S3 Lifecycle configuration policy to remove old snapshots.

19. Which of these AWS CLI commands will launch a spot fleet?
   A. `aws ec2 request-fleet --spot-fleet-request-config file://Config.json`
   B. `aws ec2 spot-fleet --spot-fleet-request-config file://Config.json`
   C. `aws ec2 launch-spot-fleet --spot-fleet-request-config /file://Config.json`
   D. `aws ec2 request-spot-fleet --spot-fleet-request-config /file://Config.json`

20. Which of the following elements is *not* something you'd include in your spot fleet request?
   A. Availability zone
   B. Target capacity
   C. Platform (the instance OS)
   D. AMI

Chapter 14

The Operational Excellence Pillar

THE AWS CERTIFIED SOLUTIONS ARCHITECT ASSOCIATE EXAM OBJECTIVES COVERED IN THIS CHAPTER MAY INCLUDE, BUT ARE NOT LIMITED TO, THE FOLLOWING:

✓ **Domain 1: Design Resilient Architecture**

- 1.2 Design highly available and/or fault-tolerant architectures
- 1.3 Design decoupling mechanisms using AWS services

Introduction

Operations include the ongoing activities required to keep your applications running smoothly and securely on AWS. Achieving operational excellence requires automating some or all of the tasks that support your organization's workloads, goals, and compliance requirements.

The key prerequisite to automating is to define your infrastructure and operations as annotated code. Unlike manual tasks, which can be time-consuming and are prone to human error, tasks embodied in code are repeatable and can be carried out rapidly. It's also easier to track code changes over time and repurpose it for testing and experimentation. For us humans, annotated code also functions as de facto documentation of your infrastructure configurations and operational workflows. In this chapter, you'll learn how to leverage the following AWS services to automate the deployment and operation of your AWS resources:

- CloudFormation
- AWS Developer Tools, which include the following:
  - CodeCommit
  - CodeDeploy
  - CodePipeline
- AWS Systems Manager
- AWS Landing Zone

CloudFormation

As you learned in Chapter 11, "The Performance Efficiency Pillar," CloudFormation uses templates to let you simultaneously deploy, configure, and document your infrastructure as code. Because CloudFormation templates are code, you can store them in a version-controlled repository just as you would any other codebase. The resources defined by a template compose a *stack*. When you create a stack, you must give it a name that's unique within the account.

One advantage of CloudFormation is that you can use the same template to build multiple identical environments. For example, you can use a single template to define and deploy two different stacks for an application, one for production and another for development. This approach would ensure both environments are as alike as possible. In this section, we'll consider two templates: `network-stack.json` and `web-stack.json`. To download these templates, visit `awscsa.github.io`.

Creating Stacks

In a CloudFormation template, you define your resources in the `Resources` section of the template. You give each resource an identifier called a *logical ID*. For example, the following snippet of the `network-stack.json` template defines a VPC with the logical ID of `PublicVPC`:

```
"Resources": {
      "PublicVPC": {
         "Type": "AWS::EC2::VPC",
         "Properties": {
            "EnableDnsSupport": "true",
            "EnableDnsHostnames": "true",
            "CidrBlock": "10.0.0.0/16"
         }
      }
   }
```

The logical ID, also sometimes called the *logical name*, must be unique within a template. The VPC created by CloudFormation will have a VPC ID, such as `vpc-0380494054677f4b8`, also known as the *physical ID*.

To create a stack named `Network` using a template stored locally, you can issue the following AWS command-line interface (CLI) command:

```
aws cloudformation create-stack --stack-name Network --template-body file://network-stack.json
```

CloudFormation can also read templates from an S3 bucket. To create a stack named using a template stored in an S3 bucket, you can issue the AWS command-line interface (CLI) command in the following format:

```
aws cloudformation create-stack --stack-name Network --template-url https://s3.amazonaws.com/cf-templates-c23z8b2vpmbb-us-east-1/network-stack.json
```

You can optionally define *parameters* in a template. A parameter lets you pass custom values to your stack when you create it, as opposed to hard-coding values into the template. For instance, instead of hard-coding the Classless Interdomain Routing (CIDR) block into a template that creates a VPC, you can define a parameter that will prompt for a CIDR when creating the stack.

> **NOTE** Stack names and logical IDs are case-sensitive.

Deleting Stacks

You can delete a stack from the web console or the AWS command-line interface (CLI). For instance, to delete the stack named `Network`, issue the following command:

```
aws cloudformation delete-stack --stack-name Network
```

If termination protection is not enabled on the stack, CloudFormation will immediately delete the stack and all resources that were created by it.

Using Multiple Stacks

You don't have to define all of your AWS infrastructure in a single stack. Instead, you can break your infrastructure across different stacks. A best practice is to organize stacks by lifecycle and ownership. For example, the network team may create a stack named `Network` to define the networking infrastructure for a web-based application. The network infrastructure would include a virtual private cloud (VPC), subnets, Internet gateway, and a route table. The development team may create a stack named `Web` to define the runtime environment, including a launch template, Auto Scaling group, application load balancer, IAM roles, instance profile, and a security group. Each of these teams can maintain their own stack.

When using multiple stacks with related resources, it's common to need to pass values from one stack to another. For example, an application load balancer in the `Web` stack needs the logical ID of the VPC in the `Network` stack. You therefore need a way to pass the VPC's logical ID from the `Network` stack to the `Web` stack. There are two ways to accomplish this: using nested stacks and exporting stack output values.

Nesting Stacks

Because CloudFormation stacks are AWS resources, you can configure a template to create additional stacks. These additional stacks are called *nested stacks*, and the stack that creates them is called the *parent stack*. To see how this works, we'll consider the `network-stack.json` and `web-stack.json` templates.

In the template for the `Web` stack (`web-stack.json`), you'd define the template for the `Network` stack as a resource, as shown by the following snippet:

```
"Resources": {
        "NetworkStack" : {
            "Type" : "AWS::CloudFormation::Stack",
            "Properties" : {
                "TemplateURL" : "https://s3.amazonaws.com/cf-templates-c23z8b2vpmbb-us-east-1/network-stack.json"
            }
        },
```

The logical ID of the stack is `NetworkStack`, and the `TemplateURL` indicates the location of the template in S3. When you create the `Web` stack, CloudFormation will automatically create the `Network` stack first.

The templates for your nested stacks can contain an `Outputs` section where you define which values you want to pass back to the parent stack. You can then reference these values in the template for the parent stack. For example, the `network-stack.json` template defines an output with the logical ID `VPCID` and the value of the VPC's physical ID, as shown in the following snippet:

```
"Outputs": {
    "VPCID": {
        "Description": "VPC ID",
        "Value": {
            "Ref": "PublicVPC"
        }
    },
```

The `Ref` intrinsic function returns the physical ID of the `PublicVPC` resource. The parent `web-stack.json` template can then reference this value using the `Fn::GetAtt` intrinsic function, as shown in the following snippet from the `Resources` section:

```
"ALBTargetGroup": {
  "Type": "AWS::ElasticLoadBalancingV2::TargetGroup",
  "Properties": {
    "VpcId": { "Fn::GetAtt" : [ "NetworkStack", "Outputs.VPCID" ] },
```

Follow the steps in Exercise 14.1 to create a nested stack.

EXERCISE 14.1

Create a Nested Stack

In this exercise, you'll create a nested stack that creates an EC2 Auto Scaling group and the supporting network infrastructure. To complete this exercise, you'll need to know the name of a Secure Shell (SSH) key pair in the region you're using.

1. Visit awscsa.github.io and click Chapter 14. Download the web-stack.json and network-stack.json CloudFormation templates.

2. Create an S3 bucket and upload the network-stack.json template to the bucket.

3. Edit the web-stack.json template and locate the NetworkStack resource. Change the TemplateURL value to the URL of the network-stack.json template in S3.

4. Upload the web-stack.json template to the same S3 bucket you used in step 2.

EXERCISE 14.1 *(continued)*

5. Run the following AWS CLI command:

   ```
   aws cloudformation create-stack --stack-name web-stack
   --template-url [URL of the web-stack.json template]
   --parameters ParameterKey=KeyName,ParameterValue=[the name of your SSH keypair]
   --capabilities CAPABILITY_NAMED_IAM
   ```

6. CloudFormation will begin to create the Web stack and the nested network stack. Enter the following command to view the stacks:

   ```
   aws cloudformation describe-stacks
   ```

7. To delete the stacks, run the following command:

   ```
   aws cloudformation delete-stack --stack-name web-stack
   ```

Exporting Stack Output Values

If you want to share information with stacks outside of a nested hierarchy, you can selectively export a stack output value by defining it in the Export field of the Output section, as follows:

```
"Outputs": {
       "VPCID": {
           "Description": "VPC ID",
           "Value": {
               "Ref": "PublicVPC"
           },
           "Export": {
               "Name": {
                   "Fn::Sub": "${AWS::StackName}-VPCID"
               }
           }
       },
```

> **NOTE** AWS::StackName is a pseudo-parameter that returns the name of the stack.

Any other template in the same account and region can then import the value using the Fn::ImportValue intrinsic function, as follows:

```
"ALBTargetGroup": {
  "Type": "AWS::ElasticLoadBalancingV2::TargetGroup",
```

```
"Properties": {
  "VpcId": { "Fn::ImportValue" : {"Fn::Sub": "${NetworkStackName}-VPCID"} },
```

Keep in mind that you can't delete a stack if another stack is referencing any of its outputs.

Stack Updates

If you need to reconfigure a resource in a stack, the best way is to modify the resource configuration in the source template and then either perform a *direct update* or create a *change set*.

Direct Update

To perform a direct update, upload the updated template. CloudFormation will ask you to enter any parameters the template requires. CloudFormation will deploy the changes immediately, modifying only the resources that changed in the template.

Change Set

If you want to understand exactly what changes CloudFormation will make, create a change set instead. You submit the updated template, and once you create the change set, CloudFormation will display a list of every resource it will add, remove, or modify. You can then choose to execute the change set to make the changes immediately, delete the change set, or do nothing. You can create multiple change sets using different templates, compare them, and then choose which one to execute. This approach is useful if you want to compare several different configurations without having to create a new stack each time.

> You can also use change sets to create new stacks.

Update Behavior

How CloudFormation updates a resource depends on the update behavior of the resource's property that you're updating. Update behaviors can be one of the following:

Update with No Interruption   The resource is updated with no interruption and without changing its physical ID. For example, changing the IAM instance profile attached to an EC2 instance doesn't require an interruption to the instance.

Update with Some Interruption   The resource encounters a momentary interruption but retains its physical ID. This happens when, for example, you change the instance type for an EBS-backed EC2 instance.

Replacement   CloudFormation creates a new resource with a new physical ID. It then points dependent resources to the new resource and deletes the original resource. For example, if you change the availability zone for an instance defined in a template, CloudFormation will create a new instance in the availability zone defined in the updated template and then delete the original instance.

Preventing Updates to Specific Resources

To prevent specific resources in a stack from being modified by a stack update, you can create a *stack policy* when you create the stack. If you want to modify a stack policy or apply one to an existing stack, you can do so using the AWS CLI. You can't remove a stack policy.

A stack policy follows the same format as other resource policies and consists of the same `Effect`, `Action`, `Principal`, `Resource`, and `Condition` elements. The `Effect` element functions the same as in other resource policies, but the `Action`, `Principal`, `Resource`, and `Condition` elements differ as follows:

> Action   The action must be one of the following:
>
>> Update:Modify   Allows updates to the specific resource only if the update will modify and not replace or delete the resource
>>
>> Update:Replace   Allows updates to the specific resource only if the update will replace the resource
>>
>> Update:Delete   Allows updates to the specific resource only if the update will delete the resource
>>
>> Update:*   Allows all update actions
>
> Principal   The principal must always be the wildcard (*). You can't specify any other principal.
>
> Resource   This element specifies the logical ID of the resource to which the policy applies. You must prefix it with the text `LogicalResourceId/`.
>
> Condition   This element specifies the resource types, such as `AWS::EC2::VPC`. You can use wildcards to specify multiple resource types within a service, such as `AWS::EC2::*` to match all EC2 resources. If you use a wildcard, you must use the `StringLike` condition. Otherwise, you can use the `StringEquals` condition.

The following stack policy document named `stackpolicy.json` allows all stack updates except for those that would replace the `PublicVPC` resource. Such an update would include changing the VPC CIDR.

```
{
  "Statement" : [
    {
      "Effect" : "Allow",
      "Action" : "Update:*",
      "Principal": "*",
      "Resource" : "*"
    },
```

```
    {
      "Effect" : "Deny",
      "Action" : "Update:Replace",
      "Principal": "*",
      "Resource" : "LogicalResourceId/PublicVPC",
      "Condition" : {
        "StringLike" : {
            "ResourceType" : ["AWS::EC2::VPC"]
        }
      }
    }
  ]
}
```

> **NOTE** A stack policy cannot prevent a principal from updating a resource directly, deleting a resource, or deleting an entire stack. If you need to control which principals can modify a stack or resource, you should use IAM policies.

CloudFormation doesn't preemptively check whether an update will violate a stack policy. If you attempt to update a stack in such a way that's prevented by the stack policy, CloudFormation will still attempt to update the stack. The update will fail only when CloudFormation attempts to perform an update prohibited by the stack policy. Therefore, when updating a stack, you must verify that the update succeeded; don't just start the update and walk away.

Overriding Stack Policies

You can temporarily override a stack policy when doing a direct update. When you perform a direct update, you can specify a stack policy that overrides the existing one. CloudFormation will apply the updated policy during the update. After the update is complete, CloudFormation will revert to the original policy.

CodeCommit

Git (git-scm.com) is a free and open source version control system invented by Linus Torvalds to facilitate collaborative software development projects. However, people often use Git to store and track a variety of file-based assets, including source code, scripts, documents, and binary files. Git stores these files in a *repository*, colloquially referred to as a *repo*. The AWS CodeCommit service hosts private Git-based repositories for version-controlled files, including code, documents, and binary files. CodeCommit provides advantages over other private Git repositories, including the following:

- Encryption at rest using an AWS-managed key stored in KMS
- Encryption in transit using HTTPS or SSH
- No size limitation for repositories
- 2 GB size limitation for a single file
- Integration with other AWS services

A Git repository and S3 have some things in common. They both store files and provide automatic versioning, allowing you to revert to a previous version of a deleted or overwritten file. But Git tracks changes to individual files using a process called *differencing*, allowing you to see what changed in each file, who changed it, and when. It also allows you to create additional *branches*, which are essentially snapshots of an entire repository. For example, you could take an existing source code repository, create a new branch, and experiment on that branch without causing any problems in the original or master branch.

Creating a Repository

You create a repository in CodeCommit using the AWS Management Console or the AWS CLI. When you create a new repository in CodeCommit, it is always empty. You can add files to the repository in three ways:

- Upload files using the CodeCommit management console.
- Use Git to clone the empty repository locally, add files to it, commit those changes, and then push those changes to CodeCommit.
- Use Git to push an existing local repository to CodeCommit.

Repository Security

CodeCommit uses IAM policies to control access to repositories. To help you control access to your repositories, AWS provides three managed policies. Each of the following policies allows access to all CodeCommit repositories using both the AWS Management Console and Git.

AWSCodeCommitFullAccess   This policy provides unrestricted access to CodeCommit. This is the policy you'd generally assign to repository administrators.

AWSCodeCommitPowerUser   This policy provides near full access to CodeCommit repositories but doesn't allow the principal to delete repositories. This is the policy you'd assign to users who need both read and write access to repositories.

AWSCodeCommitReadOnly   This policy grants read-only access to CodeCommit repositories.

If you want to restrict access to a specific repository, you can copy one of these policies to your own customer-managed policy and specify the ARN of the repository.

Interacting with a Repository Using Git

Most users will interact with a CodeCommit repository via the Git command-line interface, which you can download from git-scm.com. Many integrated development environments (IDEs) such as Eclipse, IntelliJ, Visual Studio, and Xcode provide their own user-friendly Git-based tools.

Only IAM principals can interact with a CodeCommit repository. AWS recommends generating a Git username and password for each IAM user from the IAM Management Console. You can generate up to two Git credentials per user. Since CodeCommit doesn't use resource-based policies, it doesn't allow anonymous access to repositories.

If you don't want to configure IAM users or if you need to grant repository access to a federated user or application, you can grant repository access to a role. You can't assign Git credentials to a role, so instead you must configure Git to use the AWS CLI Credential Helper to obtain temporary credentials that Git can use.

Git-based connections to CodeCommit must be encrypted in transit using either HTTPS or SSH. The easiest option is HTTPS, since it requires inputting only your Git credentials. However, if you can't use Git credentials—perhaps because of security requirements—you can use SSH. If you go this route, you must generate public and private SSH keys for each IAM user and upload the public key to AWS. The user must also have permissions granted by the IAMUserSSHKeys AWS-managed policy. Follow the steps in Exercise 14.2 to create your own CodeCommit repository.

EXERCISE 14.2

Create and Interact with a CodeCommit Repository

In this exercise, you'll create a CodeCommit repository and learn how to use Git to work with it.

1. On the same machine on which you installed the AWS CLI, download Git from git-scm.com and install it.

2. Use the following AWS CLI command to create a repository named myrepo:

   ```
   aws codecommit create-repository --repository-name myrepo
   ```

   You should see output listing the URL and ARN of the repository, like so:

   ```
   {
       "repositoryMetadata": {
           "repositoryName": "myrepo",
           "cloneUrlSsh": "ssh://git-codecommit.us-east-1.amazonaws.com/v1/repos/myrepo",
           "lastModifiedDate": 1540919473.55,
           "repositoryId": "f0d6111a-db12-461c-8bc4-a3727c0e8481",
           "cloneUrlHttp": "https://git-codecommit.us-east-1.amazonaws.com/v1/repos/myrepo",
   ```

EXERCISE 14.2 *(continued)*

```
            "creationDate": 1540919473.55,
            "Arn": "arn:aws:codecommit:us-east-1:xxxxxxxxxxxx:myrepo",
            "accountId": "xxxxxxxxxxxx"
        }
    }
```

3. In the IAM Management Console, create a new IAM policy by importing the AWSCodeCommitPowerUser policy. For the CodeCommit service permissions, specify as the resource the repository's ARN from step 2. Give the policy a name of your choice, such as **CodeCommit-myrepo-PowerUser**.

4. Create a new IAM user with programmatic access. Assign the user the policy you created in step 3.

5. In the IAM Management Console, click Users on the menu on the left and then click the name of the user you just created.

6. Select the Security Credentials tab.

7. Under HTTPS Git Credentials For AWS CodeCommit, click the Generate button. Note the username and password that IAM generates.

8. At a command prompt, issue the following command to clone the myrepo repository:

    ```
    git clone https://git-codecommit.us-east-1.amazonaws.com/v1/repos/myrepo
    ```

    This is the same URL you saw in step 2.

9. Git will prompt you for the username and password IAM generated in step 7. Enter these and you should see output indicating that Git successfully cloned the repository, as follows:

    ```
        Cloning into 'myrepo'...
    Username for 'https://git-codecommit.us-east-1.amazonaws.com': user-at-xxxxxxxxxxxx
    Password for 'https://user-at-xxxxxxxxxxxx@git-codecommit.us-east-1.amazonaws.com':
        warning: You appear to have cloned an empty repository.
        Checking connectivity... done.
    ```

10. Descend into the local repository by typing **cd myrepo**.

11. Before you can add files to the local repository, you'll need to set up Git by configuring a name and email address. These can be anything you want, but Git does require them to be set as shown in the following two commands (replace the values in quotes to a username and email address that you prefer):

    ```
    git config user.name "Ben Piper"
    git config user.email "ben@benpiper.com"
    ```

12. Create a file to add to the repo by entering `echo test > file.txt`.
13. Instruct Git to stage the file for addition to the local repository by typing `git add file.txt`.
14. Commit the changes to the repository by typing `git commit -m "Add file .txt"`. The text in quotes can be whatever you want but should be a description of the changes you've made. You should see output similar to the following:

    ```
    [master (root-commit) bc914ea] Add file.txt
     1 file changed, 1 insertion(+)
     create mode 100644 file.txt
    ```

15. Push the changes up to the CodeCommit repository using the command `git push`. Git will prompt you for your Git credentials. Once you've entered them successfully, Git will display output similar to the following:

    ```
    Counting objects: 3, done.
    Writing objects: 100% (3/3), 218 bytes | 0 bytes/s, done.
    Total 3 (delta 0), reused 0 (delta 0)
    To https://git-codecommit.us-east-1.amazonaws.com/v1/repos/myrepo
     * [new branch]      master -> master
    ```

16. Go to the CodeCommit Management Console and click Repositories on the menu on the left.
17. Click the myrepo repository. You should see the file named `file.txt`.

CodeDeploy

CodeDeploy is a service that can deploy applications to EC2 or on-premises instances. You can use CodeDeploy to deploy binary executables, scripts, web assets, images, and anything else you can store in an S3 bucket or GitHub repository. You can also use CodeDeploy to deploy Lambda functions. CodeDeploy offers a number of advantages over manual deployments, including the following:

- Simultaneous deployment to multiple instances or servers
- Automatic rollback if there are deployment errors
- Centralized monitoring and control of deployments
- Deployment of multiple applications at once

The CodeDeploy Agent

The CodeDeploy agent is a service that runs on your Linux or Windows instances and performs the hands-on work of deploying the application onto an instance. You can install the CodeDeploy agent on an EC2 instance at launch using a user data script or you can bake it into an AMI. You can also use AWS Systems Manager to install it automatically. You'll learn about Systems Manager later in this chapter.

Deployments

To deploy an application using CodeDeploy, you must create a *deployment* that defines the compute platform, which can be EC2/on-premises or Lambda, and the location of the application source files. CodeDeploy currently supports only deployments from S3 or GitHub. CodeDeploy doesn't automatically perform deployments. If you want to automate deployments, you can use CodePipeline, which we'll cover later in this chapter.

Deployment Groups

Prior to creating a deployment, you must create a *deployment group* to define which instances CodeDeploy will deploy your application to. A deployment group can be based on an EC2 Auto Scaling group, EC2 instance tags, or on-premises instance tags.

Deployment Types

When creating a deployment group for EC2 or on-premises instances, you must specify a *deployment type*. CodeDeploy gives you the following two deployment types to give you control over how you deploy your applications.

In-Place Deployment

With an in-place deployment, you deploy the application to existing instances. In-place deployments are useful for initial deployments to instances that don't already have the application. These instances can be stand-alone or, in the case of EC2 instances, part of an existing Auto Scaling group.

On each instance, the application is stopped and upgraded (if applicable) and then restarted. If the instance is behind an elastic load balancer, the instance is deregistered before the application is stopped and then reregistered to the load balancer after the deployment to the instance succeeds. Although an elastic load balancer isn't required to perform an in-place deployment, having one already set up allows CodeDeploy to prevent traffic from going to instances that are in the middle of a deployment.

> **NOTE** You must use the in-place deployment type for on-premises instance deployments.

Blue/Green Deployment

A blue/green deployment is used to upgrade an existing application with minimal interruption. With Lambda deployments, CodeDeploy deploys a new version of a Lambda function and automatically shifts traffic to the new version. Lambda deployments always use the blue/green deployment type.

In a blue/green deployment against EC2 instances, the existing instances in the deployment group are left untouched. A new set of instances is created, to which CodeDeploy deploys the application.

Blue/green deployments require an existing Application, Classic, or Network Load balancer. CodeDeploy registers the instances to the load balancer's target group after a successful deployment. At the same time, instances in the original environment are deregistered from the target group.

Note that if you're using an Auto Scaling group, CodeDeploy will create a new Auto Scaling group with the same configuration as the original. CodeDeploy will not modify the minimum, maximum, or desired capacity settings for an Auto Scaling group. You can choose to terminate the original instances or keep them running. You may choose to keep them running if you need to keep them available for testing or forensic analysis.

Deployment Configurations

When creating your deployment group, you must also select a *deployment configuration*. The deployment configuration defines the number of instances CodeDeploy simultaneously deploys to, as well as how many instances the deployment must succeed on for the entire deployment to be considered successful. The effect of a deployment configuration differs based on the deployment type. There are three preconfigured deployment configurations you can choose from: OneAtATime, HalfAtATime, and AllAtOnce.

OneAtATime

For both in-place and blue/green deployments, if the deployment group has more than one instance, CodeDeploy must successfully deploy the application to one instance before moving on to the next one. The overall deployment succeeds if the application is deployed successfully to all but the last instance. For example, if the deployment succeeds to the first two instances in a group of three, the entire deployment will succeed. If the deployment fails to any instance but the last one, the entire deployment fails.

If the deployment group has only one instance, the overall deployment succeeds only if the deployment to the instance succeeds. For blue/green deployments, CodeDeploy reroutes traffic to each instance as deployment succeeds on the instance. If CodeDeploy is unable to reroute traffic to any instance except the last one, the entire deployment fails.

HalfAtATime

For in-place and blue/green deployments, CodeDeploy will deploy to up to half of the instances in the deployment group before moving on to the remaining instances. The entire deployment succeeds only if deployment to at least half of the instances succeeds. For blue/green deployments, CodeDeploy must be able to reroute traffic to at least half of the new instances for the entire deployment to succeed.

AllAtOnce

For in-place and blue/green deployments, CodeDeploy simultaneously deploys the application to as many instances as possible. If the application is deployed to at least one instance, the entire deployment succeeds. For blue/green deployments, the entire deployment succeeds if CodeDeploy reroutes traffic to at least one new instance.

Custom Deployment Configurations

You can also create custom deployment configurations. This approach is useful if you want to customize how many instances CodeDeploy attempts to deploy to simultaneously. The deployment must complete successfully on these instances before CodeDeploy moves onto the remaining instances in the deployment group. Hence, the value you must specify when creating a custom deployment configuration is called the number of *healthy instances*. The number of healthy instances can be a percentage of all instances in the group or a number of instances.

Lifecycle Events

An instance deployment is divided into lifecycle events, which include stopping the application (if applicable), installing prerequisites, installing the application, and validating the application. During some of these lifecycle events, you can have the agent execute a *lifecycle event hook*, which is a script of your choosing. The following are all the lifecycle events during which you can have the agent automatically run a script:

ApplicationStop   You can use this hook to stop an application gracefully prior to an in-place deployment. You can also use it to perform cleanup tasks. This event occurs prior to the agent copying any application files from your repository. This event doesn't occur on original instances in a blue/green deployment, nor does it occur the first time you deploy to an instance.

BeforeInstall   This hook occurs after the agent copies the application files to a temporary location on the instance but before it copies them to their final location. If your application files require some manipulation, such as decryption or the insertion of a unique identifier, this would be the hook to use.

AfterInstall   After the agent copies your application files to their final destination, this hook performs further needed tasks, such as setting file permissions.

ApplicationStart   You use this hook to start your application. For example, on a Linux instance running an Apache web server, this may be as simple as running a script that executes the `systemctl httpd start` command.

`ValidateService`   With this event, you can check that the application is working as expected. For instance, you may check that it's generating log files or that it's established a connection to a backend database. This is the final hook for in-place deployments that don't use an elastic load balancer.

`BeforeBlockTraffic`   With an in-place deployment using an elastic load balancer, this hook occurs first, before the instance is unregistered.

`AfterBlockTraffic`   This hook occurs after the instance is unregistered from an elastic load balancer. You can use this hook to wait for user sessions or in-process transfers to complete.

`BeforeAllowTraffic`   For deployments using an elastic load balancer, this event occurs after the application is installed and validated. You can use this hook to perform any tasks needed to warm up the application or otherwise prepare it to accept traffic.

`AfterAllowTraffic`   This is the final event for deployments using an elastic load balancer.

Notice that not all of these lifecycle events can occur on all instances in a blue/green deployment. The `BeforeBlockTraffic` event, for example, wouldn't occur on a replacement instance since it makes no sense for CodeDeploy to unregister a replacement instance from a load balancer during a deployment.

Each script run during a lifecycle event must complete successfully before CodeDeploy will allow the deployment to advance to the next event. By default, the agent will wait one hour for the script to complete before it considers the instance deployment failed. You can optionally set the timeout to a lower value, as shown in the following section.

The Application Specification File

The application specification (AppSpec) file defines where the agent should copy the application files onto your instance and what scripts it should run during the deployment process. You must place the file in the root of your application repository and name it `appspec.yml`. It consists of the following five sections:

`Version`   Currently the only allowed version of the AppSpec file is 0.0.

`OS`   Because the CodeDeploy agent works only on Linux and Windows, you must specify one of these as the operating system.

`Files`   This section specifies one or more source and destination pairs identifying the files or directories to copy from your repository to the instance.

`Permissions`   This section optionally sets ownership, group membership, file permissions, and Security-Enhanced Linux (SELinux) context labels for the files after they're copied to the instance. This applies to Amazon Linux, Ubuntu, and RedHat Enterprise Linux instances only.

Hooks   This section is where you define the scripts the agent should run at each lifecycle event. You must specify the name of the lifecycle event followed by a tuple containing the following:

> Location   This must be a full path to an executable.
>
> Timeout   You can optionally set the time in seconds that the agent will wait for the script to execute before it considers the instance deployment failed.
>
> Run As   For Amazon Linux and Ubuntu instances, this lets you specify the user the script should run as.

Note that you can specify multiple locations, timeouts, and script tuples under a single lifecycle event. Keep in mind that the total timeouts for a single lifecycle event can't exceed one hour. The following is a sample `appspec.yml` file:

```
version: 0.0
os: linux
files:
  - source: /index.html
    destination: /var/www/html/
hooks:
  BeforeInstall:
    - location: scripts/install_dependencies
      timeout: 300
      runas: root
```

Triggers and Alarms

You can optionally set up triggers to generate an SNS notification for certain deployment and instance events, such as when a deployment succeeds or fails. You can also configure your deployment group to monitor up to 10 CloudWatch alarms. If an alarm exceeds or falls below a threshold you define, the deployment will stop.

Rollbacks

You can optionally have CodeDeploy *roll back*, or revert, to the last successful revision of an application if the deployment fails or if a CloudWatch alarm is triggered during deployment. Despite the name, rollbacks are actually new deployments.

CodePipeline

CodePipeline lets you automate the different *stages* of your software development and release process. These stages are often implemented using *continuous integration* (CI) and *continuous delivery* (CD) workflows or *pipelines*. Continuous integration (CI) and continuous delivery (CD) are different but related concepts.

Continuous Integration

Continuous integration is a method whereby developers use a version control system such as Git to regularly submit or check in their changes to a common repository. This first stage of the pipeline is called the source stage.

Depending on the application, a build system may compile the code or build it into a binary file, such as an executable, AMI, or container image. This is called the *build stage*. One goal of CI is to ensure that the code developers are adding to the repository works as expected and meets the requirements of the application. Thus, the build stage may also include unit tests, such as verifying that a function given a certain input returns the correct output. This way, if a change to an application causes something to break, the developer can learn of the error and fix it early. Not all applications require a build stage. For example, a web-based application using an interpreted language like PHP doesn't need to be compiled.

Continuous Delivery

Continuous delivery incorporates elements of the CI process but also deploys the application to production. A goal of CD is to allow frequent updates to an application while minimizing the risk of failure. To do this, CD pipelines usually include a test stage. As with the build stage, the actions performed in the test stage depend on the application. For example, testing a web application may include deploying it to a test web server and verifying that the web pages display the correct content. On the other hand, testing a Linux executable that you plan to release publicly may involve deploying it to test servers running a variety of Linux distributions and versions. Of course, you always want to run such tests in a separate, non-production VPC.

The final stage is deployment, in which the application is deployed to production. Although CD can be fully automated without requiring human intervention, it's common to require manual approval before releasing an application to production. You can also schedule releases to occur regularly or during opportune times such as maintenance windows.

Because continuous integration and continuous delivery pipelines overlap, you'll often see them combined as the term *CI/CD pipeline*. Keep in mind that even though a CI/CD pipeline includes every stage from source to deployment, that doesn't mean you have to deploy to production every time you make a change. You can add logic to require a manual approval before deployment. Or you can disable transitions from one stage to the next. For instance, you may disable the transition from the test stage to the deployment stage until you're actually ready to deploy.

Creating the Pipeline

Every CodeDeploy pipeline must include at least two stages and can have up to 10. Within each stage, you must define at least one task or *action* to occur during the stage. An action can be one of the following types:

- Source
- Build
- Test
- Approval
- Deploy
- Invoke

CodePipeline integrates with other AWS and third-party *providers* to perform the actions. You can have up to 20 actions in the same stage, and they can run sequentially or in parallel. For example, during your testing stage you can have two separate test actions that execute concurrently. Note that different action types can occur in the same stage. For instance, you can perform build and test actions in the same stage.

Source

The source action type specifies the source of your application files. The first stage of a pipeline must include at least one source action and can't include any other types of actions. Valid providers for the source type are CodeCommit, S3, or GitHub.

If you specify CodeCommit or S3, you must also specify the ARN of a repository or bucket. AWS can use CloudWatch events to detect when a change is made to the repository or bucket. Alternatively, you can have CodePipeline periodically poll for changes.

To add a GitHub repository as a source, you'll have to grant CodePipeline permission to access your repositories. Whenever you update the repository, GitHub creates a webhook that notifies CodePipeline of the change.

Build

Not all applications require build actions. An interpreted language such as those used in shell scripts and declarative code such as CloudFormation templates doesn't require compiling. However, even noncompiled languages may benefit from a build stage that analyzes the code for syntax errors and style conventions.

The build action type can use AWS CodeBuild as well as third-party providers CloudBees, Jenkins, Solano CI, and TeamCity. AWS CodeBuild is a managed build service that lets you compile source code and perform unit tests. CodeBuild offers on-demand build environments for a variety of programming languages, saving you from having to create and manage your own build servers.

Test

The test action type can also use AWS CodeBuild as a provider. For testing against smartphone platforms, AWS DeviceFarm offers testing services for Android iOS and web applications. Other supported providers are BlazeMeter, Ghost Inspector, HPE StormRunner Load, Nouvola, and Runscope.

Approval

The approval action type includes only one action: manual approval. When pipeline execution reaches this action, it awaits manual approval before continuing to the next stage. If there's no manual approval within seven days, the action is denied and pipeline execution halts. You can optionally send an SNS notification, which includes a link to approve or deny the request and may include a URL for the approver to review.

Deploy

For deployment, CodePipeline offers integrations with CodeDeploy, CloudFormation, Elastic Container Service, Elastic Beanstalk, OpsWorks Stacks, Service Catalog, and XebiaLabs. Recall that CodeDeploy doesn't let you specify a CodeCommit repository as a source for your application files. But you can specify CodeCommit as the provider for the source action and CodeDeploy as the provider for the deploy action.

Invoke

If you want to run a custom Lambda function as part of your pipeline, you can invoke it by using the invoke action type. For example, you can write a function to create an EBS snapshot, perform application testing, clean up unused resources, and so on.

At least one stage must have an action that's not a source action type.

Artifacts

When you create a pipeline, you must specify an S3 bucket to store the files used during different stages of the pipeline. CodePipeline compresses these files into a zip file called an *artifact*. Different actions in the pipeline can take an artifact as an input, generate it as an output, or both.

The first stage in your pipeline must include a source action specifying the location of your application files. When your pipeline runs, CodePipeline compresses the files to create a source artifact.

If the second stage of your pipeline is a build stage, CodePipeline then unzips the source artifact and passes the contents along to the build provider. The build provider uses this as an input artifact. The build provider yields its output; let's say it's a binary file. CodePipeline takes that file and compresses it into another zip file, called an *output artifact*.

This process continues throughout the pipeline. When creating a pipeline, you must specify an IAM service role for CodePipeline to assume. It uses this role to obtain permissions to the S3 bucket. The bucket must exist in the same region as the pipeline. You can use the same bucket for multiple pipelines, but each pipeline can use only one bucket for artifact storage.

AWS Systems Manager

AWS Systems Manager, formerly known as EC2 Systems Manager and Simple Systems Manager (SSM), lets you automatically or manually perform *actions* against your AWS resources and on-premises servers.

From an operational perspective, Systems Manager can handle many of the maintenance tasks that often require manual intervention or writing scripts. For on-premises and EC2 instances, these tasks include upgrading installed packages, taking an inventory of installed software, and installing a new application. For your AWS resources, such tasks may include creating an AMI golden image from an EBS snapshot, attaching IAM instance profiles, or disabling public read access to S3 buckets. Systems Manager provides the following two capabilities:

- Actions
- Insights

Actions

Actions let you automatically or manually perform actions against your AWS resources, either individually or in bulk. These actions must be defined in *documents*, which are divided into three types:

- **Automation**—Actions you can run against your AWS resources
- **Command**—Actions you run against your Linux or Windows instances
- **Policy**—Defines metadata to collect, such as software inventory and network configurations

Automation

Automation enables you to perform actions against your AWS resources in bulk. For example, you can restart multiple EC2 instances, update CloudFormation stacks, and patch AMIs. Automation provides granular control over how it carries out its individual actions. It can perform the entire automation task in one fell swoop, or it can perform one step at a time, enabling you to control precisely what happens and when. Automation also offers rate control, so you can specify as a number or a percentage how many resources to target at once.

Run Command

While automation enables you to automate tasks against your AWS resources, run commands let you execute tasks on your managed instances that would otherwise require logging in or using a third-party tool to execute a custom script.

Systems Manager accomplishes this via an agent installed on your EC2 and on-premises *managed instances*. The Systems Manager agent is installed on all Windows Server and Amazon Linux AMIs.

> **NOTE** By default, Systems Manager doesn't have permissions to do anything on your instances. You first need to apply an IAM instance profile role that contains the permissions in the `AmazonEC2RoleforSSM` policy.

AWS offers a variety of preconfigured command documents for Linux and Windows instances; for example, the `AWS-InstallApplication` document installs software on Windows, and the `AWS-RunShellScript` document allows you to execute arbitrary shell scripts against Linux instances. Other documents include tasks such as restarting a Windows service or installing the CodeDeploy agent.

You can target instances by tag or select them individually. As with automation, you optionally may use rate limiting to control how many instances you target at once.

Session Manager

Session Manager lets you achieve interactive Bash and PowerShell access to your Linux and Windows instances, respectively, without having to open inbound ports on a security group or a network ACL or even having your instances in a public subnet. You don't need to set up a bastion host or worry about SSH keys. All Linux versions and Windows Server 2008 R2 through 2016 are supported.

You open a session using the web console or AWS CLI. You must first install the Session Manager plug-in on your local machine to use the AWS CLI to start a session. The Session Manager SDK has libraries for developers to create custom applications that connect to instances. This is useful if you want to integrate an existing configuration management system with your instances without opening ports in a security group or network ACL.

Connections made via Session Manager are secured using TLS 1.2. Session Manager can keep a log of all logins in CloudTrail and store a record of commands run within a session in an S3 bucket.

Patch Manager

Patch Manager helps you automate the patching of your Linux and Windows instances. You can individually choose instances to patch, patch according to tags, or create a *patch group*. A patch group is a collection of instances with the tag key `Patch Group`. For example, if you wanted to include some instances in the `Webservers` patch group, you'd assign tags to each instance with the tag key of `Patch Group` and the tag value of `Webservers`. Keep in mind that the tag key is case-sensitive.

Patch Manager uses *patch baselines* to define which available patches to install, as well as whether the patches will be installed automatically or require approval. AWS offers default baselines that differ according to operating system but include patches that are classified as security-related, critical, important, or required. The patch baselines for all operating systems except Ubuntu automatically approve these patches after seven days. This is called an *auto-approval delay*.

For more control over which patches get installed, you can create your own custom baselines. Each custom baseline contains one or more approval rules that define the operating system, the classification and severity level of patches to install, and an auto-approval delay.

You can also specify approved patches in a custom baseline configuration. For Windows baselines, you can specify knowledgebase and security bulletin IDs. For Linux baselines, you can specify CVE IDs or full package names. If a patch is approved, it will be installed during a maintenance window that you specify. Alternatively, you can forego a maintenance window and patch your instances immediately. Patch Manager executes the `AWS-RunPatchBaseline` document to perform patching.

State Manager

Patch Manager can help ensure your instances are all at the same patch level, but State Manager is a configuration management tool that ensures your instances have the software you want them to have and are configured in the way you define. More generally, State Manager can automatically run command and policy documents against your instances, either one time only or on a schedule. For example, you may want to install antivirus software on your instances and then take a software inventory.

To use State Manager, you must create an *association* that defines the command document to run, any parameters you want to pass to it, the target instances, and the schedule. Once you create an association, State Manager will immediately execute it against the target instances that are online. Thereafter, it will follow the schedule.

There is currently only one policy document you can use with State Manager: `AWS-GatherSoftwareInventory`. This document defines what specific metadata to collect from your instances. Despite the name, in addition to collecting software inventory, you can also have it collect network configurations; file information; CPU information; and, for Windows, registry values.

Insights

Insights aggregate health, compliance, and operational details about your AWS resources into a single area of AWS Systems Manager. Some insights are categorized according to *AWS resource groups*, which are collections of resources in an AWS region. You define a resource group based on one or more tag keys and optionally tag values. For example, you can apply the same tag key to all resources related to a particular application—EC2 instances, S3 buckets, EBS volumes, security groups, and so on. Insights are categorized, as we'll cover next.

Built-In Insights

Built-in insights are monitoring views that Systems Manager makes available to you by default. Built-in insights include the following:

AWS Config Compliance   This insight shows the total number of resources in a resource group that are compliant or noncompliant with AWS Config rules, as well as compliance by resource. It also shows a brief history of configuration changes tracked by AWS Config.

CloudTrail Events   This insight displays each resource in the group, the resource type, and the last event that CloudTrail recorded against the resource.

Personal Health Dashboard   The Personal Health Dashboard contains alerts when AWS experiences an issue that may impact your resources. For example, some service APIs occasionally experience increased latency. It also shows you the number of events that AWS resolved within the last 24 hours.

Trusted Advisor Recommendations   The AWS Trusted Advisor tool can check your AWS environment for optimizations and recommendations for cost optimization, performance, security, and fault tolerance. It will also show you when you've exceeded 80 percent of your limit for a service.

Business and Enterprise support customers get access to all Trusted Advisor checks. All AWS customers get the following security checks for free:

- Public access to an S3 bucket, particularly upload and delete access
- Security groups with unrestricted access to ports that normally should be restricted, such as TCP port 1433 (MySQL) and 3389 (Remote Desktop Protocol)
- Whether you've created an IAM user
- Whether multifactor authentication is enabled for the root user
- Public access to an EBS or RDS snapshot

Inventory Manager

The Inventory Manager collects data from your instances, including operating system and application versions. Inventory Manager can collect data for the following:
- Operating system name and version
- Applications and filenames, versions, and sizes
- Network configuration, including IP and MAC addresses
- Windows updates, roles, services, and registry values
- CPU model, cores, and speed

You choose which instances to collect data from by creating a regionwide *inventory association* by executing the `AWS-GatherSoftwareInventory` policy document. You can choose all instances in your account or select instances manually or by tag. When you choose all instances in your account, it's called a *global inventory association*, and new instances you create in the region are automatically added to it. Inventory collection occurs at least every 30 minutes.

When you configure the Systems Manager agent on an on-premises server, you specify a region for inventory purposes. To aggregate metadata for instances from different regions and accounts, you may configure Resource Data Sync in each region to store all inventory data in a single S3 bucket.

Compliance

Compliance insights show how the patch and association status of your instances stacks up against the rules you've configured. Patch compliance shows the number of instances that have the patches in their configured baseline, as well as details of the specific patches installed. Association compliance shows the number of instances that have had an association successfully executed against them.

AWS Landing Zone

Deploying your resources to different AWS accounts helps you improve the security and scalability of your AWS environment. Recall that the root user for an AWS account has full access to the account. By having your resources spread out across multiple accounts, you reduce your risk footprint. AWS Landing Zone can automatically set up a multiaccount environment according to AWS best practices. Landing Zone uses the concept of an Account Vending Machine (AVM) that allows users to create new AWS accounts that are preconfigured with a network and a security baseline. Landing Zone begins with four core accounts, each containing different services and performing different roles:

AWS Organizations Account In this account, you create and manage other member accounts. This account contains an S3 bucket with member account configuration settings.

Shared Services Account   This account contains a shared services VPC and is designed for hosting services that other accounts can access.

Log Archive Account   This account aggregates CloudTrail and AWS Config log files.

Security Account   This account contains GuardDuty, cross-account roles, and SNS topics for security notifications.

Summary

Automation is certainly nothing new. System administrators have been writing scripts for decades to automate common (and often mundane) tasks. But in the cloud, automation can be extended to infrastructure deployments as well. Because of the flexibility of cloud infrastructure, the line between infrastructure—which has traditionally been static—and operations is becoming more blurred. In the cloud, deploying infrastructure and operating it are essentially the same thing.

Despite this, there are some differences to keep in mind. Because the infrastructure belongs to AWS, you have to use the AWS web console, CLI, or SDKs to configure it. You can automate this task by using CloudFormation, or you can write your own scripts. Conversely, when it comes to configuring the operating system and applications running on instances, you can write scripts and use your own tooling, or you can use the services AWS provides, namely, the AWS Developer Tools and Systems Manager. The combination of scripts, tools, and services you use is up to you, but as an AWS architect, you need to understand all options available to you and how they all fit together.

Exam Essentials

Know how to use CloudFormation stacks.   A stack is a collection of AWS resources created by a CloudFormation template. These resources can be created, updated, and deleted as a group. Stacks can also pass information to each other through output exports and nested stacks.

Understand the use cases for CodeCommit and Git.   CodeCommit offers Git-based repositories for storing files. It provides automatic versioning and differencing, allowing multiple IAM users to collaborate on software development projects, document creation, and more.

Be able to decide whether to use a blue/green or in-place deployment.   In a blue/green deployment you deploy an application to a new set of instances and then route traffic to them, resulting in minimal downtime. An in-place deployment is appropriate when you're initially deploying an application to existing instances. It's also a good option when you don't want to create new instances but are able to safely take your existing application offline to update it.

Know how CodePipeline works. CodePipeline lets you create workflows that define the different stages of software development, all the way from coding to deployment. It can integrate with other AWS and third-party tools to perform builds and unit testing.

Understand the features of Systems Manager. Systems Manager lets you automatically or manually take actions against your AWS resources and instances in bulk or individually. It also gives you insights into the state of your resources, such as compliance status and software inventory.

Review Questions

1. When using CloudFormation to provision multiple stacks of related resources, by which of the following should you organize your resources into different stacks? (Choose two.)
   A. Cost
   B. S3 bucket
   C. Lifecycle
   D. Ownership

2. Which of the following resource properties are good candidates for definition as parameters in a CloudFormation template? (Choose two.)
   A. AMI ID
   B. EC2 key pair name
   C. Stack name
   D. Logical ID

3. You want to use nested stacks to create an EC2 Auto Scaling group and the supporting VPC infrastructure. These stacks do not need to pass any information to stacks outside of the nested stack hierarchy. Which of the following must you add to the template that creates the Auto Scaling group?
   A. An Export field to the Output section
   B. A resource of the type AWS::EC2::VPC
   C. A resource of the type AWS::CloudFormation::Stack
   D. Fn::ImportValue

4. You need to update a stack that has a stack policy applied. What must you do to verify the specific resources CloudFormation will change before updating the stack?
   A. Create a change set.
   B. Perform a direct update.
   C. Update the stack policy.
   D. Override the stack policy.

5. You've granted a developer's IAM user permissions to read and write to a CodeCommit repository using Git. What information should you give the developer to access the repository as an IAM user?
   A. IAM username and password
   B. Access key and secret key
   C. Git username and password
   D. SSH public key

6. You need grant access to a specific CodeCommit repository to only one IAM user. How can you do this?
   A. Specify the repository's clone URL in an IAM policy.
   B. Generate Git credentials only for the user.
   C. Specify the user's ARN in a repository policy.
   D. Specify the repository's ARN in an IAM policy.

7. You need to store text-based documentation for your data center infrastructure. This documentation changes frequently, and auditors need to be able to see how the documents change over time. The documents must also be encrypted at rest. Which service should you use to store the documents and why?
   A. CodeCommit, because it offers differencing
   B. S3, because it offers versioning
   C. S3, because it works with customer-managed KMS keys
   D. CodeCommit, because it works with customer-managed KMS keys

8. Which command will download a CodeCommit repository?
   A. `aws codecommit get-repository`
   B. `git clone`
   C. `git push`
   D. `git add`

9. You need to deploy an application using CodeDeploy. Where must you place your application files so that CodeDeploy can deploy them?
   A. An EBS snapshot
   B. A CodeCommit repository
   C. A self-hosted Git repository
   D. An S3 bucket

10. Which deployment type requires an elastic load balancer?
    A. In-place instance deployment
    B. Blue/green instance deployment
    C. Blue/green Lambda deployment
    D. In-place Lambda deployment

11. You want to use CodeDeploy to perform an in-place upgrade of an application running on five instances. You consider the entire deployment successful if the deployment succeeds even on only one instance. Which preconfigured deployment configuration should you use?
    A. OneAtATime
    B. HalfAtATime
    C. AllAtOnce
    D. OnlyOne

12. You want CodeDeploy to run a shell script that performs final checks against your application after allowing traffic to it and before declaring the deployment successful. Which lifecycle event hook should you use?
    A. ValidateService
    B. AfterAllowTraffic
    C. BeforeAllowTraffic
    D. AllowTraffic

13. The build stage of your software development pipeline compiles Java source code into a binary JAR file that can be deployed to a web server. CodePipeline compresses this file and puts it in an S3 bucket. What's the term for this compressed file?
    A. An artifact
    B. A provider
    C. An asset
    D. A snapshot

14. You're designing an automated continuous integration pipeline and want to ensure developers don't accidentally trigger a release to production when checking in code. What are two ways to accomplish this? (Choose two).
    A. Create a separate bucket in S3 to store artifacts for deployment.
    B. Implement an approval action before the deploy stage.
    C. Disable the transition to the deploy stage.
    D. Don't allow developers access to the deployment artifact bucket.

15. You have CloudFormation templates stored in a CodeCommit repository. Whenever someone updates a template, you want a new CloudFormation stack automatically deployed. How should you design a CodePipeline pipeline to achieve this? (Choose all that apply).
    A. Use a source action with the CodeCommit provider.
    B. Use a build action with the CloudFormation provider.
    C. Use a deploy action with the CodeCommit provider.
    D. Use a deploy action with the CloudFormation provider.
    E. Create a two-stage pipeline.
    F. Create a three-stage pipeline.
    G. Create a single-stage pipeline.

16. How many stages can you have in a pipeline?
    A. 1
    B. 10
    C. 20
    D. 21

17. You need to manually take EBS snapshots of several hundred volumes. Which type of Systems Manager document enables you to do this?
    A. Command
    B. Automation
    C. Policy
    D. Manual

18. You want to use Systems Manager to perform routine administration tasks and collect software inventory on your EC2 instances running Amazon Linux. You already have an instance profile attached to these instances. Which of the following should you do to enable you to use Systems Manager for these tasks?
    A. Add the permissions from the `AmazonEC2RoleforSSM` managed policy to the role you're using for the instance profile.
    B. Manually install the Systems Manager agent.
    C. Use Session Manager to install the Systems Manager agent.
    D. Modify the instance security groups to allow access from Systems Manager.

19. You've configured Patch Manager to patch your Windows instances every Saturday. The custom patch baseline you're using has a seven-day auto-approval delay for security-related patches. On this Monday, a critical security patch was released, and you want to push it to your instances as soon as possible. You also want to take the opportunity to install all other available security-related packages. How can you accomplish this? (Choose two).
    A. Execute the `AWS-RunPatchBaseline` document.
    B. Add the patch to the list of approved patches in the patch baseline.
    C. Change the maintenance window to occur every Monday at midnight.
    D. Set the patch baseline's auto-approval delay to zero days.

20. You've installed the Systems Manager agent on an Ubuntu instance and ensured the correct instance profile is applied. But Systems Manager Insights don't display the current network configuration. Which of the following must you do to be able to automatically collect and view the network configuration for this and future instances in the same region?
    A. Make sure the instance is running.
    B. Create a global inventory association.
    C. Execute the `AWS-GatherSoftwareInventory` policy document against the instance.
    D. Execute the `AWS-SetupManagedInstance` automation document against the instance.

Appendix

Answers to Review Questions

Chapter 1: Introduction to Cloud Computing and AWS

1. B. Elastic Beanstalk takes care of the ongoing underlying deployment details for you, allowing you to focus exclusively on your code. Lambda will respond to trigger events by running code a single time, Auto Scaling will ramp up existing infrastructure in response to demand, and Route 53 manages DNS and network routing.

2. A. CloudFront maintains a network of endpoints where cached versions of your application data are stored to provide quicker responses to user requests. Route 53 manages DNS and network routing, Elastic Load Balancing routes incoming user requests among a cluster of available servers, and Glacier provides high-latency, low-cost file storage.

3. D. Elastic Block Store provides virtual block devices (think: storage drives) on which you can install and run filesystems and data operations. It is not normally a cost-effective option for long-term data storage.

4. A, C. AWS IAM lets you create user accounts, groups, and roles and assign them rights and permissions over specific services and resources within your AWS account. Directory Service allows you to integrate your resources with external users and resources through third-party authentication services. KMS is a tool for generating and managing encryption keys, and SWF is a tool for coordinating application tasks. Amazon Cognito can be used to manage authentication for your application users, but not your internal admin teams.

5. C. DynamoDB provides a NoSQL (nonrelational) database service. Both are good for workloads that can be more efficiently run without the relational schema of SQL database engines (like those, including Aurora, that are offered by RDS). KMS is a tool for generating and managing encryption keys.

6. D. EC2 endpoints will always start with an `ec2` prefix followed by the region designation (`eu-west-1` in the case of Ireland).

7. A. An availability zone is an isolated physical data center within an AWS region. Regions are geographic areas that contain multiple availability zones, subnets are IP address blocks that can be used within a zone to organize your networked resources, and there can be multiple data centers within an availability zone.

8. B. VPCs are virtualized network environments where you can control the connectivity of your EC2 (and RDS, etc.) infrastructure. Load Balancing routes incoming user requests among a cluster of available servers, CloudFront maintains a network of endpoints where cached versions of your application data are stored to provide quicker responses to user requests, and AWS endpoints are URIs that point to AWS resources within your account.

9. C. The AWS service level agreement tells you the level of service availability you can realistically expect from a particular AWS service. You can use this information when assessing your compliance with external standards. Log records, though they can offer important historical performance metrics, probably won't be enough to prove compliance. The AWS Compliance Programs page will show you only which regulatory programs can be satisfied with AWS resources, not whether a particular configuration will meet their demands.

The AWS Shared Responsibility Model outlines who is responsible for various elements of your AWS infrastructure. There is no AWS Program Compliance tool.

10. B. The AWS Command Line Interface (CLI) is a tool for accessing AWS APIs from the command-line shell of your local computer. The AWS SDK is for accessing resources programmatically, the AWS Console works graphically through your browser, and AWS Config is a service for editing and auditing your AWS account resources.

11. A. Unlike the Basic and Developer plans (which allow access to a support associate to no or one user, respectively), the Business plan allows multiple team members.

Chapter 2: Amazon Elastic Compute Cloud and Amazon Elastic Block Store

1. A, C. Many third-party companies maintain official and supported AMIs running their software on the AWS Marketplace. AMIs hosted among the community AMIs are not always official and supported versions. Since your company will need multiple such instances, you'll be better off automating the process by bootstrapping rather than having to configure the software manually each time. The Site-to-Site VPN tool doesn't use OpenVPN.

2. B, C. The VM Import/Export tool handles the secure and reliable transfer for a virtual machine between your AWS account and local data center. A successfully imported VM will appear among the private AMIs in the region you selected. Direct S3 uploads and SSH tunnels are not associated with VM Import/Export.

3. D. AMIs are specific to a single AWS region and cannot be deployed into any other region. If your AWS CLI or its key pair was not configured properly, your connection would have failed completely. A public AMI being unavailable because it's "updating" is theoretically possible but unlikely.

4. A. Only Dedicated Host tenancy offers full isolation. Shared tenancy instances will often share hardware with operations belonging to other organizations. Dedicated instance tenancy instances may be hosted on the same physical server as other instances within your account.

5. A, E. Reserve instances will give you the best price for instances you know will be running 24/7, whereas on-demand makes the most sense for workloads that will run at unpredictable times but can't be shut down until they're no longer needed. Load balancing controls traffic routing and, on its own, has no impact on your ability to meet changing demand. Since the m5.large instance type is all you need to meet normal workloads, you'll be wasting money by running a larger type 24/7.

6. B. Spot market instances can be shut down with only a minimal (two-minute) warning, so they're not recommended for workloads that require reliably predictable service. Even if your AMI can be relaunched, the interrupted workload will still be lost. Static S3 websites don't run on EC2 infrastructure in the first place.

7. A. You can edit or even add or remove security groups from running instances and the changes will take effect instantly. Similarly, you can associate or release an elastic IP address to/from a running instance. You can change an instance type as long as you shut down the instance first. But the AMI can't be changed; you'll need to create an entirely new instance.

8. B. The first of two (and *not* three) strings in a resource tag is the *key*—the group to which the specific resource belongs. The second string is the *value*, which identifies the resource itself. If the key looks too much like the value, it can cause confusion.

9. D. Provisioned-IOPS SSD volumes are currently the only type that comes close to 20,000 IOPS. In fact, they can deliver up to 64,000 IOPS.

10. B, C, E. Options B, C, and E are steps necessary for creating and sharing such an image. When an image is created, a snapshot is automatically created from which an AMI is built. You do not, however, create a snapshot from an image. The AWS Marketplace contains only public images: hopefully, no one will have uploaded your organization's private image there!

11. A, C. The fact that instance volumes are physically attached to the host server and add nothing to an instance cost is a benefit. The data on instance volumes is ephemeral and will be lost as soon as the instance is shut down. There is no way to set termination protection for instance volumes because they're dependent on the lifecycle of their host instances.

12. C, D. By default, EC2 uses the standard address blocks for private subnets, so all private addresses will fall within these ranges: 10.0.0.0 to 10.255.255.255, 172.16.0.0 to 172.31.255.255, and 192.168.0.0 to 192.168.255.255.

13. A, B, D. Ports and source and destinations addresses are considered by security group rules. Security group rules do not take packet size into consideration. Since a security group is directly associated with specific objects, there's no need to reference the target address.

14. A, D. IAM roles define how resources access other resources. Users cannot authenticate as an instance role, nor can a role be associated with an instance's internal system process.

15. B, D. NAT instances and NAT gateways are AWS tools for safely routing traffic between private and public subnets and from there, out to the Internet. An Internet gateway connects a VPC with the Internet, and a virtual private gateway connects a VPC with a remote site over a secure VPN. A stand-alone VPN wouldn't normally be helpful for this purpose.

16. D. The client computer in an encrypted operation must always use the private key to authenticate. For EC2 instances running Windows, you retrieve the password you'll use for the GUI login using your private key.

17. B. Placement groups allow you to specify where your EC2 instances will live. Load balancing directs external user requests between multiple EC2 instances, Systems Manager provides tools for monitoring and managing your resources, and Fargate is an interface for administering Docker containers on Amazon ECS.

18. A. Lambda can be used as such a trigger. Beanstalk launches and manages infrastructure for your application that will remain running until you manually stop it, ECS manages Docker containers but doesn't necessarily stop them when a task is done, and Auto Scaling can *add* instances to an already running deployment to meet demand.

19. C. VM Import/Export will do this. S3 buckets are used to store an image, but they're not directly involved in the import operation. Snowball is a physical high-capacity storage device that Amazon ships to your office for you to load data and ship back. Direct Connect uses Amazon partner providers to build a high-speed connection between your servers and your AWS VPC.

20. B. You can modify a launch template by creating a new version of it; however, the question indicates that the Auto Scaling group was created using a launch configuration. You can't modify a launch configuration. Auto Scaling doesn't use CloudFormation templates.

21. A. Auto Scaling strives to maintain the number of instances specified in the desired capacity setting. If the desired capacity setting isn't set, Auto Scaling will attempt to maintain the number of instances specified by the minimum group size. Given a desired capacity value of 5, there should be five healthy instances. If you manually terminate two of them, Auto Scaling will create two new ones to replace them. Auto Scaling will not adjust the desired capacity or minimum group size.

22. B, C. Scheduled actions can adjust the minimum and maximum group sizes and the desired capacity on a schedule, which is useful when your application has a predictable load pattern. To add more instances in proportion to the aggregate CPU utilization of the group, implement step scaling policies. Target tracking policies adjust the desired capacity of a group to keep the threshold of a given metric near a predefined value. Simple scaling policies simply add more instances when a defined CloudWatch alarm triggers, but the number of instances added is not proportional to the value of the metric.

23. B. Automation documents let you perform actions against your AWS resources, including taking EBS snapshots. Although called automation documents, you can still manually execute them. A command document performs actions within a Linux or Windows instance. A policy document works only with State Manager and can't take an EBS snapshot. There's no manual document type.

Chapter 3: AWS Storage

1. A, C. Storage Gateway and EFS provide the required read/write access. S3 can be used to share files, but it doesn't offer low-latency access—and its eventual consistency won't work well with filesystems. EBS volumes can be used only for a single instance at a time.

2. D. In theory, at least, there's no limit to the data you can upload to a single bucket or to all the buckets in your account or to the number of times you upload (using the PUT command). By default, however, you are allowed only 100 S3 buckets per account.

3. A. HTTP (web) requests must address the s3.amazonaws.com domain along with the bucket and filenames.

4. C. A prefix is the name common to the objects you want to group, and a slash character (/) can be used as a delimiter. The bar character (|) would be treated as part of the name rather than as a delimiter. Although DNS names can have prefixes, they're not the same as prefixes in S3.

5. A, C. Client-side encryption occurs before an object reaches the bucket (i.e., before it comes to rest in the bucket). Only AWS KMS-Managed Keys provide an audit trail. AWS End-to-End managed keys doesn't exist as an AWS service.

6. A, B, E. S3 server access logs don't report the source bucket's current size. They don't track API calls—that's something covered by AWS CloudTrail.

7. C, E. The S3 guarantee only covers the physical infrastructure owned by AWS. Temporary service outages are related to "availability" and not "durability."

8. A. One Zone-IA data is heavily replicated but only within a single availability zone, whereas Reduced Redundancy data is only lightly replicated.

9. B. The S3 Standard-IA (Infrequent Access) class is guaranteed to be available 99.9 percent of the time.

10. D. S3 can't guarantee instant consistency across their infrastructure for changes to existing objects, but there aren't such concerns for newly created objects.

11. C. Object versioning must be manually enabled for each object to prevent older versions of the object from being deleted.

12. A. S3 lifecycle rules can incorporate specifying objects by prefix. There's no such thing as a lifecycle template.

13. A. Glacier offers the least expensive and most highly resilient storage within the AWS ecosystem. Reduced Redundancy is not resilient and, in any case, is no longer recommended. S3 One Zone and S3 Standard are relatively expensive.

14. B, C. ACLs are a legacy feature that isn't as flexible as IAM or S3 bucket polices. Security groups are not used with S3 buckets. KMS is an encryption key management tool and isn't used for authentication.

15. D. In this context, a principal is an identity to which bucket access is assigned.

16. B. The default expiry value for a presigned URL is 3,600 seconds (one hour).

17. A, D. The AWS Certificate Manager can (when used as part of a CloudFront distribution) apply an SSL/TLS encryption certificate to your website. You can use Route 53 to associate a DNS domain name to your site. EC2 instances and RDS database instances would never be used for static websites. You would normally not use KMS for a static website—websites are usually meant to be public and encrypting the website assets with a KMS key would make it impossible for clients to download them.

18. B. As of this writing, a single Glacier archive can be no larger than 40 TB.

19. C. Direct Connect can provide fast network connections to AWS, but it's very expensive and can take up to 90 days to install. Server Migration Service and Storage Gateway aren't meant for moving data at such scale.

20. A. FSx for Lustre and Elastic File System are primarily designed for access from Linux file systems. EBS volumes can't be accessed by more than a single instance at a time.

Chapter 4: Amazon Virtual Private Cloud

1. A. The allowed range of prefix lengths for a VPC CIDR is between /16 and /28 inclusive. The maximum possible prefix length for an IP subnet is /32, so /56 is not a valid length.

2. C. A secondary CIDR may come from the same RFC 1918 address range as the primary, but it may not overlap with the primary CIDR. 192.168.0.0/24 comes from the same address range (192.168.0.0–192.168.255.255) as the primary and does not overlap with 192.168.16.0/24; 192.168.0.0/16 and 192.168.16.0/23 both overlap with 192.168.16.0/24; and 172.31.0.0/16 is not in the same range as the primary CIDR.

3. A, D. Options A and D (10.0.0.0/24 and 10.0.0.0/23) are within the VPC CIDR and leave room for a second subnet; 10.0.0.0/8 is wrong because prefix lengths less than /16 aren't allowed; and 10.0.0.0/16 doesn't leave room for another subnet.

4. B. Multiple subnets may exist in a single availability zone. A subnet cannot span availability zones.

5. A. Every ENI must have a primary private IP address. It can have secondary IP addresses, but all addresses must come from the subnet the ENI resides in. Once created, the ENI cannot be moved to a different subnet. An ENI can be created independently of an instance and later attached to an instance.

6. D. Each VPC contains a default security group that can't be deleted. You can create a security group by itself without attaching it to anything. But if you want to use it, you must attach it to an ENI. You also attach multiple security groups to the same ENI.

7. A. An NACL is stateless, meaning it doesn't track connection state. Every inbound rule must have a corresponding outbound rule to permit traffic, and vice versa. An NACL is attached to a subnet, whereas a security group is attached to an ENI. An NACL can be associated with multiple subnets, but a subnet can have only one NACL.

8. D. An Internet gateway has no management IP address. It can be associated with only one VPC at a time and so cannot grant Internet access to instances in multiple VPCs. It is a logical VPC resource and not a virtual or physical router.

9. A. The destination 0.0.0.0/0 matches all IP prefixes and hence covers all publicly accessible hosts on the Internet. ::/0 is an IPv6 prefix, not an IPv4 prefix. An Internet gateway is the target of the default route, not the destination.

10. A. Every subnet is associated with the main route table by default. You can explicitly associate a subnet with another route table. There is no such thing as a default route table, but you can create a default route within a route table.

11. A. An instance must have a public IP address to be directly reachable from the Internet. The instance may be able to reach the Internet via a NAT device. The instance won't necessarily receive the same private IP address because it was automatically assigned. The instance will be able to reach other instances in the subnet because a public IP is not required.

12. B. Assigning an EIP to an instance is a two-step process. First you must allocate an EIP, and then you must associate it with an ENI. You can't allocate an ENI, and there's no such thing as an instance's primary EIP. Configuring the instance to use an automatically assigned public IP must occur at instance creation. Changing an ENI's private IP to match an EIP doesn't actually assign a public IP to the instance, because the ENI's private address is still private.

13. A. Internet-bound traffic from an instance with an automatically assigned public IP will traverse an Internet gateway that will perform NAT. The source address will be the instance's public IP. An instance with an automatically assigned public IP cannot also have an EIP. The NAT process will replace the private IP source address with the public IP. Option D, 0.0.0.0, is not a valid source address.

14. A. The NAT device's default route must point to an Internet gateway, and the instance's default route must point to the NAT device. No differing NACL configurations between subnets are required to use a NAT device. Security groups are applied at the ENI level. A NAT device doesn't require multiple interfaces.

15. D. A NAT gateway is a VPC resource that scales automatically to accommodate increased bandwidth requirements. A NAT instance can't do this. A NAT gateway exists in only one availability zone. There are not multiple NAT gateway types. A NAT instance is a regular EC2 instance that comes in different types.

16. A. An Internet gateway performs NAT for instances that have a public IP address. A route table defines how traffic from instances is forwarded. An EIP is a public IP address and can't perform NAT. An ENI is a network interface and doesn't perform NAT.

17. A. The source/destination check on the NAT instance's ENI must be disabled to allow the instance to receive traffic not destined for its IP and to send traffic using a source address that it doesn't own. The NAT instance's default route must point to an Internet gateway as the target. You can't assign a primary private IP address after the instance is created.

18. A. You cannot route through a VPC using transitive routing. Instead, you must directly peer the VPCs containing the instances that need to communicate. A VPC peering connection uses the AWS internal network and requires no public IP address. Because a peering connection is a point-to-point connection, it can connect only two VPCs. A peering connection can be used only for instance-to-instance communication. You can't use it to share other VPC resources.

19. A, D. Each peered VPC needs a route to the CIDR of its peer; therefore, you must create two routes with the peering connection as the target. Creating only one route is not

sufficient to enable bidirectional communication. Additionally, the instances' security groups must allow for bidirectional communication. You can't create more than one peering connection between a pair of VPCs.

20. C. Interregion VPC peering connections aren't available in all regions and support a maximum MTU of 1,500 bytes. You can use IPv4 across an inter-region peering connection but not IPv6.

21. B. VPN connections are always encrypted.

22. A, C, D. VPC peering, transit gateways, and VPNs all allow EC2 instances in different regions to communicate using private IP addresses. Direct Connect is for connecting VPCs to on-premises networks, not for connecting VPCs together.

23. B. A transit gateway route table can hold a blackhole route. If the transit gateway receives traffic that matches the route, it will drop the traffic.

24. D. Tightly coupled workloads include simulations such as weather forecasting. They can't be broken down into smaller, independent pieces, and so require the entire cluster to function as a single supercomputer.

Chapter 5: Database Services

1. A, C. Different relational databases use different terminology. A row, record, and tuple all describe an ordered set of columns. An attribute is another term for column. A table contains rows and columns.

2. C. A table must contain at least one attribute or column. Primary and foreign keys are used for relating data in different tables, but they're not required. A row can exist within a table, but a table doesn't need a row in order to exist.

3. D. The SELECT statement retrieves data from a table. INSERT is used for adding data to a table. QUERY and SCAN are commands used by DynamoDB, which is a nonrelational database.

4. B. Online transaction processing databases are designed to handle multiple transactions per second. Online analytics processing databases are for complex queries against large data sets. A key/value store such as DynamoDB can handle multiple transactions per second, but it's not a relational database. There's no such thing as an offline transaction processing database.

5. B. Although there are six database engines to choose from, a single database instance can run only one database engine. If you want to run more than one database engine, you will need a separate database instance for each engine.

6. B, C. MariaDB and Aurora are designed as binary drop-in replacements for MySQL. PostgreSQL is designed for compatibility with Oracle databases. Microsoft SQL Server does not support MySQL databases.

7. C. InnoDB is the only storage engine Amazon recommends for MySQL and MariaDB deployments in RDS and the only engine Aurora supports. MyISAM is another storage engine that works with MySQL but is not compatible with automated backups. XtraDB is another storage engine for MariaDB, but Amazon no longer recommends it. The PostgreSQL database engine uses its own storage engine by the same name and is not compatible with other database engines.

8. A, C. All editions of the Oracle database engine support the bring-your-own-license model in RDS. Microsoft SQL Server and PostgreSQL only support the license-included model.

9. B. Memory-optimized instances are EBS optimized, providing dedicated bandwidth for EBS storage. Standard instances are not EBS optimized and top out at 10,000 Mbps disk throughput. Burstable performance instances are designed for development and test workloads and provide the lowest disk throughput of any instance class. There is no instance class called storage optimized.

10. A. MariaDB has a page size of 16 KB. To write 200 MB (204,800 KB) of data every second, it would need 12,800 IOPS. Oracle, PostgreSQL, or Microsoft SQL Server, which all use an 8 KB page size, would need 25,600 IOPS to achieve the same throughput. When provisioning IOPS, you must specify IOPS in increments of 1,000, so 200 and 16 IOPS—which would be woefully insufficient anyway—are not valid answers.

11. A. General-purpose SSD storage allocates three IOPS per gigabyte, up to 10,000 IOPS. Therefore, to get 600 IOPS, you'd need to allocate 200 GB. Allocating 100 GB would give you only 300 IOPS. The maximum storage size for gp2 storage is 16 TB, so 200 TB is not a valid value. The minimum amount of storage you can allocate depends on the database engine, but it's no less than 20 GB, so 200 MB is not valid.

12. C. When you provision IOPS using io1 storage, you must do so in a ratio no greater than 50 IOPS for 1 GB. Allocating 240 GB of storage would give you 12,000 IOPS. Allocating 200 GB of storage would fall short, yielding just 10,000 IOPS. Allocating 12 TB would be overkill for the amount of storage required.

13. A. A read replica only services queries and cannot write to a database. A standby database instance in a multi-AZ deployment does not accept queries. Both a primary and a master database instance can service queries and writes.

14. D. Multi-AZ deployments using Oracle, PostgreSQL, MariaDB, MySQL, or Microsoft SQL Server replicate data synchronously from the primary to a standby instance. Only a multi-AZ deployment using Aurora uses a cluster volume and replicates data to a specific type of read replica called an Aurora replica.

15. A. When you restore from a snapshot, RDS creates a new instance and doesn't make any changes to the failed instance. A snapshot is a copy of the entire instance, not just a copy of the individual databases. RDS does not delete a snapshot after restoring from it.

16. B. The ALL distribution style ensures every compute node has a complete copy of every table. The EVEN distribution style splits tables up evenly across all compute nodes. The KEY distribution style distributes data according to the value in a specified column. There is no distribution style called ODD.

17. D. The dense compute type can store up to 326 TB of data on magnetic storage. The dense storage type can store up to 2 PB of data on solid state drives. A leader node coordinates communication among compute nodes but doesn't store any databases. There is no such thing as a dense memory node type.

18. **A, B.** In a nonrelational database, a primary key is required to uniquely identify an item and hence must be unique within a table. All primary key values within a table must have the same data type. Only relational databases use primary keys to correlate data across different tables.

19. **B.** An order date would not be unique within a table, so it would be inappropriate for a partition (hash) key or a simple primary key. It would be appropriate as a sort key, as DynamoDB would order items according to the order date, which would make it possible to query items with a specific date or within a date range.

20. **A.** A single strongly consistent read of an item up to 4 KB consumes one read capacity unit. Hence, reading 11 KB of data per second using strongly consistent reads would consume three read capacity units. Were you to use eventually consistent reads, you would need only two read capacity units, as one eventually consistent read gives you up to 8 KB of data per second. Regardless, you must specify a read capacity of at least 1, so 0 is not a valid answer.

21. **B.** The dense storage node type uses fast SSDs, whereas the dense compute node uses slower magnetic storage. The leader node doesn't access the database but coordinates communication among compute nodes. KEY is a data distribution strategy Redshift uses, but there is no such thing as a key node.

22. **D.** When you create a table, you can choose to create a global secondary index with a different partition and hash key. A local secondary index can be created after the table is created, but the partition key must be the same as the base table, although the hash key can be different. There is no such thing as a global primary index or eventually consistent index.

23. **B.** NoSQL databases are optimized for queries against a primary key. If you need to query data based only on one attribute, you'd make that attribute the primary key. NoSQL databases are not designed for complex queries. Both NoSQL and relational databases can store JSON documents, and both database types can be used by different applications.

24. **D.** A graph database is a type of nonrelational database that discovers relationships among items. A document-oriented store is a nonrelational database that analyzes and extracts data from documents. Relational databases can enforce relationships between records but don't discover them. A SQL database is a type of relational database.

Chapter 6: Authentication and Authorization—AWS Identity and Access Management

1. **C.** Although each of the other options represents possible concerns, none of them carries consequences as disastrous as the complete loss of control over your account.

2. **B.** The * character does, indeed, represent global application. The Action element refers to the kind of action requested (list, create, etc.), the Resource element refers to the particular AWS account resource that's the target of the policy, and the Effect element refers to the way IAM should react to a request.

3. A, B, C. Unless there's a policy that explicitly allows an action, it will be denied. Therefore, a user with no policies or with a policy permitting S3 actions doesn't permit EC2 instance permissions. Similarly, when two policies conflict, the more restrictive will be honored. The AdministratorAccess policy opens up nearly all AWS resources, including EC2. There's no such thing as an IAM action statement.

4. B, C. If you don't perform any administration operations with regular IAM users, then there really is no point for them to exist. Similarly, without access keys, there's a limit to what a user will be able to accomplish. Ideally, all users should use MFA and strong passwords. The AWS CLI is an important tool, but it isn't necessarily the most secure.

5. D. The top-level command is `iam`, and the correct subcommand is `get-access-key-last-used`. The parameter is identified by `--access-last-key-id`. Parameters (not subcommands) are always prefixed with `--` characters.

6. B. IAM groups are primarily about simplifying administration. It has no direct impact on resource usage or response times and only an indirect impact on locking down the root user.

7. C. X.509 certificates are used for encrypting SOAP requests, not authentication. The other choices are all valid identities within the context of an IAM role.

8. A. AWS CloudHSM provides encryption that's FIPS 140-2 compliant. Key Management Service manages encryption infrastructure but isn't FIPS 140-2 compliant. Security Token Service is used to issue tokens for valid IAM roles, and Secrets Manager handles secrets for third-party services or databases.

9. B. AWS Directory Service for Microsoft Active Directory provides Active Directory authentication within a VPC environment. Amazon Cognito provides user administration for your applications. AWS Secrets Manager handles secrets for third-party services or databases. AWS Key Management Service manages encryption infrastructure.

10. A. Identity pools provide temporary access to defined AWS services to your application users. Sign-up and sign-in is managed through Cognito user pools. KMS and/or CloudHSM provide encryption infrastructure. Credential delivery to databases or third-party applications is provided by AWS Secrets Manager.

11. A, D, E. Options A, D, and E are appropriate steps. Your IAM policies will be as effective as ever, even if outsiders know your policies. Since even an account's root user would never have known other users' passwords, there's no reason to change them.

12. B. IAM policies are global—they're not restricted to any one region. Policies *do*, however, require an action (like create buckets), an effect (allow), and a resource (S3).

13. B, C. IAM roles require a defined trusted entity and at least one policy. However, the relevant actions are defined by the policies you choose, and roles themselves are uninterested in which applications use them.

14. D. STS tokens are used as temporary credentials to external identities for resource access to IAM roles. Users and groups would not use tokens to authenticate, and policies are used to define the access a token will provide, not the recipient of the access.

15. C. Policies must be written in JSON format.

16. B, D. The correct Resource line would read "Resource": "*". And the correct Action line would read "Action": "*". There is no "Target" line in an IAM policy. "Permit" is not a valid value for "Effect".

17. B. User pools provide sign-up and sign-in for your application's users. Temporary access to defined AWS services to your application users is provided by identity pools. KMS and/or CloudHSM provide encryption infrastructure. Credential delivery to databases or third-party applications is provided by AWS Secrets Manager.

18. C, D. An AWS managed service takes care of all underlying infrastructure management for you. In this case, that will include data replication and software updates. On-premises integration and multi-AZ deployment are important infrastructure features, but they're not unique to "managed" services.

19. B, C, D. Options B, C, and D are all parts of the key rotation process. In this context, key usage monitoring is only useful to ensure that none of your applications is still using an old key that's set to be retired. X.509 certificates aren't used for access keys.

20. A. You attach IAM roles to services in order to give them permissions over resources in other services within your account.

Chapter 7: CloudTrail, CloudWatch, and AWS Config

1. B, D. Creating a bucket and subnet are API actions, regardless of whether they're performed from the web console or AWS CLI. Uploading an object to an S3 bucket is a data event, not a management event. Logging into the AWS console is a non-API management event.

2. C. Data events include S3 object-level activity and Lambda function executions. Downloading an object from S3 is a read-only event. Uploading a file to an S3 bucket is a write-only event and hence would not be logged by the trail. Viewing an S3 bucket and creating a Lambda function are management events, not data events.

3. C. CloudTrail stores 90 days of event history for each region, regardless of whether a trail is configured. Event history is specific to the events occurring in that region. Because the trail was configured to log read-only management events, the trail logs would not contain a record of the trail's deletion. They might contain a record of who viewed the trail, but that would be insufficient to establish who deleted it. There is no such thing as an IAM user log.

4. B. CloudWatch uses dimensions to uniquely identify metrics with the same name and namespace. Metrics in the same namespace will necessarily be in the same region. The data point of a metric and the timestamp that it contains are not unique and can't be used to uniquely identify a metric.

5. C. Basic monitoring sends metrics every five minutes, whereas detailed monitoring sends them every minute. CloudWatch can store metrics at regular or high resolution, but this affects how the metric is timestamped, rather than the frequency with which it's delivered to CloudWatch.

6. A. CloudWatch can store high-resolution metrics at subminute resolution. Therefore, updating a metric at 15:57:08 and again at 15:57:37 will result in CloudWatch storing two separate data points. Only if the metric were regular resolution would CloudWatch overwrite an earlier data point with a later one. Under no circumstances would CloudWatch ignore a metric update.

7. D. Metrics stored at one-hour resolution age out after 15 months. Five-minute resolutions are stored for 63 days. One-minute resolution metrics are stored for 15 days. High-resolution metrics are kept for 3 hours.

8. A. To graph a metric's data points, specify the Sum statistic and set the period equal to the metric's resolution, which in this case is five minutes. Graphing the Sum or Average statistic over a one-hour period will not graph the metric's data points but rather the Sum or Average of those data points over a one-hour period. Using the Sample count statistic over a five-minute period will yield a value of 1 for each period, since there's only one data point per period.

9. B. CloudWatch uses a log stream to store log events from a single source. Log groups store and organize log streams but do not directly store log events. A metric filter extracts metrics from logs but doesn't store anything. The CloudWatch agent can deliver logs to CloudWatch from a server but doesn't store logs.

10. A, D. Every log stream must be in a log group. The retention period setting of a log group controls how long CloudWatch retains log events within those streams. You can't manually delete log events individually, but you can delete all events in a log stream by deleting the stream. You can't set a retention period on a log stream directly.

11. A, C. CloudTrail will not stream events greater than 256 KB in size. There's also a normal delay, typically up to 15 minutes, before an event appears in a CloudWatch log stream. Metric filters have no bearing on what log events get put into a log stream. Although a misconfigured or missing IAM role would prevent CloudTrail from streaming logs to CloudWatch, the question indicates that some events are present. Hence, the IAM role is correctly configured.

12. B, D. If an EBS volume isn't attached to a running instance, EBS won't generate any metrics to send to CloudWatch. Hence, the alarm won't be able to collect enough data points to alarm. The evaluation period can be no more than 24 hours, and the alarm was created two days ago, so the evaluation period has elapsed. The data points to monitor don't have to cross the threshold for CloudWatch to determine the alarm state.

13. B. To have CloudWatch treat missing data as exceeding the threshold, set the Treat Missing Data As option to Breaching. Setting it to Not Breaching will have the opposite effect. Setting it to As Missing will cause CloudWatch to ignore the missing data and behave as if those evaluation periods didn't occur. The Ignore option causes the alarm not to change state in response to missing data. There's no option to treat missing data as Not Missing.

14. C, D. CloudWatch can use the Simple Notification Service to send a text message. CloudWatch refers to this as a Notification action. To reboot an instance, you must use an EC2 action. The Auto Scaling action will not reboot an instance. SMS is not a valid CloudWatch alarm action.

15. A. The recover action is useful when there's a problem with an instance that requires AWS involvement to repair, such as a hardware failure. The recover action migrates the same instance to a new host. Rebooting an instance assumes the instance is running and entails the instance remaining on the same host. Recovering an instance does not involve restoring any data from a snapshot, as the instance retains the same EBS volume(s).

16. B. If CloudTrail were logging write-only management events in the same region as the instance, it would have generated trail logs containing the deletion event. Deleting a log stream containing CloudTrail events does not delete those events from the trail logs stored in S3. Deleting an EC2 instance is not an IAM event. If AWS Config were tracking changes to EC2 instances in the region, it would have recorded a timestamped configuration item for the deletion, but it would not include the principal that deleted the instance.

17. B, C, D. The delivery channel must include an S3 bucket name and may specify an SNS topic and the delivery frequency of configuration snapshots. You can't specify a CloudWatch log stream.

18. D. You can't delete configuration items manually, but you can have AWS Config delete them after no less than 30 days. Pausing or deleting the configuration recorder will stop AWS Config from recording new changes but will not delete configuration items. Deleting configuration snapshots, which are objects stored in S3, will not delete the configuration items.

19. C, D. CloudWatch can graph only a time series. METRICS()/AVG(m1) and m1/m2 both return a time series. AVG(m1)-m1 and AVG(m1) return scalar values and can't be graphed directly.

20. B. Deleting the rule will prevent AWS Config from evaluating resources configurations against it. Turning off the configuration recorder won't prevent AWS Config from evaluating the rule. It's not possible to delete the configuration history for a resource from AWS Config. When you specify a frequency for periodic checks, you must specify a valid frequency, or else AWS Config will not accept the configuration.

21. B. EventBridge can take an action in response to an event, such as an EC2 instance launch. CloudWatch Alarms can take an action based only on a metric. CloudTrail logs events but doesn't generate any alerts by itself. CloudWatch Metrics is used for graphing metrics.

Chapter 8: The Domain Name System and Network Routing: Amazon Route 53 and Amazon CloudFront

1. A. Option A is the correct answer. Name servers resolve IP addresses from domain names, allowing clients to connect to resources. Domain registration is performed by domain name registrars. Routing policies are applied through record sets within hosted zones.

2. C. A domain is a set of resources identified by a single domain name. FQDN stands for fully qualified domain name. Policies for resolving requests are called routing policies.

3. D. The rightmost section of an FQDN address is the TLD. `aws.` would be a subdomain or host, `amazon.` is the SLD, and `amazon.com/documentation/` points to a resource stored at the web root of the domain server.

4. A. CNAME is a record type. TTL, record type, and record data are all configuration elements, not record types.

5. C. An A record maps a hostname to an IPv4 address. NS records identify name servers. SOA records document start of authority data. CNAME records define one hostname as an alias for another.

6. A, C, D. Route 53 provides domain registration, health checks, and DNS management. Content delivery network services are provided by CloudFront. Secure and fast network connections to a VPC can be created using AWS Direct Connect.

7. C. Geolocation can control routing by the geographic origin of the request. The simple policy sends traffic to a single resource. Latency sends content using the fastest origin resource. Multivalue can be used to make a deployment more highly available.

8. A. Latency selects the available resource with the lowest latency. Weighted policies route among multiple resources by percentage. Geolocation tailors request responses to the end user's location but isn't concerned with response speed. Failover incorporates backup resources for higher availability.

9. B. Weighted policies route among multiple resources by percentage. Failover incorporates backup resources for higher availability. Latency selects the available resource with the lowest latency. Geolocation tailors request responses to the end user's location.

10. D. Failover incorporates backup resources for higher availability. Latency selects the available resource with the lowest latency. Weighted policies route among multiple resources by percentage. Geolocation tailors request responses to the end user's location.

11. A, D. Public and private hosting zones are real options. Regional, hybrid, and VPC zones don't exist (although private zones do map to VPCs).

12. A, B. To transfer a domain, you'll need to make sure the domain isn't set to locked. You'll also need an authorization code that you'll provide to Route 53. Copying name server addresses is necessary only for managing domains that are hosted on but not registered with Route 53. CNAME record sets are used to define one hostname as an alias for another.

13. B. You can enable remotely registered domains on Route 53 by copying name server addresses into the remote registrar-provided interface (not the other way around). Making sure the domain isn't set to locked and requesting authorization codes are used to *transfer a domain* to Route 53, not just to manage the routing. CNAME record sets are used to define one hostname as an alias for another.

14. C. You specify the web page that you want used for testing when you configure your health check. There is no default page. Remote SSH sessions would be impossible for a number of reasons and wouldn't definitively confirm a running resource in any case.

15. A. Geoproximity is about precisely pinpointing users, whereas geolocation uses geopolitical boundaries.

16. A, D. CloudFront is optimized for handling heavy download traffic and for caching website content. Users on a single corporate campus or accessing resources through a VPN will not benefit from the distributed delivery provided by CloudFront.

17. C. API Gateway is used to generate custom client SDKs for your APIs to connect your back-end systems to mobile, web, and server applications or services.

18. A. Choosing a price class offering limited distribution is the best way to reduce costs. Non-HTTPS traffic can be excluded (thereby saving some money) but not through the configuration of an SSL certificate (you'd need further configuration). Disabling Alternate Domain Names or enabling Compress Objects Automatically won't reduce costs.

19. C. Not every CloudFront distribution is optimized for low-latency service. Requests of an edge location will only achieve lower latency after copies of your origin files are already cached. Therefore, a response to the first request might not be fast because CloudFront still has to copy the file from the origin server.

20. B. RTMP distributions can manage content only from S3 buckets. RTMP is intended for the distribution of video content.

Chapter 9: Simple Queue Service and Kinesis

1. C, D. After a consumer grabs a message, the message is not deleted. Instead, the message becomes invisible to other consumers for the duration of the visibility timeout. The message is automatically deleted from the queue after it's been in there for the duration of the retention period.

2. B. The default visibility timeout for a queue is 30 seconds. It can be configured to between 0 seconds and 12 hours.

3. D. The default retention period is 4 days but can be set to between 1 minute and 14 days.

4. B. You can use a message timer to hide a message for up to 15 minutes. Per-queue delay settings apply to all messages in the queue unless you specifically override the setting using a message timer.

5. B. A standard queue can handle up to 120,000 in-flight messages. A FIFO queue can handle up to about 20,000. Delay and short are not valid queue types.

6. A. FIFO queues always deliver messages in the order they were received. Standard queues usually do as well, but they're not guaranteed to. LIFO, FILO, and basic aren't valid queue types.

7. C. Standard queues may occasionally deliver a message more than once. FIFO queues will not. Using long polling alone doesn't result in duplicate messages.

8. B. Short polling, which is the default, may occasionally fail to deliver messages. To ensure delivery of these messages, use long polling.

9. D. Dead-letter queues are for messages that a consumer is unable to process. To use a dead-letter queue, you create a queue of the same type as the source queue, and set the `maxReceiveCount` to the maximum number of times a message can be received before it's moved to the dead-letter queue.

10. C. If the retention period for the dead-letter queue is 10 days, and a message is already 6 days old when it's moved to the dead-letter queue, it will spend at most 4 days in the dead-letter queue before being deleted.

11. B. Kinesis Video Streams is designed to work with time-indexed data such as RADAR images. Kinesis ML doesn't exist.

12. A, C. You can't specify a retention period over 7 days, so your only option is to create a Kinesis Data Firehose delivery stream that receives data from the Kinesis Data Stream and sends the data to an S3 bucket.

13. C. Kinesis Data Firehose requires you to specify a destination for a delivery stream. Kinesis Video Streams and Kinesis Data Streams use a producer-consumer model that allows consumers to subscribe to a stream. There is no such thing as Kinesis Data Warehouse.

14. B. The Amazon Kinesis Agent can automatically stream the contents of a file to Kinesis. There's no need to write any custom code or move the application to EC2. The CloudWatch Logs Agent can't send logs to a Kinesis Data Stream.

15. C. SQS and Kinesis Data Streams are similar. But SQS is designed to temporarily hold a small message until a single consumer processes it, whereas Kinesis Data Streams is designed to provide durable storage and playback of large data streams to multiple consumers.

16. B, C. You should stream the log data to Kinesis Data Streams and then have Kinesis Data Firehose consume the data and stream it to Redshift.

17. C. Kinesis is for streaming data such as stock feeds and video. Static websites are not streaming data.

18. B. Shards determine the capacity of a Kinesis Data Stream. A single shard gives you writes of up to 1 MB per second, so you'd need two shards to get 2 MB of throughput.

19. A. Shards determine the capacity of a Kinesis Data Stream. Each shard supports 2 MB of reads per second. Because consumers are already receiving a total of 3 MB per second, it implies you have at least two shards already configured, supporting a total of 4 MB per second. Therefore, to support 5 MB per second you need to add just one more shard.

20. A. Kinesis Data Firehose is designed to funnel streaming data to big data applications, such as Redshift or Hadoop. It's not designed for videoconferencing.

Chapter 10: The Reliability Pillar

1. C. Availability of 99.95 percent translates to about 22 minutes of downtime per month, or 4 hours and 23 minutes per year. Availability of 99.999 percent is less than 30 seconds of downtime per month, but the question calls for the minimum level of availability. Availability of 99 percent yields more than 7 hours of downtime per month, whereas 99.9 percent is more than 43 minutes of downtime per month.

2. A. The EC2 instances are redundant components, so to calculate their availability, you multiply the component failure rates and subtract the product from 100 percent. In this case, 100% – (10% × 10%) = 99%. Because the database represents a hard dependency, you multiply the availability of the EC2 instances by the availability of the RDS instance, which is 95 percent. In this case, 99% × 95% = 94.05%. A total availability of 99 percent may seem intuitive, but because the redundant EC2 instances have a hard dependency on the RDS instance, you must multiple the availabilities together. A total availability of 99.99 percent is unachievable since it's well above the availability of any of the components.

3. B. DynamoDB offers 99.99 percent availability and low latency. Because it's distributed, data is stored across multiple availability zones. You can also use DynamoDB global tables to achieve even higher availability: 99.999 percent. Multi-AZ RDS offerings can provide low latency performance, particularly when using Aurora, but the guaranteed availability is capped at 99.95 percent. Hosting your own SQL database isn't a good option because, although you could theoretically achieve high availability, it would come at the cost of significant time and effort.

4. B, D. One cause of application failures is resource exhaustion. By scoping out large enough instances and scaling out to make sure you have enough of them, you can prevent failure and thus increase availability. Scaling instances in may help with cost savings but won't help availability. Storing web assets in S3 instead of hosting them from an instance can help with performance but won't have an impact on availability.

5. B. You can modify a launch template by creating a new version of it; however, the question indicates that the Auto Scaling group was created using a launch configuration. You can't modify a launch configuration. Auto Scaling doesn't use CloudFormation templates.

6. A. Auto Scaling strives to maintain the number of instances specified in the desired capacity setting. If the desired capacity setting isn't set, Auto Scaling will attempt to maintain the number of instances specified by the minimum group size. Given a desired capacity of 5, there should be five healthy instances. If you manually terminate two of them, Auto Scaling will create two new ones to replace them. Auto Scaling will not adjust the desired capacity or minimum group size.

7. A, D, E. Auto Scaling monitors the health of instances in the group using either ELB or EC2 instance and system checks. It can't use Route 53 health checks. Dynamic scaling policies can use CloudWatch Alarms, but these are unrelated to checking the health of instances.

8. B, C. Scheduled actions can adjust the minimum and maximum group sizes and the desired capacity on a schedule, which is useful when your application has a predictable load pattern. To add more instances in proportion to the aggregate CPU utilization of the group, implement step scaling policies. Target tracking policies adjust the desired capacity of a group to keep the threshold of a given metric near a predefined value. Simple scaling policies simply add more instances when a defined CloudWatch alarm triggers, but the number of instances added is not proportional to the value of the metric.

9. A, D. Enabling versioning protects objects against data corruption and deletion by keeping before and after copies of every object. The Standard storage class replicates objects across multiple availability zones in a region, guarding against the failure of an entire zone. Bucket policies may protect against accidental deletion, but they don't guard against data corruption. Cross-region replication applies to new objects, not existing ones.

10. C. The Data Lifecycle Manager can automatically create snapshots of an EBS volume every 12 or 24 hours and retain up to 1,000 snapshots. Backing up files to EFS is not an option because a spot instance may terminate before the cron job has a chance to complete. CloudWatch Logs doesn't support storing binary files.

11. D. Aurora allows you to have up to 15 replicas. MariaDB, MySQL, and PostgreSQL allow you to have only up to five.

12. B. When you enable automated snapshots, RDS backs up database transaction logs about every five minutes. Configuring multi-AZ will enable synchronous replication between the two instances, but this is useful for avoiding failures and is unrelated to the time it takes to recover a database. Read replicas are not appropriate for disaster recovery because data is copied to them asynchronously, and there can be a significant delay in replication, resulting in an RPO of well over five minutes.

13. A, C. AWS sometimes adds additional availability zones to a region. To take advantage of a new zone, you'll need to be able to add a new subnet in it. You also may decide later that you may need another subnet or tier for segmentation or security purposes. RDS doesn't require a separate subnet. It can share the same subnet with other VPC resources. Adding a secondary CIDR to a VPC doesn't require adding another subnet.

14. A, D. Fifty EC2 instances, each with two private IP addresses, would consume 100 IP addresses in a subnet. Additionally, AWS reserves five IP addresses in every subnet. The subnet therefore must be large enough to hold 105 IP addresses. 172.21.0.0/25 and 10.0.0.0/21 are sufficiently large. 172.21.0.0/26 allows room for only 63 IP addresses. 10.0.0.0/8 is large enough, but a subnet prefix length must be at least /16.

15. A, D. Direct Connect offers consistent speeds and latency to the AWS cloud. Because Direct Connect bypasses the public Internet, it's more secure. For speeds, you can choose 1 Gbps or 10 Gbps, so Direct Connect wouldn't offer a bandwidth increase over using the existing 10 Gbps Internet connection. Adding a Direct Connect connection wouldn't have an effect on end-user experience, since they would still use the Internet to reach your AWS resources.

16. B. When connecting a VPC to an external network, whether via a VPN connection or Direct Connect, make sure the IP address ranges don't overlap. In-transit encryption, though useful for securing network traffic, isn't required for proper connectivity. IAM policies restrict API access to AWS resources, but this is unrelated to network connectivity. Security groups are VPC constructs and aren't something you configure on a data center firewall.

17. A, C. CloudFormation lets you provision and configure EC2 instances by defining your infrastructure as code. This lets you update the AMI easily and build a new instance from it as needed. You can include application installation scripts in the user data to automate the build process. Auto Scaling isn't appropriate for this scenario because you're going to sometimes terminate and re-create the instance. Dynamic scaling policies are part of Auto Scaling,

18. D. By running four instances in each zone, you have a total of 12 instances in the region. If one zone fails, you lose four of those instances and are left with eight. Running eight or 16 instances in each zone would allow you to withstand one zone failure, but the question asks for the minimum number of instances. Three instances per zone would give you nine total in the region, but if one zone fails, you'd be left with only six.

19. C. Availability of 99.99 percent corresponds to about 52 minutes of downtime per year; 99 percent, 99.9 percent, and 99.95 percent entail significantly more downtime.

20. A, C. Because users access a public domain name that resolves to an elastic load balancer, you'll need to update the DNS record to point to the load balancer in the other region. You'll also need to fail the database over to the other region so that the read replica can become the primary. Load balancers are not cross-region, so it's not possible to point the load balancer in one region to instances in another. Restoring the database isn't necessary because the primary database instance asynchronously replicates data to the read replicas in the other region.

Chapter 11: The Performance Efficiency Pillar

1. A, B, D. ECUs, vCPUs, and the Intel AES-NI encryption set are all instance type parameters. Aggregate cumulative cost per request has nothing to do with EC2 instances but is a common key performance indicator (KPI). Read replicas are a feature used with database engines.

2. A, B, C. A launch configuration pointing to an EC2 AMI and an associated load balancer are all, normally, essential to an Auto Scaling operation. Passing a startup script to the instance at runtime may not be necessary, especially if your application is already set up as part of your AMI. OpsWorks stacks are orchestration automation tools and aren't necessary for successful Auto Scaling.

3. B. Defining a capacity metric, minimum and maximum instances, and a load balancer are all done during Auto Scaling configuration. Only the AMI is defined by the launch configuration.

4. A. Elastic Container Service is a good platform for microservices. Lambda functions executions are short-lived (having a 15-minute maximum) and wouldn't work well for this kind of deployment. Beanstalk operations aren't ideal for microservices. ECR is a repository for container images and isn't a deployment platform on its own.

5. D. RAID optimization is an OS-level configuration and can, therefore, be performed only from within the OS.

6. C. Cross-region replication can provide both low-latency and resilience. CloudFront and S3 Transfer Acceleration deliver low latency but not resilience. RAID arrays can deliver both, but only on EBS volumes.

7. A. S3 Transfer Acceleration makes use of CloudFront locations. Neither S3 Cross-Region Replication nor EC2 Auto Scaling uses CloudFront edge locations, and the EBS Data Transfer Wizard doesn't exist (although perhaps it should).

8. B. Scalability is managed automatically by RDS, and there is no way for you to improve it through user configurations. Indexes, schemas, and views should be optimized as much as possible.

9. D, E. Automated patches, out-of-the-box Auto Scaling, and updates are benefits of a managed service like RDS, not of custom-built EC2-based databases.

10. B, D. Integrated enhanced graphics and Auto Scaling can both help here. Amazon Lightsail is meant for providing quick and easy compute deployments. Elasticsearch isn't likely to help with a graphics workload. CloudFront can help with media transfers, but not with graphics processing.

11. C. The network load balancer is designed for any TCP-based application and preserves the source IP address. The application load balancer terminates HTTP and HTTPS connections, and it's designed for applications running in a VPC, but it doesn't preserve the source IP address. The Classic load balancer works with any TCP-based application but doesn't preserve the source IP address. There is no such thing as a Dynamic load balancer.

12. A, B, D. The CloudFormation wizard, prebuilt templates, and JSON formatting are all useful for CloudFormation deployments. CloudDeploy and Systems Manager are not good sources for CloudFormation templates.

13. A. There is no default node name in a CloudFormation configuration—nor is there a node of any sort.

14. B, E. Chef and Puppet are both integrated with AWS OpsWorks. Terraform, SaltStack, and Ansible are not directly integrated with OpsWorks.

15. A, C. Dashboards and SNS are important elements of resource monitoring. There are no tools named CloudWatch OneView or AWS Config dashboards.

16. A, B. Advance permission from AWS is helpful only for penetration testing operations. A complete record of your account's resource configuration changes would make sense in the context of AWS Config, but not CloudWatch. Service Catalog helps you audit your resources but doesn't contribute to ongoing event monitoring.

17. D. Config is an auditing tool. CloudTrail tracks API calls. CloudWatch monitors system performance. CodePipeline is a continuous integration/continuous deployment (CI/CD) orchestration service.

18. B, C. ElastiCache executions can use either Redis or Memcached. Varnish and Nginx are both caching engines but are not integrated into ElastiCache.

19. A, D. Redis is useful for operations that require persistent session states and or greater flexibility. If you're after speed, Redis might not be the best choice; in many cases, Memcached will provide faster service. Redis configuration has a rather steep learning curve.

20. B. Read replicas based on the Oracle database are not possible.

Chapter 12: The Security Pillar

1. A, C. A password policy can specify a minimum password length but not a maximum. It can prevent a user from reusing a password they used before but not one that another user has used. A password policy can require a password to contain numbers. It can also require administrator approval to reset an expired password.

2. B. The Condition element lets you require MFA to grant the permissions defined in the policy. The Resource and Action elements define what those permissions are but not the conditions under which those permissions are granted. The Principal element is not used in an identity-based policy.

3. A, D. IAM keeps five versions of every customer managed policy. When CloudTrail is configured to log global management events, it will record any policy changes in the request parameters of the `CreatePolicyVersion` operation. There is no such thing as a policy snapshot. CloudTrail data event logs will not log IAM events.

4. B. When an IAM user assumes a role, the user gains the permissions assigned to that role but loses the permissions assigned to the IAM user. The `RunInstances` action launches a new instance. Because the role can perform the `RunInstances` action in the `us-east-1` region, the user, upon assuming the role, can create a new instance in the `us-east-1` region but cannot perform any other actions. `StartInstances` starts an existing instance but doesn't launch a new one.

5. A. Granting a user access to use a KMS key to decrypt data requires adding the user to the key policy as a key user. Adding the user as a key administrator is insufficient to grant this access, as is granting the user access to the key using an IAM policy. Adding the user to a bucket policy can grant the user permission to access encrypted objects in the bucket but doesn't necessarily give the user the ability to decrypt those objects.

6. C. VPC flow logs record source IP address information for traffic coming into your VPC. DNS query logs record the IP addresses of DNS queries, but those won't necessarily be the same IP addresses accessing your application. Because users won't directly connect to your RDS instance, RDS logs won't record their IP addresses. CloudTrail logs can record the source IP address of API requests but not connections to an EC2 instance.

7. C, D. Athena lets you perform advanced SQL queries against data stored in S3. A metric filter can increment based on the occurrence of a value in a CloudWatch log group but can't tell you the most frequently occurring IP address.

8. A. The Behavior finding type is triggered by an instance sending abnormally large amounts of data or communicating on a protocol and port that it typically doesn't. The Backdoor finding type indicates that an instance has resolved a DNS name associated with a command-and-control server or is communicating on TCP port 25. The Stealth finding type is triggered by weakening password policies or modifying a CloudTrail configuration. The ResourceConsumption finding type is triggered when an IAM user launches an EC2 instance when they've never done so.

9. A, C. The AWS Config timeline will show every configuration change that occurred on the instance, including the attachment and detachment of security groups. CloudTrail management event logs will also show the actions that detached and attached the security group. Although AWS Config rules use Lambda functions, the Lambda logs for AWS managed rules are not available to you. VPC flow logs capture traffic ingressing a VPC, but not API events.

10. D. The Security Best Practices rules package has rules that apply to only Linux instances. The other rules contain rules for both Windows and Linux instances.

11. C, D. You can use an IAM policy or SQS access policy to restrict queue access to certain principals or those coming from a specified IP range. You cannot use network access control lists or security groups to restrict access to a public endpoint.

12. A, C. HTTPS traffic traverses TCP port 443, so the security group should allow inbound access to this protocol and port. HTTP traffic uses TCP port 80. Because users need to reach the ALB but not the instances directly, the security group should be attached to the ALB. Removing the Internet gateway would prevent users from reaching the ALB as well as the EC2 instances directly.

13. B. A security group to restrict inbound access to authorized sources is sufficient to guard against a UDP-based DDoS attack. Elastic load balancers do not provide UDP listeners, only TCP. AWS Shield is enabled by default and protects against those UDP-based attacks from sources that are allowed by the security group.

14. A, C. WAF can block SQL injection attacks against your application, but only if it's behind an application load balancer. It's not necessary for the EC2 instances to have an elastic IP address. Blocking access to TCP port 3306, which is the port that MySQL listens on for database connections, may prevent direct access to the database server but won't prevent a SQL injection attack.

15. B, D. Both WAF and Shield Advanced can protect against HTTP flood attacks, which are marked by excessive or malformed requests. Shield Advanced includes WAF at no charge. Shield Standard does not offer protection against Layer 7 attacks. GuardDuty looks for signs of an attack but does not prevent one.

16. A, D. You can revoke and rotate both a customer-managed CMK and a customer-provided key at will. You can't revoke or rotate an AWS-managed CMK or an S3-managed key.

17. C, D. Customer-managed customer master keys (CMKs) can be rotated at will, whereas AWS-managed CMKs are rotated only once a year. RDS and DynamoDB let you use a customer-managed CMK to encrypt data. RedShift is not designed for highly transactional databases and is not appropriate for the application. KMS stores and manages encryption keys but doesn't store application data.

18. B, D. To encrypt data on an unencrypted EBS volume, you must first take a snapshot. The snapshot will inherit the encryption characteristics of the source volume, so an unencrypted EBS volume will always yield an unencrypted snapshot. You can then simultaneously encrypt the snapshot as you copy it to another region.

19. B. You can enable encryption on an EFS filesystem only when you create it; therefore, the only option to encrypt the data using KMS is to create a new EFS filesystem and copy the data to it. A third-party encryption program can't use KMS keys to encrypt data. Encrypting the EBS volume will encrypt the data stored on the volume, but not on the EFS filesystem.

20. A, D. You can install an ACM-generated certificate on a CloudFront distribution or application load balancer. You can't export the private key of an ACM-generated certificate, so you can't install it on an EC2 instance. AWS manages the TLS certificates used by S3.

21. C. Security Hub checks the configuration of your AWS services against AWS best practices.

Chapter 13: The Cost Optimization Pillar

1. C. The Free Tier provides free access to *basic* levels of AWS services for a new account's first *year*.

2. A. Standard provides the most replicated and quickest-access service and is, therefore, the most expensive option. Storage rates for Standard-Infrequent and One Zone-Infrequent are lower than Standard but are still more expensive than Glacier.

3. B. Cost Explorer provides usage and spending data. Organizations lets you combine multiple AWS accounts under a single administration. TCO Calculator lets you compare the costs of running an application on AWS versus locally.

4. D. Cost Explorer provides usage and spending data, but without the ability to easily incorporate Redshift and QuickSight that Cost and Usage Reports offers. Trusted Advisor checks your account for best-practice compliance. Budgets allows you to set alerts for problematic usage.

5. A, B, D. As efficient as Organizations can be, so does the threat they represent grow. There is no such thing as a specially hardened organization-level VPC. Security groups don't require any special configuration.

6. B, C. Trusted Advisor monitors your EC2 instances for lower than 10 percent CPU and network I/O below 5 MB on four or more days. Trusted Advisor doesn't monitor Route 53 hosted zones or the status of S3 data transfers. Proper OS-level configuration of your EC2 instances is your responsibility.

7. B. The Pricing Calculator is the most direct tool for this kind of calculation. TCO Calculator helps you compare costs of on-premises to AWS deployments. Trusted Advisor checks your account for best-practice compliance. Cost and Usage Reports helps you analyze data from an existing deployment.

8. A. Monitoring of EBS volumes for capacity is not within the scope of budgets.

9. A, B. Tags can take up to 24 hours to appear and they can't be applied to legacy resources. You're actually allowed only two free budgets per account. Cost allocation tags are managed from the Cost Allocation Tags page.

10. D. The most effective approach would be to run three reserve instances 12 months/year and purchase three scheduled reserve instances for the summer. Spot instances are not appropriate because they shut down automatically. Since it's possible to schedule an RI to launch within a recurring block of time, provisioning other instance configurations for the summer months will be wasteful.

11. C. Interruption polices are relevant to spot instances, not reserved instances. Payment options (All Upfront, Partial Upfront, or No Upfront), reservation types (Standard or Convertible RI), and tenancy (Default or Dedicated) are all necessary settings for RIs.

12. C. No Upfront is the most expensive option. The more you pay up front, the lower the overall cost. There's no option called Monthly.

13. B, D. Containers are more dense and lightweight. Containers do tend to launch more quickly than EC2 instances and do make it easy to replicate server environments, but those are not primarily cost savings.

14. B. Standard reserve instances make the most sense when they need to be available 24/7 for at least a full year, with even greater savings over three years. Irregular or partial workdays are not good candidates for this pricing model.

15. D. A spot instance pool is made up of unused EC2 instances. There are three request types: Request, Request And Maintain, and Reserve For Duration. A spot instance interruption occurs when the spot price rises above your maximum. A spot fleet is a group of spot instances launched together.

16. A. A spot instance interruption occurs when the spot price rises above your maximum. Workload completions and data center outages are never referred to as interruptions. Spot requests can't be manually restarted.

17. B. Target capacity represents the maximum instances you want running. A spot instance pool contains unused EC2 instances matching a particular set of launch specifications. Spot maximum and spot cap sound good but aren't terms normally used in this context.

18. A. The EBS Lifecycle Manager can be configured to remove older EBS snapshots according to your needs. Creating a script is possible, but it's nowhere near as simple and it's not tightly integrated with your AWS infrastructure. There is no "EBS Scheduled Reserve Instance" but there is an "EC2 Scheduled Reserve Instance." Tying a string? Really? EBS snapshots are stored in S3, but you can't access the buckets that they're kept in.

19. D. The command is `request-spot-fleet`. The `--spot-fleet-request-config` argument points to a JSON configuration file.

20. C. The availability zone, target capacity, and AMI are all elements of a complete spot fleet request.

Chapter 14: The Operational Excellence Pillar

1. C, D. It's a best practice to organize stacks by lifecycle (e.g., development, test, production) and ownership (e.g., network team, development team). You can store templates for multiple stacks in the same bucket, and there's no need to separate templates for different stacks into different buckets. Organizing stacks by resource cost doesn't offer any advantage since the cost is the same regardless of which stack a resource is in.

2. A, B. Parameters let you input custom values into a template when you create a stack. The purpose of parameters is to avoid hard-coding those values into a template. An AMI ID and EC2 key pair name are values that likely would not be hard-coded into a template. Although you define the stack name when you create a stack, it is not a parameter that you define in a template. The logical ID of a resource must be hard-coded in the template.

3. C. When using nested stacks, the parent stack defines a resource of the type `AWS::CloudFormation::Stack`, which points to the template used to generate the nested stack. Because of this, there's no need to define a VPC resource directly in the template that creates the parent stack. There is also no need to export stack output values because the nested stacks do not need to pass any information to stacks outside of the nested stack hierarchy. For this same reason, you don't need to use the `Fn::ImportValue` intrinsic function, since it is used to import values exported by another stack.

4. A. A change set lets you see the changes CloudFormation will make before updating the stack. A direct update doesn't show you the changes before making them. There's no need to update or override the stack policy before using a change set to view the changes that CloudFormation would make.

5. C. To use Git to access a repository as an IAM user, the developer must use a Git username and password generated by IAM. Neither an AWS access key and secret key combination nor an IAM username and password will work. Although SSH is an option, the developer would need a private key. The public key is what you'd provide to IAM.

6. D. You can allow repository access for a specific IAM user by using an IAM policy that specifies the repository ARN as the resource. Specifying the repository's clone URL would not work, since the resource must be an ARN. Generating Git credentials also would not work, because the user still needs permissions via IAM. There is no such thing as a repository policy.

7. A. CodeCommit offers differencing, allowing you (and the auditors) to see file-level changes over time. CodeCommit offers at-rest encryption using AWS-managed KMS keys but not customer-managed keys. S3 offers versioning and at-rest encryption, but not differencing.

8. B. The `git clone` command clones or downloads a repository. The `git push` command pushes or uploads changes to a repository. The `git add` command stages files for commit to a local repository but doesn't commit them or upload them to CodeCommit. The `aws codecommit get-repository` command lists the metadata of a repository, such as the clone URL and ARN, but doesn't download the files in it.

9. D. CodeDeploy can deploy from an S3 bucket or GitHub repository. It can't deploy from any other Git repository or an EBS snapshot.

10. B. A blue/green instance deployment requires an elastic load balancer (ELB) in order to direct traffic to the replacement instances. An in-place instance deployment can use an ELB but doesn't require it. A blue/green Lambda deployment doesn't use an ELB because ELB is for routing traffic to instances. There's no such thing as an in-place Lambda deployment.

11. C. The AllAtOnce deployment configuration considers the entire deployment to have succeeded if the application is deployed successfully to at least one instance. HalfAtATime and OneAtATime require the deployment to succeed on multiple instances. There's no preconfigured deployment configuration called OnlyOne.

12. B. The `AfterAllowTraffic` lifecycle event occurs last in any instance deployment that uses an elastic load balancer. `ValidateService` and `BeforeAllowTraffic` occur before CodeDeploy allowing traffic to the instances. `AllowTraffic` is a lifecycle event, but you can't hook into it to run a script.

13. A. CodePipeline stores pipeline artifacts in an S3 bucket. An artifact can serve as an input to a stage, an output from a stage, or both. A provider is a service that performs an action, such as building or testing. An asset is a term that often refers to the supporting files for an application, such as images or audio. S3 doesn't offer snapshots, but it does offer versioning for objects.

14. B, C. You can implement an approval action to require manual approval before transitioning to the deploy stage. Instead of or in addition to this, you can disable the transition to the deploy stage, which would require manually enabling the transition to deploy to production. Because CodePipeline uses one bucket for all stages of the pipeline, you can't create a separate bucket for the deploy stage. Even if you could, disallowing developers access to that bucket would not prevent a deployment, since CodePipeline obtains its permission to the bucket by virtue of its IAM service role.

15. A, D, E. A pipeline must consist of at least two stages. The first stage must contain only source actions. Since the templates are stored in CodeCommit, it must be the provider for the source action. The second stage of the pipeline should contain a deploy action with a CloudFormation provider, since it's the service that creates the stack. There's no need for a build stage, because CloudFormation templates are declarative code that don't need to be compiled. Hence, the pipeline should only be two stages. CodeCommit is not a valid provider for the deploy action.

16. B. A pipeline can have anywhere from two to 10 stages. Each stage can have one to 20 actions.

17. B. Automation documents let you perform actions against your AWS resources, including taking EBS snapshots. Although they're called automation documents, you can still manually execute them. A command document performs actions within a Linux or Windows instance. A policy document works only with State Manager and can't take an EBS snapshot. There's no manual document type.

18. A. The `AmazonEC2RoleforSSM` managed policy contains permissions allowing the Systems Manager agent to interact with the Systems Manager service. There's no need to install the agent because Amazon Linux comes with it preinstalled. There's also no need to open inbound ports to use Systems Manager.

19. A, D. Setting the patch baseline's auto-approval delay to 0 and then running the `AWS-RunPatchBaseline` document would immediately install all available security patches. Adding the patch to the list of approved patches would approve the specific patch for installation but not any other security updates released within the preceding seven days. Changing the maintenance window to occur Monday at midnight wouldn't install the patch until the following Monday.

20. A, B. Creating a global inventory association will immediately run the `AWS-GatherSoftwareInventory` policy document against the instance, collecting both network configuration and software inventory information. State Manager will execute the document against future instances according to the schedule you define. Simply running the `AWS-GatherSoftwareInventory` policy document won't automatically gather configuration information for future instances. Of course, an instance must be running in order for the Systems Manager agent to collect data from it. The `AWS-SetupManagedInstance` document is an automation document and thus can perform operations on AWS resources and not tasks within an instance.

Index

A

accelerated computing instances, 25
access control list (ACL) rules, 67–69
access keys, 170–172
actions
   AWS Systems Manager, 47–49
   CloudWatch, 200–201
   operational excellence and, 374–377
agents (CloudWatch), 196–197
alarms
   CloudWatch, 198–201
   operational excellence and, 370
alias records, 216
ALL distribution, 150
AllAtOnce deployment configuration, 368
Amazon Machine Images (AMIs), 23
anomaly detection, 198
answers
   to Assessment Test, xxxiii–xxxv
   to review questions, 386–413
API Gateway, 10
application instances, health checks against, 41–42, 256
application load balancer (ALB), 41, 255–256, 268, 284
Application management category, 7, 9, 10
application problems, detecting, 184
application specification (AppSpec) file, 369–370
approval action type, 373
architecture
   cloud computing, 4–5
   S3 service, 61–64
archive, 71
artifacts, 373–374
Assessment Test, xxvii–xxxii, xxxiii–xxxv

Athena (Amazon), 315–317
attributes
   in database tables, 134, 135
   in DynamoDB, 154–155
Aurora (Amazon), 139, 145, 147
authentication tools, 173–175, 178
Auto Scaling
   about, 7, 38–39, 53
   actions, 201
   configuring, 53
   cost optimization and, 347
   DynamoDB, 156–157
   Elastic Compute Cloud (EC2), 276–278, 295
   exercises, 277–278
   groups, 40–42, 255–256
   launch configurations, 39
   launch templates, 39–40
   options, 42–46, 256–260
   review questions, 297–300
auto-approval delay, 48, 376
automatic snapshots, 148–149
automation, 47, 375
availability
   calculating, 248–253, 265, 266, 267, 268
   of data, 302
   defined, 248
   designing for, 264–267
   increasing, 252–253
   monitoring with Amazon Route 53, 219
   review questions, 269–272
availability zone (AZ), 88–89
AWS, 13–15
AWS Budgets, 337–338
AWS Certificate Manager (ACM), 224
AWS Certified Solutions Architect-Associate Exam, xxi–xxii

AWS Cloud, 6–10
AWS CloudHSM, 175
AWS Config
    about, 9, 184–185, 202, 206
    auditing resource configurations
        with, 317–318
    configuration history, 203
    configuration items, 203
    configuration recorder, 203
    configuration snapshots, 203–204
    monitoring changes, 204–205
    review questions, 397–399. 207–210
AWS Config Compliance, 49, 377
AWS Cost and Usage Reports, 339

B

backups
    databases and, 148
    DynamoDB, 158
base table, 157
Basic plan, 14
behaviors, configuring for instances, 28
billing, 23
blackhole routes, 122
blue/green deployment, 367
bring your own license (BYOL), 139–140
build action type, 372–373
built-in insights, 49–50, 377
burstable performance instance classes, 141
Business plan, 15

C

caching, 291–293, 296
Center for Internet Security Benchmarks, 321
centralized router, 116–117
Chef, 288–289
child table, 136
Classless Inter-Domain Routing (CIDR), 68, 84–87, 87–88, 127

CLI examples, 51–52, 75–76, 176–177, 225–226
client-side encryption, 63
cloud computing
    Exam Essentials, 16
    exercises, 16
    review questions, 17–19, 386–387
    virtualization and, 4–6
CloudFormation
    about, 9, 286–288
    exercises, 287–288
    operational excellence and, 354–361, 379
CloudFront (Amazon)
    about, 8, 223–225, 227
        about, 212
        Amazon S3 and, 282
    review questions, 228–231
cloud-native applications, availability and, 250–251, 268
CloudTrail
    about, 9, 184–185, 206, 313–314
    data events, 186
    event history, 186
    events, 50, 377
    log file integrity validation, 189
    logs, 197
    Logs, 314
    management events, 185–186
    review questions, 207–210, 397–399
    trails, 186–188
CloudWatch
    about, 9, 184–185, 189–190, 206
    agents, 196–197
    alarms, 198–201, 206
    Amazon EventBridge, 201–202, 206
    exercises, 290–291
    logs, 195–196, 206, 314
    metrics, 190–195
    review questions, 207–210, 397–399
cluster groups, 28
CodeCommit, 361–365, 379
CodeDeploy, 365–370
CodeDeploy Agent, 366
CodePipeline, 371–374, 380

Cognito (Amazon), 174
cold HDD, 33, 144
columns, in database tables, 135
Command Line Interface (CLI), 14
Common Vulnerabilities and Exposures
   (CVE), 49, 321
Community AMIs, 23
compliance, 51
compliance insights, 378
composite primary key, 153
compression, 294
compute, cost optimization and, 342–347
Compute category, 6, 7–8
compute nodes, 149
compute optimized instances, 25
confidentiality, of data, 302
configuration history, 203
configuration items, 203
configuration recorder, 203, 205
configuration snapshots, 203–204
consumer, 239
content delivery network (CDN), 223
continuous delivery (CD), 371–372
continuous integration (CI), 371
control plane operations, 185
core AWS services, 7–10
Cost Explorer, 338–339
cost optimization
   about, 336
   Auto Scaling, 347
   AWS Budgets, 337–338
   AWS Cost and Usage Reports, 339
   AWS Organizations, 339–340
   AWS Total Cost of Ownership
     Calculator, 342
   AWS Trusted Advisor, 340
   cloud computing and, 6
   compute, 342–347
   Cost Explorer, 338–339
   EC2 reserved instances, 343–344, 348
   EC2 spot instances, 344–345, 348
   Elastic Block Store Lifecycle
     Manager, 347
   maximizing server density, 343
   monitoring tools, 338–339
   online calculator tools, 340
   planning, tracking, and controlling costs,
     336–342, 348
   review questions, 349–352, 409–411
   Savings Plans, 344
   Simple Monthly Calculator, 341–342
customer-managed policy, 304

D

data
   distribution styles of, 150
   eventually consistent, 65–66
   missing, 200
   optimizing operations, 291–294
   optimizing transfers, 77
   querying, 137, 152
   querying in nonrelational (NoSQL)
     databases, 152
   reading in DynamoDB, 157–158
   at rest, 325
   storing, 137, 151
   storing in nonrelational (NoSQL)
     databases, 151
   in transit, 326–327
data encryption
   about, 324
   Amazon S3, 325
   data at rest, 325
   data in transit, 326–327
   Elastic Block Store (EBS), 325–326
   Elastic File System (EFS), 326
   Macie, 327
data events (CloudTrail), 186, 187
Data Firehose, 239–240
data points, 190, 198, 199
Data Streams, 238–240
data warehouse, 138
Database category, 7, 9
database engines, 138–139
Database Migration Service (DMS), 150
databases

about, 134, 282–283, 295
Amazon RedShift, 149–151
Amazon Relational Database Services (RDS), 138–149
DynamoDB, 153–158
instance classes, 140–141
Nonrelational (NoSQL), 151–152, 159
option groups, 140
relational, 134–138, 159
resiliency of, 262
review questions, 161–164, 393–395
DataSync, 74–75
deactivating unused access keys, 171
dead-letter queues, 237
dedicated connection, 123–124
default route, 95–96
default security group, 100–101
delay queues, 235
delimiters, 61
delivery channel, 203
deploy action type, 373
deployment
about, 366, 379
configurations for, 367–368
groups for, 366
types of, 366–367
destinations, in rules, 99
Detective (Amazon), 322
detective controls
Amazon Athena Logs, 315–317
Amazon Detective, 322
Amazon GuardDuty, 318–321, 328
Amazon Inspector, 321–322, 328
auditing resource configurations with AWS Config, 317–318
CloudTrail, 313–314
CloudWatch Logs, 314–315
Security Hub, 323
Developer plan, 15
dimension, 190
Direct Connect, 8, 123–124, 263–264
Directory Service, 10
DNS management, 217

document data type, 154–155
documents, actions defined in, 47
domain layers, 214
Domain Name System (DNS)
about, 212
alias records, 216
domain layers, 214
domain registration, 214, 226
domains/domain names, 213, 226
fully qualified domain names (FDQN), 214–215
name servers, 213
namespaces, 212–213
record types, 215–216, 227
review questions, 228–231, 399–401
zones/zone files, 215
domain registration, 214, 217, 226
domains/domain names, 213, 226
dynamic scaling policy option, in Auto Scaling, 42–43, 257
DynamoDB, 9, 153–158

E

EBS general-purpose SSD, 32
EBS-provisioned IOPS SSD, 32
Elastic Beanstalk, 8
Elastic Block Store (EBS). *See also* Elastic Compute Cloud (EC2)
about, 8, 325–326
for data backup and recovery, 261–262, 268
exercises, 325–326
Lifecycle Manager, 347
RAID-Optimized volumes, 278–280, 295
review questions, 387–391
storage, 278, 295
volume features, 33
volumes, 32–34
Elastic Compute Cloud (EC2)
about, 7, 22
accessing instances, 35–36
actions, 201

Auto Scaling, 38–46, 53, 253, 276–278, 295
  exercises, 27–28
  instance types, 275–276, 295
  instances, 22–32
  reserved instances, 343–344, 348
  review questions, 54–58, 297–300, 387–389
  securing instances, 36–38
  serverless workloads, 278, 295
  spot instances, 344–345, 348
  storage volumes, 32–34
  Systems Manager (*See* Systems Manager)
elastic fabric adapter (EFA), 125–126
Elastic File System (EFS), 73, 261, 268, 326
elastic IP addresses, 107–109
Elastic Load Balancing, 7
elastic network interface (ENI), 91–93, 127
ElastiCache (Amazon), 292–293
elasticity, cloud computing and, 5
encryption, 62–63
endpoint addresses, 12
enhanced networking, 93
Enterprise plan, 15
envelope key, 63
environments, configuring for instances, 25–27
ephemeral ports, 105
EVEN distribution, 150
event buses, 201
event history (CloudTrail), 186
EventBridge (Amazon), 201–202
exercises
  access keys, 171–172
  administrator privileges, 305–306
  Auto Scaling, 277–278
  AWS Budgets, 338
  AWS resources, 52
  blackhole routes, 122–123
  cloud computing, 16
  CloudFormation, 287–288
  CloudFront, 224–225
  CloudTrail logs, 197
  CloudTrail trails, 187–188

CloudWatch, 197, 290–291
CodeCommit, 363–365
DynamoDB tables, 156
EDS database instances, 143
Elastic Block Store (EBS), 325–326
Elastic Compute Cloud (EC2), 27–28
elastic IP addresses, 107–109
elastic network interface (ENI), 92–93
graphing metrics in CloudWatch, 193
health check, 219–220
IAM policies, 169
inbound rules, 103–104
instances, 30, 34, 51
Internet gateways, 96–97
ISM groups, 172–173
launch templates, 40, 254–255
lifecycle costs, 73
lifecycle management, 67
load balancer, 285–286
nested stacks, 357–358
predesigned URLs, 69
read replicas, 145, 146
roles, 311–313
root users, 169
Route 53, 218–219, 219–220, 221–222
route tables, 96–97
routing policies, 221–222
security group, 100–101
Simple Monthly Calculator, 341–342
Simple Storage Service (S3), 62, 281, 282
spot fleets, 345–347
static website hosting, 70
subnets, 89–90
Transit Gateways, 117–122
versioning, 67
virtual private cloud (VPC), 85–87, 315
expiration, of metrics in CloudWatch, 191
external connectivity, 263–264

F

failover routing, 221
firewall, stateful, 99–100
first-in, first-out (FIFO) queues, 236

foreign key, 136
FSx (Amazon), 73
fully qualified domain names (FDQN), 214–215

G

gateways
    Direct Connect, 124
    NAT, 113
general purpose instances, 24
general-purpose SSD, 142–143
geolocation routing, 221
geoproximity routing, 222
Git, interacting with repositories using, 363–365, 379
Glacier, 71–73, 77
Global Accelerator, 109
global inventory association, 50
global secondary index (GSI), 157
global tables, 158
graphing metrics in CloudWatch, 192–193
groups
    Auto Scaling, 255–256
    Identity and Access Management (IAM), 172–173
GuardDuty (Amazon), 318–321

H

HalfAtATime deployment configuration, 367–368
hash keys, 153–154
health check, 219–220, 256
help forums, 15
high-performance computing, 125–126
high-resolution metrics, 191
hosted connection, 124
hostnames, 213
hot partition, 154
hybrid cloud networking, 115–124

I

IAM roles, 37
identities, 166–173
Identity and Access Management (IAM)
    about, 9, 166, 302
    access keys, 170–172
    authentication tools, 173–175, 178
    CLI example, 176–177
    fine-grained authorization, 303–305
    groups, 172–173
    identities, 166–173
    permissions boundaries, 305–306
    policies, 167–168, 177
    protecting credentials, 303
    review questions, 179–182, 395–397
    roles, 173, 306–313
    root accounts, 168–170, 178
    user accounts, 168–170
identity pools, 174
implicit router, 94
implied router, 94
inbound rules
    exercises, 103–104
    network access control lists, 102–104
    security groups, 98–99
infrastructure automation
    about, 286
    CloudFormation, 286–288
    third-party solutions, 288–289
infrastructure configurations, optimizing, 289–291
inline policy, 305
in-place deployment, 366
input/output operations per second (IOPS), 141–142
insights
    about, 49
    built-in, 49–50
    compliance, 51
    Inventory Manager, 50
    operational excellence and, 377–378
Inspector (Amazon), 321–322

instance store volumes, 34
instances
    defined, 5
    exercises, 30, 34, 51
    NAT, 113
    pricing of, 29–30
    profiles for, 307–311
    terminating, 30
instances (EC2)
    about, 22, 275–276, 295
    accessing, 35–36
    configuring behavior of, 28
    configuring environments for, 25–27
    launching, 53
    lifecycle of, 30
    pricing, 53
    provisioning, 23–27
    resource tags, 30–31
    securing, 36–38
    service limits, 31–32
    types of, 24–25
integrity, of data, 302
Internet gateways, 93–94, 96–97
inventory, maintaining, 184–185
Inventory Manager, 50, 378
invoke action type, 373
IP addresses, 91
IP prefix, 84
IPv6 CIDR blocks, 85, 91
isolated VPCs, 117
items, in DynamoDB, 154–155

J

JavaScript Object Notation (JSON), 67–68

K

Key Management Service (KMS), 9, 175
key pairs, 38
Kinesis (Amazon)
    about, 237
    Data Firehose, 239–240
    Data Streams, 238–240
    review questions, 241–244, 401–402
    video streams, 237–238, 240

L

Lambda, 7, 251–252, 314
Landing Zone, 378–379
latency-based routing, 220–221
launch configurations, 39, 253, 268
launch templates, 39–40, 254
leader node, 149
licensing considerations, 139–140
lifecycle events, operational excellence
    and, 368–369
lifecycle management, 66, 67
Link Layer Discovery Protocol (LLDP), 117
load balancing, network optimization and,
    284–286, 295
load testing, 289–290
local route, 95
local secondary index (LSI), 158
Log Archive Account, 379
log file integrity validation, 189
logging
    about, 63–64
    data events, 188
    events, 184
    management events, 187, 188
logs
    CloudTrail, 197
    CloudWatch, 195–196, 197
loosely coupled workloads, 125, 234

M

m out of n alarm, 199
M4/M5 instances, 24
Macie, 327
magnetic storage, 144
main route table, 94
maintenance items, 149
managed instances, 47

Managed Microsoft 4D, 174
managed policy, 304
management events, 185–186
manual scaling option, in Auto Scaling, 42, 256–257
MariaDB, 138, 147
Marketplace AMIs, 23
master database instances, 145
maximizing server density, 343
memory optimized instances, 25, 141
message timers, 235
metrics (CloudWatch), 190–195
Microsoft SQL Server, 139, 147
missing data, 200
monitoring
    changes in AWS Config, 204–205
    CloudWatch, 190–191
    cost optimization and, 338–339
multi-AZ deployment, 146–147
multicast, 122
multi-master cluster, 147–148
multiple database tables, 135–136
multivalue answer routing, 221
MySQL, 138, 147

N

name servers, 213
namespaces, 212–213
Neptune (Amazon), 152
nested stacks, 356–358
network access control lists (NACLs), 37, 101–106, 127, 323
network address translation (NAT), 109–110, 128
network address translation (NAT) devices, 37–38, 110–113
network boundaries, protecting, 323–324
network load balancer, 284
network optimization, load balancing and, 284–286, 295
Network Reachability, 321
Networking category, 6, 8, 93
Nonrelational (NoSQL) databases, 151–152, 159

O

objects
    accessing, 67–71
    lifecycle of, 66–67, 77
    working with large, 61–62
OneAtATime deployment configuration, 367
online analytic processing (OLAP), 137, 138
online calculator tools, 340
online resources, 14
online transaction processing (OLTP), 137
operational excellence
    about, 354
    actions, 374–377
    alarms, 370
    application specification (AppSpec) file, 369–370
    artifacts, 373–374
    AWS Landing Zone, 378–379
    AWS Systems Manager, 374–378, 380
    CloudFormation, 354–361, 379
    CodeCommit, 361–365, 379
    CodeDeploy, 365–370
    CodeDeploy Agent, 366
    CodePipeline, 371–374, 380
    continuous delivery (CD), 371–372
    continuous integration (CI), 371
    creating pipelines, 372–373
    creating repositories, 362
    creating stacks, 355
    deleting stacks, 356
    deployment configurations, 367–368
    deployment groups, 366
    deployment types, 366–367
    deployments, 366, 379
    insights, 377–378
    interacting with repositories using Git, 363–365, 379
    lifecycle events, 368–369
    overriding stack policies, 361
    preventing updates to resources, 360–361
    review questions, 381–384, 411–413
    rollbacks, 370
    security of repositories, 362
    stack updates, 359

triggers, 370
using multiple stacks, 356–359
OpsWorks, 288–289
Oracle, 139, 147
Organizations, 339–340
Organizations Account, 378
outbound rules
    network access control lists, 105
    security groups, 99

P

ParallelCluster, 126
parameters, 355
parent stack, 356
parent table, 136
partition groups, 28
partition key, 153–154
partitioning, 138, 293–294
patch baselines, 48, 376
patch group, 48
Patch Manager, 48–49, 376
peering, 122
performance
    about, 274
    infrastructure automation, 286–289
    optimizing data operations, 291–294
    optimizing for core AWS services, 274–286
    optimizing infrastructure configurations, 289–291
    review questions, 405–407
    tracking, 184
permissions policy, 305
Personal Health Dashboard, 50, 377
pipelines, creating, 372–373
placement groups, 28
planning, tracking, and controlling costs, 336–342, 348
platform architecture, 10–12, 16
policies (IAM), 167–168, 177
polling (SQS), 236
port address translation (PAT), 111
PostgreSQL, 139, 147
prefixes, 61, 66, 84

pricing, of instances, 29–30
primary database instance, 146
primary private IP address, 91
principals, 166
Private AMIs, 23
private subnet, 95
private virtual interface, 124
producer, 239
projected attributes, 157
provisioned IOPS SSD, 144
provisioned throughput, 155
provisioning instances (EC2), 23–27
public IP addresses, 106–107
public subnet, 95
public virtual interface, 124
publisher, 200
Puppet, 288–289

Q

querying data, 137
queues
    Simple Queue Service (SQS), 234–235, 240
    types of, 235–236
Quick Start AMIs (Amazon), 23

R

RAID-Optimized volumes, 278–280, 295
read capacity units (RCUs), 155
read replicas, 145–146
records, 215–216, 227. *See also* attributes
recovery point objective (RPO), 148
recovery time objective (RTO), 148
RedShift (Amazon), 149–151
RedShift Spectrum, 150
regions, AWS, 11, 25
regular-resolution metrics, 191
Relational Database Service (RDS), 9, 138–149, 314
relational databases, 134–138, 135, 159
reliability

about, 12, 248
AWS Service Level Agreement, 13
AWS Shared Responsibility Model, 12–13
review questions, 269–272, 403–405
repositories
   creating, 362
   interacting with using Git, 363–365, 379
   security of, 362
request type, 345
reserve instance, 29
reserved capacity, 157
resiliency. *See* reliability
resilient networks, creating, 263–264
Resolver, Amazon Route 53, 223
resource groups (AWS), 49, 377
resource records, 215
resource tags, 30–31
resources
   exercises, 52
   organizing, 77
retention period, 148, 235
review questions
   Auto Scaling, 297–300
   availability, 269–272
   AWS Config, 207–210, 397–399
   cloud computing, 17–19, 386–387
   CloudFront (Amazon), 228–231
   CloudTrail, 207–210, 397–399
   CloudWatch, 207–210, 397–399
   cost optimization and, 349–352, 409–411
   databases, 161–164, 393–395
   Domain Name System (DNS), 228–231, 399–401
   Elastic Block Store (EBS), 387–391
   Elastic Compute Cloud (EC2), 54–58, 297–300, 387–389
   Identity and Access Management (IAM), 179–182, 395–397
   Kinesis (Amazon), 241–244, 401–402
   operational excellence, 381–384, 411–413
   performance, 405–407
   reliability, 269–272, 403–405
   Route 53 (Amazon), 228–231
   security, 329–333, 407–409
   Simple Queue Service (SQS), 241–244, 401–402

   Simple Storage Service (S3), 78–81
   Virtual Private Cloud, 129–132, 391–393
roles
   assuming, 311–313
   exercises, 311–313
   IAM, 306–313
   Identity and Access Management (IAM), 173
rollbacks, operational excellence and, 370
root accounts, 168–170, 178
Route 53 (Amazon)
   about, 8, 212, 216–217
   availability monitoring, 219
   CLI example, 225–226
   DNS management, 217
   domain registration, 217
   exercises, 218–219, 219–220, 221–222
   queries, 314
   Resolver, 223
   review questions, 228–231
   routing policies, 220–222, 227
   Traffic Flow, 222–223
route propagation, 116
route tables
   about, 94–97, 127
   configuring to use NAT devices, 112
   exercises, 96–97
router, centralized, 116–117
routing, 220–222, 227
rules (EventBridge), 202, 205
Run command
   about, 47–48
   AWS Systems Manager and, 375
running instances, accessing, 53
Runtime Behavior Analysis, 321

S

Savings Plans, 344
scalability, cloud computing and, 5
scalar data type, 154
scaling horizontally, 145–146
scaling vertically, 145

scheduled actions, in Auto Scaling,
    45–46, 259–260
secondary CIDR blocks, 85
secondary indexes, 157
secondary private IP address, 91
second-level domain (SLD), 214
Secrets Manager, 175
security
    about, 302
    AWS Identity and Access Management
        (IAM), 302–313
    data encryption, 324–327
    detecting problems, 184
    detective controls, 313–323
    protecting network boundaries, 323–324
    of repositories, 362
    review questions, 329–333, 407–409
    of S3 data, 77
Security Account, 379
Security and identity category, 7, 9–10
security groups
    about, 36–37, 323
    Amazon Virtual Private Cloud, 98–101,
        127
    configuring, 53
    exercises, 100–101
    using network access control
        lists with, 106
Security Hub, 323
SELECT statement, 137
serverless applications, building with
    Lambda, 251–252
serverless workloads, 278, 295
server-side encryption, 63
service categories, 6–7
Service Level Agreement (SLA), 13
service limits, 31–32
service-level protection, 313
Session Manager, 48, 375–376
set data type, 154
sharding, 138, 293–294
Shared Responsibility Model, 12–13
shared services, isolated VPCs with, 117
Shared Services Account, 379

shared tenancy, 26–27
Shield, 324, 328
Simple Monthly Calculator, 73, 341–342
Simple Notification Service (SNS), 10, 200
Simple Queue Service (SQS)
    about, 10, 234
    dead-letter queues, 237
    polling, 236
    queue types, 235–236
    queues, 234–235, 240
    review questions, 241–244, 401–402
simple scaling policy option, in Auto Scaling,
    43, 257–258
Simple Storage Service (S3)
    about, 8, 60, 325
    accessing objects, 67–71
    architecture, 61–64
    CloudFront and, 282
    for data backup and recovery, 261, 268
    durability and availability, 64–66
    exercises, 62, 281, 282
    Glacier, 71–73, 77
    object lifecycle, 66–67, 77
    organizing resources, 77
    Reduced Redundancy Storage (RRS), 64
    review questions, 78–81
    S3 cross-region replication, 280–281
    Transfer Acceleration, 281–282
    Transfer Acceleration Speed Comparison
        tool (website), 62
Simple Storage Service (S3) Glacier, 8
Simple Workflow (SWF), 10
Single Sign-On (SSO), 174
single-master cluster, 147
single-root input/output virtualization
    (SR-IOV), 93
slash notation, 84
Snowball, 74
software development kits (SDKs), 14
source action type, 372
sources, in rules, 99
spot fleet, 345–347
spot instance interruption, 345
spot instance pool, 345

spot price, 344
spread groups, 28
stacks
   creating, 355
   deleting, 356
   nested, 356–358
   overriding policies, 361
   parent, 356
   updates to, 359
   using multiple, 356–359
standard instance classes, 140
standard queues, 236
standby database instance, 146
State Manager, 49, 376–377
stateful firewall, 99–100
static threshold, 198
static websites, 70, 77
step scaling policy, in Auto Scaling, 44, 258–259
storage. *See also* Simple Storage Service (S3)
   Amazon Elastic File System (EFS), 73
   Amazon FSx, 73
   AWS DataSync, 74–75
   AWS Snowball, 74
   AWS Storage Gateway, 73–74
   CLI example, 75–76
   of data, 137
   databases and, 141–144
   Elastic Block Store (EBS), 278, 295
   pricing for, 71
Storage category, 6, 8–9
Storage Gateway, 9, 73–74
storage optimized instances, 25
storage volumes, 32–34, 53
Structured Query Language (SQL), 137
subnets, 87–91, 127
support plans, 14–15, 16
Systems Manager
   about, 46–47, 374–378, 380
   actions, 47–49
   insights, 49–51

T

T2/T3 instances, 24
target tracking policy, in Auto Scaling, 44–45, 259
targets (Amazon EventBridge), 202
technical support, 14
tenancy, 26–27
test action type, 373
third-party automation solutions, 288–289
threshold, 198
throughput capacity, 155–157
throughput-optimized HDD, 33, 144
tightly coupled workloads, 125
top-level domain (TLD), 214
Total Cost of Ownership (TCO) Calculator, 6, 342
traditional applications, availability and, 249–250, 268
Traffic Flow, 222–223
trailing slash (/), 36
trails (CloudTrail), 186–188
Transfer Acceleration, 62
Transit Gateway, 115–123
transit virtual interface, 124
transparent data encryption (TDE), 140
triggers, operational excellence and, 370
trust policy, 308
Trusted Advisor, 252, 340
Trusted Advisor Recommendation, 50, 377
trusted entity, 173
tuples. *See* attributes

U

URLs, predesigned, 69
user accounts, 168–170
user pools, 174

V

vault, 71
versioning, 66, 67
video streams, 237–238, 240
virtual interfaces, 124
Virtual Private Cloud (VPC)
   about, 8, 12, 26, 84
   AWS Global Accelerator, 109
   Classless Inter-Domain Routing (CIDR) blocks, 84–87, 127
   design considerations for, 263
   elastic IP addresses, 107–109
   elastic network interface (ENI), 91–93, 127
   exercises, 85–87, 315
   high-performance computing, 125–126
   hybrid cloud networking, 115–124
   Internet gateways, 93–94
   logs, 314
   network access control list (NACL) functions, 101–106, 127
   network address translation (NAT), 109–110, 128
   network address translation devices, 110–113
   public IP addresses, 106–107
   review questions, 129–132, 391–393
   route tables, 94–97, 127
   security groups, 98–101, 127
   subnets, 87–91, 127
   VPC peering connection, 114, 128
virtual private networks (VPNs), 115
virtualization
   cloud computing and, 4–6
   single-root input/output virtualization (SR-IOV), 93
visibility timeout, 235
visualization, 290–291
VPC peering connection, 114, 128

W

Web Application Firewall (WAF), 323–324, 328
weighted routing, 220
Well-Architected (website), 15
whitelisting, 98
write capacity units (WCUs), 155

Z

zones/zone files, 215

Online Test Bank

Register to gain one year of FREE access to the online interactive test bank to help you study for your AWS Certified Solutions Architect Associate certification exam—included with your purchase of this book! All of the chapter review questions and the practice tests in this book are included in the online test bank so you can practice in a timed and graded setting.

Register and Access the Online Test Bank

To register your book and get access to the online test bank, follow these steps:

1. Go to bit.ly/SybexTest (this address is case sensitive)!
2. Select your book from the list.
3. Complete the required registration information, including answering the security verification to prove book ownership. You will be emailed a pin code.
4. Follow the directions in the email or go to www.wiley.com/go/sybextestprep.
5. Find your book on that page and click the "Register or Login" link with it. Then enter the pin code you received and click the "Activate PIN" button.
6. On the Create an Account or Login page, enter your username and password, and click Login or, if you don't have an account already, create a new account.
7. At this point, you should be in the test bank site with your new test bank listed at the top of the page. If you do not see it there, please refresh the page or log out and log back in.

SYBEX
A Wiley Brand